THE SHIFT AND SHAPE OF SPECTRAL LINES

THE SHIFT AND SHAPE OF SPECTRAL LINES

BY

ROBERT G. BREENE, Jr.

PERGAMON PRESS

NEW YORK · OXFORD · LONDON · PARIS

1961

PERGAMON PRESS INC.
122 East 55th Street, New York 22, N.Y.
1404 New York Avenue N. W., Washington 5 D. C.
Statler Center 640, 900 Wilshire Boulevard
Los Angeles 17, California

PERGAMON PRESS LTD.
Headington Hill Hall, Oxford
4 & 5 Fitzroy Square, London, W. 1

PERGAMON PRESS S. A. R. L.
24 Rue des Écoles, Paris Ve

PERGAMON PRESS G. m. b. H.
Kaiserstrasse 75, Frankfurt am Main

Library of Congress Card No. 60-14951

PRINTED IN THE GERMAN DEMOCRATIC REPUBLIC

By Paul Dünnhaupt, Köthen (Anhalt)

TO A FINE
SOLDIER AND GENTLEMAN
MY FATHER

CONTENTS

INTRODUCTION

IN RECENT years man has generated tremendous enthusiasm for the development and operation of missiles and space vehicles designed, among other things, for re-entry into the atmospheres of this and other planets. In preparing for these attempts he has been and is busily engaged in experiments, some confined to the mundane limits of the laboratory, others in the actual environments of the re-entering vehicles. The high temperatures encountered in the gases surrounding the actual vehicles in the latter case or the simulating models in the former require something new in the way of probes for the determination of temperatures, densities, and the like. It is as one of these probes that the breadths and shapes of spectral lines are gaining a "practical" importance and hence popularity which they have not hitherto enjoyed. Of more ancient vintage is the application of line broadening to stellar study and description. After the same fashion that the width and shape serve as a probe of the space vehicle environment, so they may and have served as a probe of the atomic environment provided by the interior of a distant star. Whichever the domain of interest of the reader, it is to be hoped that what follows will be of some assistance to him.

The first chapter deals only with the historical background of line broadening but is deserving of more than simply pedagogical concern. Certainly the very early work of Michelson (1895) and Rayleigh (1889) still forms the basis for everything we do in this field.

Chapters 2 and 3 deal with the basic theory of line broadening. The Interruption Theory and the Statistical Theory are, as we shall see, the limiting theories for all types of spectral line broadening. The former is the limit for low densities and frequencies near line center while the latter is, of course, the high density, line wing limit. Matters of adiabaticity and the applicability of a classical path assumption are certainly of basic interest and are discussed as the need for such discussion arises. With this basis laid in these two chapters we devote the remaining chapters to special cases of the general theory. It is true, for example, that the same type treatment is applied to the case of Resonance Broadening as is applied to, say, broadening by van der Waals forces. However, the special attention which must and has been paid to self broadening renders its consideration as a special case justifiable. In like manner the tremendous increase in interest in plasmas of recent years more than justifies our specific attention to Stark broadening in Chapter 6. The similar remarks which could be made concerning the last two chapters are obvious.

References to and discussions of experimental work have been deliberately avoided. Of course it is to be understood that the theoretical work detailed is supported and justified by experimental work not detailed. This being so it was felt that consideration of such work could be dispensed with. With regard to experiment the reader is referred to a recent review article by Ch'en and Takeo, referenced in the bibliography, which provides a great deal of such information.

The author wishes to express his appreciation to Professors E. Lindholm and L. Spitzer, Jr. for clarifying points in certain of their valuable contributions to the field prior to the publication of the first edition. It is also a pleasure to acknowledge with many thanks the translational help rendered prior to the first edition publication by Mr. James Gough, then of the American Meteorological Society staff. The author is greatly indebted to Professor Henry Margenau whose contributions to this field have been so vast, for the continuing assistance which has been so graciously rendered whenever requested.

CHAPTER 1

EARLY LINE BROADENING THEORY

THE YEAR 1895 may well be chosen as the starting point for our study of the development of spectral intensity although some work had, of course, been done previous to this time. In November of that year Michelson published an article in the Astrophysical Journal in which he made the first attempt to bring together and consider all those factors which "... account for the finite width of the spectral lines of a substance emitting approximately homogeneous radiation..."[155]

1.1. The causes of line broadening (1895)

Michelson began his consideration of line broadening effects by a summarizing of the hypotheses which had been advanced previous to that time to account for these phenomena, and a verbatim restatement may be of interest here.

"1. As a consequence of Kirchhoff's law 'the ratio of brightness of two immediately contiguous portions of a discontinuous bright-line spectrum constantly decreases, if the number of luminous strata is multiplied or if the coefficient of absorption of the single stratum is increased, until the value is reached which, for the same wavelength and the same temperature, corresponds to the ratio in the continuous spectrum of a body completely opaque for a given thickness'.[235]

"2. The direct modification of the period of the vibrating atoms in consequence of presence (sic) of neighboring molecules.

"3. The exponential diminution in amplitude of the vibrations due to communication of energy to the surrounding medium or to other causes.

"4. The change in wavelength due to the Doppler effect of the component of the velocity of the vibrating atom in the line of sight.

To these causes of line broadening Michelson added the following:

"5. The limitation of the number of regular vibrations by more or less abrupt changes of phase amplitude or plane of vibration caused by collisions.

"6. The possible variations in the properties of the atoms within such narrow limits as to escape detection by other than spectroscopic observations."

Of these six possible contributing factors, Michelson considered only the fourth and fifth as of any real importance in line broadening, although he looked on the other factors as of minor import rather than non-existent.

As will be apparent at a later stage in the development, we can still agree with Michelson today that the fourth and fifth factors are of great import in the broadening of spectral lines.

Thus, Michelson concluded that the Doppler effect — which will be shown to be dependent on the temperature — and the effect of collisions — which will prove primarily dependent on the pressure* — are chiefly responsible for the finite width of spectrum lines.

Before considering the first development of the effect of pressure on the broadening of a spectral line as given by Michelson, it will perhaps be well to go back to the year 1889 and the first comprehensive consideration of the effect of temperature on the broadening of a spectral line. In that year Lord Rayleigh published an article[170] which dealt with the effect of molecular motion — the velocity being dependent on the gas temperature — on the radiation emitted by the moving molecule.

1.2. The Doppler effect in line broadening

Let us preface the detailed consideration of this first step in the development of the theory of line broadening and line shape by a qualitative and premature — insofar as the evolution of the theory is concerned — consideration of this spectral line which we have so far glibly taken for granted.

We are not interested at this particular time in the mechanics of the occurrence of this line, but only roughly in what it is. A collection of molecules† in the gaseous state will absorb or emit approximately homogeneous electromagnetic radiation of a number of frequencies depending on the nature of the molecule. Consider one of these frequencies. One might intuitively expect this radiation to be truly homogeneous, that is, of one unique frequency, but, as Michelson noted, this is not the case.

First, let us define the word intensity of emission (absorption) as the energy emitted (absorbed) per unit area per unit time. Then if one makes a plot of frequency vs. intensity for this "almost homogeneous" radiation, one obtains a distribution of intensities over a small frequency range giving rise to a spectral line of definite shape.

We are desirous of eventually obtaining this intensity distribution, but Lord Rayleigh, in considering this problem, was primarily interested in an expression for the brightness in the fringes which are produced by an interferometer. Thus, the expression at which we shall arrive will differ slightly from that of Lord Rayleigh.

First, let us consider the effect which the motion of a molecule will have on the radiation emitted by it.

* This is not strictly true, since the effects of collisions will be connected with the *density* of particles. Thus, for example, from the perfect gas law ($pv = NkT$) the effects of collisions are also somewhat dependent on temperature.

† A molecule may be monatomic and the possessor of one atom or polyatomic and the possessor of many.

In Fig. 1.1 a molecule whose velocity vector, v, makes an angle ϑ with the line of sight of an observer at 0 has been depicted. The molecule is emitting radiation of frequency ν and it is with the effect of the molecular motion on this frequency, that the Doppler effect is concerned. Since the velocity of the emitted radiation is c and the wavelength λ, c/λ waves would reach the observer each second were the molecule at rest, but since it possesses the velocity component ξ in the line of sight an additional ξ/λ waves will reach the observer each second. Thus, the frequency of the radiation at 0 has been perturbed by the molecular velocity from a value c/λ to:

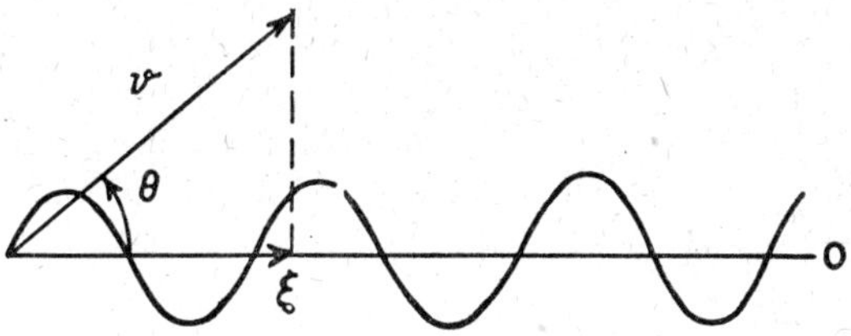

FIG. 1.1. The physical conditions requiste ior the Doppler Effect. The observer is at 0. The emitter velocity vector makes the angle θ with the line of sight at this observer

$$\nu' = \frac{c}{\lambda} + \frac{\xi}{\lambda} = \nu\left(\frac{c+\xi}{c}\right)$$

or:

$$\left.\begin{aligned} \lambda' &= \frac{c}{\nu'} = \lambda\left(\frac{c}{c+\xi}\right) = \lambda\left(\frac{1}{1+\xi/c}\right) \\ &= \lambda\left(1 - \frac{\xi}{c} + \left(\frac{\xi}{c}\right)^2 - \cdots\right) = \lambda\left(1 - \frac{\xi}{c}\right) \end{aligned}\right\} \qquad (1.1)$$

in which the binomial expansion and the reasonable assumption $\xi \ll c$ have been utilized. In these expressions ν' and λ' have been used for the perturbed frequency and wavelength respectively. Thus, a moving emitter would have its radiation effected in a manner as given by Eq. (1.1).

Ebert[36] had investigated this effect on the emitted radiation under the assumption that all the molecules of the gas moved with the same velocity and had found the predicted line widths much greater than experiment showed them to be. This discrepancy which, if explained, would have dealt "... the dynamical theory of gases ... a heavy blow, from which it could only with difficulty recover"[170] was corrected by Rayleigh who substituted Maxwell's velocity distribution for Ebert's constant velocity.

The distribution of the velocities in the line of sight is given by the Maxwellian distribution as $e^{-\beta^2\xi}$. Utilizing this distribution in conjunction with the Doppler effect, Rayleigh arrived at the following expression for the intensity in the fringes of an interferometer:

$$I'\,\mathrm{d}\nu = 4\sin^2\frac{z\pi\Delta}{\lambda}\left(1 + \frac{\xi}{c}\right) I\,\mathrm{d}\nu \qquad (1.2)$$

I gives the distribution of intensity in the incident radiation (the true spectral line) while the multiplicative factor alters this to give the intensity distribution,

I', in the interferometer fringe. It is with I that we are concerned, and this is given by the Maxwellian distribution where

$$\beta = \frac{m}{2\,kT} \tag{1.3}$$

In Eq. (1.3) m is the molecular mass, k is the Boltzmann constant, and T is the temperature in degrees Kelvin.

Now rewrite Eq. (1.1) as:

$$\xi = \frac{c}{\lambda}(\lambda - \lambda')$$

or:

$$\xi^2 = \frac{c^2}{\lambda^2}(\lambda' - \lambda)^2 = \lambda^2\, c^2 \left(\frac{1}{\lambda} - \frac{1}{\lambda'}\right)^2 = \lambda^2 (\Delta\nu)^2 \tag{1.4}$$

In Eq. (1.4) λ may now be considered as the wavelength corresponding to the maximum intensity of the line (line center), and $\Delta\nu$ as the frequency increment between the line center and the frequency whose displacement corresponds to the line of sight velocity ξ.

Thus, if one arbitrarily equates the line center intensity to one, the Doppler effect alone produces a distribution of intensities over the spectral line which may be represented by the expression:

$$I = e^{-\frac{m\lambda^2 (\Delta\nu)^2}{2\,kT}} \tag{1.5}$$

Eq. (1.5) illustrates the dependence as determined by Rayleigh of the intensity on the emitter mass, line frequency, frequency separation from line center, and, perhaps most important, temperature.

This equation then is Lord Rayleigh's solution to Michelson's Point 4. Michelson himself provided the mathematics of the fifth contributing factor, which we shall designate Interruption Broadening.

1.3. An application of the Fourier analysis to line broadening

"A group of perfectly homogeneous trains of waves . . . incident on a prism"[155] may be represented by the Fourier integral:

$$\varphi(x) = \frac{1}{\pi} \int_0^\infty du \int_{-\infty}^{+\infty} \varphi(y) \cos u(x-y)\, dy$$

Since: $\cos ux \cos uv + \sin ux \sin uv = \cos u(x-v)$ this may be rewritten in the form which was utilized by Michelson:

$$\varphi(x) = \frac{1}{\pi} \int_0^\infty du\, (C \cos ux + S \sin ux) \tag{1.6a}$$

where:

$$C = \int_{-\infty}^{+\infty} \varphi(y) \cos uy\, dy \qquad S = \int_{-\infty}^{+\infty} \varphi(y) \sin uy\, dy \tag{1.6b}$$

At this point Michelson made the assumption which was perhaps the most important in his development. He assumed that emission terminated abruptly at an intermolecular collision, thus limiting the wave train to that length which could be emitted between successive collisions. If the mean free path of the molecule is l and its velocity is v, then the time between molecular collisions will be l/v. This means that the length of the wave train which is emitted between collisions is:

$$r = \frac{l}{v} c \tag{1.7}$$

The uniform wave train may now be represented by:

$$\varphi(x) = a \cos mx + b \sin mx \tag{1.8}$$

where a and b are arbitrary constants and $m = 2\pi/\lambda$. λ is the wavelength of the incident radiation. The limits of the wave train may be taken as $-r/2$ and $+r/2$, and (1.6b) may be evaluated.

$$C = a \int_{-r/2}^{+r/2} \cos my \cos uy \, dy = a \frac{\sin \frac{1}{2}(m-u) r}{m-u} + a \frac{\sin \frac{1}{2}(m+u) r}{m+u} \tag{1.9a}$$

$$S = b \int_{-r\,2}^{r/2} \sin my \sin uy \, dy = b \frac{\sin \frac{1}{2}(m-u) r}{m-u} - b \frac{\sin \frac{1}{2}(m+u) r}{m+u} \tag{1.9b}$$

It should be remembered here that:

$$\int_{-r/2}^{r/2} \cos my \sin uy \, dy = \int_{-r/2}^{r/2} \sin my \cos uy \, dy = 0$$

The terms on the right side of Eqs. (1.9a) and (1.9b) are all small except when m is in the neighborhood of u in which case the first term on the right side of each equation is much larger than the second. For this reason the second term on the right of Eqs. (1.9a) and (1.9b) will be neglected. Further, in the neighborhood of $m = u$, C and S as given by (1.9a) and (1.9b) will control the amplitude so that, since the intensity in this case is simply the square of the amplitude:*

$$\left.\begin{aligned} I &= C^2 + S^2 \\ &= \frac{a^2 + b^2}{4} \frac{\sin^2 \pi n r}{\pi^2 n^2} \end{aligned}\right\} \tag{1.10}$$

where

$$m - u = 2\pi n = 2\pi \left(\frac{1}{\lambda} - \frac{1}{\lambda'}\right)$$

or:

$$n = \frac{1}{\lambda} - \frac{1}{\lambda'} = \frac{\Delta v}{c} \tag{1.11}$$

* The classical intensity will, of course, always be proportional to the square of the amplitude. The factor of proportionality requires additional information, specifically, it requires a knowledge of the integrated line intensity, but this information has no particular relevance to the line shape.

where λ is now the line center wavelength. If Eqs. (1.7) and (1.11) are substituted into Eq. (1.10), this equation may be expressed in a more meaningful form:

$$I = d \frac{\sin^2\left(\frac{\pi(\Delta\nu)\, l}{v}\right)}{\pi^2(\Delta\nu)^2} \tag{1.12}$$

where $d = \frac{(a^2 + b^2)\, c^2}{4}$. Eq. (1.12) illustrates the dependence upon pressure through the free path, l. Gas kinetic theory yields the mean free path:

$$l = \frac{1}{\sqrt{2}\, \pi N \sigma^2} \tag{1.13}$$

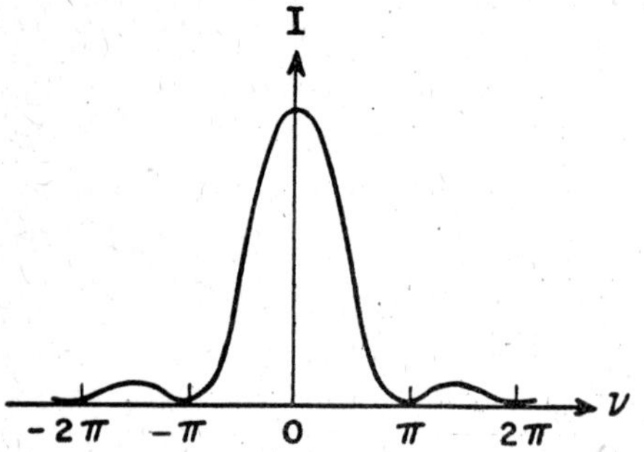

FIG. 1.2. The intensity distribution resulting from the simple application of the Fourier transform.
(After Michelson[127])

In Eq. (1.13) σ is the gas kinetic molecular diameter, a quantity which must needs arise when impenetrable spheres are chosen as molecular models. N is the number of molecules per unit of volume, and it is apparent that this is the factor through which the pressure most directly enters Eq. (1.12).

Fig. 1.2 is the curve which results from Eq. (1.12). The secondary maxima* which occur are of no import since they do not actually occur in the spectral line. It was with the central maximum that Michelson was concerned.

It is obvious that the first minimum will occur when, in Eq. (1.10), $\pi n r = \pi$, or when:

$$n = \frac{1}{r} = \frac{1}{l}\frac{v}{c} = \frac{\lambda^2}{l}\frac{v}{c} \text{ wavelengths} \tag{1.14}$$

This then would give a line width of:

$$\delta = 2\frac{\lambda^2}{l}\frac{v}{c} \tag{1.15}$$

Michelson found that Eq. (1.15) gave an order of magnitude agreement with experimental data. Somewhat better agreement was achieved by him when he took the "half-width"† as given by Eq. (1.10) — or, perhaps more correctly, as obtainable from Eq. (1.10) — and added it to the half-width which may be obtained from Eq. (1.5). As Michelson noted the problem is not herewith solved,

* For disappearance of sine see *infra*, Chap. 2.

† We now define the quantity half width. For a symmetric spectral line it is the frequency separation between the two points in a spectral line corresponding to one-half the intensity maximum in the line. For an asymmetric line it is the frequency separation between the itensity maximum and the point in the distended wing where the intensity has fallen to half maximum value.

but it will be of interest to see if "... such able contributions ... recently ... justify the prediction that a complete and satisfactory theory will be forthcoming in the near future."[155]

Let us pause briefly to consider the remaining four contributing causes as advanced by Michelson.*

To begin with, Point 3 is Michelson's manner of stating that a natural line width exists, that is, if a molecule is completely isolated from all other molecules and possesses zero velocity with respect to an observer, it will emit a spectral line having a finite width. In the classical case this is due to precisely the cause stated by Michelson as his Point 3. In the quantum mechanical case the cause may be found in a phenomenon of the type treated by Furrsow and Wlassow.† We shall not treat natural line shape.‡

Point 2 is quite interesting, for it is essentially the basis for the Statistical theory of line broadening** with which we shall deal rather extensively. This point combines with Point 5 in the Interruption theory.††

Point 1 is of no particular interest to us, and Point 6, although it shrewdly predicts isotopic spectra, is related to a "pseudo-broadening" with which we shall not concern ourselves.

It seems particularly interesting to note the correctness of the factors which Michelson listed as causing line broadening, correctness at least from the contemporary viewpoint.

The next logical step in the development would immediately appear to be the synthesis of the Doppler and Interruption broadening to arrive at some unified theory of the phenomena. This is precisely what C. Godfrey[60] attempted to do, and it is unfortunate that his results were not of too great value, since his endeavors were such a logical outgrowth of the work which has already been discussed.

1.4. The mean free path among equal spheres

Godfrey first carried out a "Fourier Analysis of m Complete Sine Waves" to arrive at Eq. (1.10) which had been given by Michelson. He then proceeded to carry out a "summation" for all molecules having definite velocities both athwart and in the line of sight. Before carrying through this procedure, it may be well to carry through the derivation for the "Mean Free Path Among Equal Spheres" which was developed by Tait.[195]

This author assumes a layer of thickness Δx "... in which quiescent spheres of diameter s ..."[125] are distributed with a density N per unit of volume. Now if spheres of diameter s are impinging on this layer in a normal direction, the fraction:

$$1 - n_1 \pi s^2 (\Delta x) \tag{1.16}$$

* See *supra*, p. 1.

† See *infra*, Chap. 5.

‡ For a complete treatment see Heitler.[208]

** See *infra*, Chap. 3.

†† See *infra*, Chaps. 1 and 2.

of the incident particles will pass through the layer without collision, since it is apparent that when the particle center separation is less than a diameter a collision will result. Tait makes the a priori assumption that "... the spheres in the very thin layer are so scattered that no one prevents another from doing its full duty* in arresting those which attempt to pass."[195]

Next the particles in the layer are considered as in a state of motion which may be described by velocity vectors of equal length, v_1 but of uniformly distributed orientation. Also, let the incident particles have a common velocity vector, v. Then the number of layer particles whose velocity vectors make an angle between ϑ and $\vartheta + d\vartheta$ with the velocity vectors of the incident particles will be given by:

$$n_1 \sin \vartheta \frac{d\vartheta}{2} \tag{1.17}$$

since this is merely the polar portion of the volume element and one-half the particles may be expected to proceed "north".

If the law of cosines is applied to the vector triangle which results from the addition of v_1 and v, a quantity which Tait calls the "relative speed" is obtained:

$$v_0 = \sqrt{v^2 + v_1^2 - 2\, v_1\, v_2 \cos \vartheta} \tag{1.18}$$

From this a quantity, the "virtual thickness" of the layer, may be defined:

$$\frac{v_0 \Delta x}{v} \tag{1.19}$$

The substitution of Eqs. (1.17) and (1.19) into Eq. (1.16) gives an expression for the number of incident particles which will penetrate the layer containing other particles whose velocity vectors have an orientation as given by Eq. (1.17):

$$1 - n_1 \pi s^2 (\Delta x)\, v_0 \sin \vartheta \frac{d\vartheta}{2v} \tag{1.20}$$

Now to find the overall fraction of particles which pass through the layer without collision, the expressions as given by Eq. (1.20) for all possible orientations of ϑ between 0 and π must be multiplied together. Each of these expressions is assumed very nearly equal to unity. Under this assumption the logarithm of the expression is given by:

$$\ln(1 - x) = -x$$

Thus, when we take the logarithm of this product, we arrive at the limit of the sum of these factors:

$$-\frac{n_1 \pi s^2 (\Delta x)}{2\, v} \int_0^{\pi} \sqrt{v^2 + v_1^2 - 2\, v\, v_1 \cos \vartheta} \sin \vartheta\, d\vartheta \tag{1.21}$$

The variable of integration may be changed from ϑ to v_0 by utilizing Eq. (1.18) to yield:

$$-\frac{n_1 \pi s^2}{2\, v^2\, v_1} \int v_0^2\, dv_0 \tag{1.21a}$$

* "The Royal Society expects every sphere to do its duty." J. I. F. King.

A straightforward application of Maxwell's law of velocity distribution, which Tait had made in a previous section of his paper, yields the number of particles in the layer whose velocity magnitudes lie between v_1 and $v_1 + dv_1$:

$$n_1 = 4\, N v_1^2\, dv_1 \sqrt{\frac{\beta^3}{\pi}}\, e^{-\beta v_1^2} \tag{1.22}$$

In Eq. (1.22) N is the particle density, and the other symbols have the meaning previously ascribed to them.

In Eq. (1.21a) the limits of integration depend on the relative magnitude of v and v_1. If $v > v_1$ the limits are $v - v_1$ and $v + v_1$. If, on the other hand, $v < v_1$ the limits become $v_1 - v$ and $v_1 + v$. Hence:

$$\left.\begin{aligned} \frac{1}{2v^2 v_1}\int v_0^2\, dv_0 &= \frac{1}{2v^2 v_1}\left[\int_{v-v_1}^{v+v_1} v_0^2\, dv_0 + \int_{v_1-v}^{v_1+v} v_0^2\, dv_0\right] \\ &= \left(1 + \frac{v_1^2}{3v^2}\right) + \left(\frac{v}{3v_1} + \frac{v_1}{v}\right) \end{aligned}\right\} \tag{1.23}$$

Eqs. (1.22) and (1.23) may be substituted into Eq. (1.21), and the result integrated over all v_1 to obtain the logarithm of the fraction of the number of incident particles of velocity v which penetrate the layer without collision.

$$-4\pi N s^2 \sqrt{\frac{\beta^3}{\pi}}\left[\int_0^v e^{-\beta v_1^2}\left(v_1^2 + \frac{v_1}{3v^2}\right) dv_1 + \int_v^\infty e^{-\beta v_1^2}\left(\frac{v v_1}{3} + \frac{v_1^3}{v}\right) dv_1\right] dx \tag{1.24}$$

It should be noted that the integral in Eq. (1.24) whose limits are 0 and v logically contains the results of integrating Eq. (1.21) in the case $v > v_1$. The second integral in Eq. (1.24) contains the results of the integration of Eq. (1.21) under the condition $v < v_1$.

1.5. A first synthesis of Doppler and interruption effects

If the bracketed expression in Eq. (1.24) is integrated by parts, the f which was utilized by Godfrey is obtained:

$$-f(\Delta x) = -4\pi N s^2 \sqrt{\frac{\beta^3}{\pi}}\left[\frac{1}{4\beta^2 v}\, e^{-\beta v^2} + \left(\frac{1}{4\beta^2 v^2} + \frac{1}{2\beta}\right)\int_0^\infty e^{-\beta v_1^2}\, dv_1\right]\Delta x \tag{1.25}$$

Recalling that this is a logarithm, the expression for the number of molecules which penetrate a distance l without collision becomes:

$$n_2 = e^{-fl} \tag{1.26}$$

From this equation it is apparent that the fraction of molecules which have free paths — distances of unimpeded penetration — between l and $l + dl$ is:

$$f\, e^{-fl}\, dl \tag{1.27}$$

These molecules will then emit wave trains of length r to $r + dr$ where r is given by Eq. (1.7) so that Eq. (1.27) may be rewritten as:

$$\frac{vf}{c}\, e^{-\frac{vfr}{c}}\, dr \tag{1.28}$$

It is at this point that Godfrey returns to Eq. (1.10) and proceeds to carry the Michelson solution a step further. Eq. (1.10) gives the intensity distribution in an Interruption broadened spectral line under the assumption that all the molecules in the emitting gas traverse free paths of the same length. Godfrey's next step was the imposition of the distribution of free paths which is given by Eq. (1.28) under the temporary restriction of definite thwart and line of sight velocities. If the constants of Eq. (1.10) are ignored, Godfrey's result is obtained:

$$\frac{fv}{n^2c}\int_0^\infty e^{-\frac{vfr}{c}} \sin^2 \pi n r \, dr \qquad (1.29)$$

It is with Eq. (1.29) that the objection of Lord Rayleigh[171] deals. Rayleigh felt that this integral favored unduly long free paths. His solution was the division of the integrand by r.* Godfrey's development will be followed, however.

If $(1 - \cos 2\pi Nr)$ is substituted for $\sin^2 \pi Nr$ in Eq. (1.29) any integral table may be consulted to obtain:

$$\frac{2\pi^2}{4\pi^2 n + v^2 f^2/c^2} \qquad (1.30)$$

Up to this point the thwart and line of sight velocities have been definite or fixed at various values. Now Godfrey wishes to add his final refinement to the theory by a consideration of these velocities.

Maxwell's velocity distribution yields $q\,e^{-\beta q^2}\,dq$ as the proportion of molecules having thwart velocities between q and $q + dq$.

Next Eq. (1.30) is integrated for a definite velocity p in the line of sight and all velocities q thwart:

$$\int_0^\infty q e^{-\beta q^2} dq \frac{1}{4\pi^2 n + \frac{v^2 f^2}{c^2}} \qquad (1.31)$$

where $v^2 = p^2 + q^2$.

The final step is the multiplication by a distribution factor for velocities in the line of sight, p, and an integration over all p. It should be apparent from previous considerations that this will bring the Doppler effect into the picture. From Eq. (1.1) it can be seen that if the Doppler perturbation is expressed in wave numbers, $\bar{\nu}$, the result is $p/c\lambda$. n in Eq. (1.31) is the Interruption perturbation in wave numbers. Let x be the total — Doppler plus Interruption — perturbation or the distance from line center to the point on the line under consideration. Then:

$$x = n + \frac{p}{c\lambda} \qquad (1.32)$$

* A consideration of Rayleigh's objection might be tempered by the fact that in voicing it he was attempting to resolve a disparity between his own and Godfrey's results. At any rate, the author feels, for reasons which will be apparent, that it is of no great moment which viewpoint is taken.

An expression for n may be obtained from this equation and substituted into Eq. (1.31). Multiplication by the line of sight velocity distribution factor and integration over all p gives:

$$\frac{1}{4\pi^2}\int_{-\infty}^{+\infty} e^{-\beta p^2}\int_0^{\infty} q e^{-\beta p^2}\frac{1}{\left(x-\frac{p}{\lambda c}\right)^2+\left(\frac{v f}{2\pi c}\right)^2}\,dp\,dq \tag{1.33}$$

Eq. (1.33) then gives the distribution of intensity over the spectral line as a function of x, the frequency separation of a point on the spectral line from the line center.

The variables of integration may now be changed from p and q to p and v where, of course, $v^2 = p^2 + q^2$.

Eq. (1.33) may be rewritten as follows:

$$\left.\begin{aligned}
\int_{-\infty}^{+\infty} dp \int_0^{\infty} f(p,q;v)\,dq &= \int_0^{\infty} dp \int_0^{\infty} f(p,q;v)\,dq + \int_{-\infty}^{0} dp\int_0^{\infty} f(p,q;v)\,dq \\
&= \int_0^{\infty} dp \int_0^{\infty} f(p,q;v)\,dq - \int_0^{-\infty} dp \int_0^{\infty} f(p,q;v)\,dq \\
&= \int_0^{\infty} dp \int_0^{\infty} dq\,[f(p,q;v) + f(-p,q;v)] \\
&= \frac{1}{4\pi^2}\int_0^{\infty} e^{-\beta p^2}\,dp\int_0^{\infty} q e^{-\beta q^2}\,dq \\
&\quad \times\left\{\frac{1}{\left(x-\frac{p}{\lambda c}\right)^2+\left(\frac{vf}{2\pi c}\right)^2}+\frac{1}{\left(x+\frac{p}{\lambda c}\right)^2+\left(\frac{vf}{2\pi c}\right)^2}\right\}
\end{aligned}\right\} \tag{1.34}$$

$f(p, q; v)$ is an even function so that the change from an integration over p from zero to minus infinity may be replaced by the integration over negative p from zero to infinity as is shown in Eq. (1.34).

During the integration over q, p is, of course, constant so that $v\,dv = q\,dq$. This leads to the result which Godfrey obtained by a slightly different integral juggle. The factor $1/4\pi^2$ has been dropped.

$$\int_0^{\infty}\int_p^{\infty} e^{-\beta v^2} v\,dp\,dv\left\{\frac{1}{\left(x-\frac{p}{\lambda c}\right)^2+\left(\frac{vf}{2\pi c}\right)^2}+\frac{1}{\left(x+\frac{p}{\lambda c}\right)^2+\left(\frac{vf}{2\pi c}\right)^2}\right\} \tag{1.35}$$

The order of integration may be changed when it is recalled that $p \leq v$.

$$I(x) \sim \int_0^{\infty}\int_0^{v} dv\,dp\,v e^{-\beta v^2}\left\{\frac{1}{\left(x-\frac{p}{\lambda c}\right)^2+\left(\frac{vf}{2\pi c}\right)^2}+\frac{1}{\left(x+\frac{p}{\lambda c}\right)^2+\left(\frac{vf}{2\pi c}\right)^2}\right\} \tag{1.36}$$

At this point Godfrey disappears into a consideration of the visibility curves of Michelson[156] (a byway down which we shall not accompany him) and finds that Eq. (1.36) is unmanageable except in the limit of zero pressure. Under this limiting condition the Doppler effect should be the only broadening factor present, so that his results should agree with those obtained by Rayleigh. Instead, he found the ten per cent discrepancy between his results and those of Rayleigh which has been mentioned previously.

Eq. (1.36) is the extent to which this synthesis of Interruption and Doppler broadening has been carried, although the beautifully straightforward and logical manner in which this synthesis emerges from the previous work renders this fact a rather unpleasant one.*

1.6. The motion of a charged particle in a radiation field

Some five years after the emergence of Eq. (1.36) and concurrently with Rayleigh's espousal of the division of Eq. (1.29) by r, H. A. Lorentz proposed a new approach to the problem of Interruption broadening. Although his approach was new, his results proved to be similar to Michelson's.

In contradistinction to Michelson's approach, Lorentz attacks the problem from the point of view of the mechanics of the absorption of radiation by molecules. He essentially explains the absorption and dispersion of light "... by means of the hypothesis that the molecules of ponderable bodies contain small bodies that are set in vibration by the periodic forces existing in a beam of light or radiant heat."[130]

Let us assume that radiation of electric field strength **E** and magnetic field strength **H** is incident on a molecular absorber. Maxwell's equations furnish us with the following relations between these two field strengths:

$$\nabla \times \mathbf{H} = \frac{1}{c}\dot{\mathbf{D}} \tag{1.37a}$$

$$\nabla \times \mathbf{E} = -\frac{1}{c}\dot{\mathbf{H}} \tag{1.37b}$$

where the electric displacement vector **D** is given by:

$$\mathbf{D} = \mathbf{E} + \mathbf{P} \tag{1.37c}$$

In Eq. (1.37c) **P** is the electric polarization. Let us define this polarization as the relative displacement of charge centers.

Each element of volume contains N molecules. We now suppose with Lorentz that each of these molecules contains a vibrating electron whose displacement from equilibrium during vibration is given by (xyz). This then will lead to the polarization:

$$P_x = Nex; \quad P_y = Ney; \quad P_z = Nez \tag{1.38}$$

since each vibrating electron has charge e.

* Weisskopf[208] later included the Doppler effect within the framework of the Fourier analysis. A rather less ponderous approach rendered this inclusion no great problem.

Now consider the forces on one of these vibrating electrons. First there will be a force due to the electric field of the incident radiation. A related force will arise due to the polarization of the surrounding medium by virtue of the electronic vibrations. The components of these forces will be:

$$e(E_x + \alpha P_x); \quad e(E_y + \alpha P_y); \quad e(E_z + \alpha P_z) \tag{1.39}$$

Since our medium is homogeneous and isotropic α is a constant.*

Next assume a linear central force to act upon the electron. The components of this force will be:

$$-fx; \quad -fy; \quad -fz \tag{1.40}$$

In Eq. (1.40) f is a constant which is dependent on the nature of the molecule. No real attempt need be made to attach further physical significance to this constant. If this were the only force acting on the electron the frequency of vibration would, of course, be related to f and the electron mass as follows:

$$\omega_0^2 = 4\pi^2 \nu_0^2 = \frac{f}{m} \tag{1.41}$$

A dissipative force which is proportional to the velocity is now introduced. The components of this force are:

$$-g\frac{dx}{dt}; \qquad -g\frac{dy}{dt}; \qquad -g\frac{dz}{dt} \tag{1.42}$$

The force whose components are given by this equation was intended to account for the absorption of the incident radiation.

If Eqs. (1.39), (1.40), and (1.42) are substituted into Newton's equation of motion, $\mathbf{F} = m\mathbf{a}$, one obtains the component equations of motion for the electron:

$$m\ddot{x} = e(E_x + \alpha P_x) - fx - g\dot{x} \tag{1.43a}$$

$$m\ddot{y} = e(E_y + \alpha P_y) - fy - g\dot{y} \tag{1.43b}$$

$$m\ddot{z} = e(E_z + \alpha P_z) - fz - g\dot{z} \tag{1.43c}$$

Now let:

$$m' = \frac{m}{Ne^2}; \qquad f' = \frac{f}{Ne^2}; \qquad g' = \frac{g}{Ne^2} \tag{1.44}$$

The substitution of Eq. (1.38) into Eqs. (1.43a) through (1.43c) and the utilization of Eq. (1.44) in the resulting expressions yields:

$$m'\ddot{P}_x = E_x + \alpha P_x - f'P_x - g'\dot{P}_x \tag{1.45a}$$

$$m'\ddot{P}_y = E_y + \alpha P_y - f'P_y - g'\dot{P}_y \tag{1.45b}$$

$$m'\ddot{P}_z = E_z + \alpha P_z - f'P_z - g'\dot{P}_z \tag{1.45c}$$

* We recall that in the most extreme case of non-isotropy the polarizability tensor α of nine components replaces this constant.

There is surely a sinusoidal time dependence associated with E and P so that the substitution of:

$$E_x = E_{0x}\,e^{i\omega t};\quad P_x = P_{0x}\,e^{i\omega t};\quad \text{etc.} \tag{1.46}$$

into Eqs. (1.45) is not an unreasonable one.

Also let:

$$\xi = f' - \alpha - m'\,\omega^2;\quad \eta = \omega g' \tag{1.47}$$

and let the substitution indicated by Eqs. (1.46) and (1.47) be made into Eqs. (1.45) to obtain:

$$E_{0x} = (\xi + i\eta)\,P_{0x} \tag{1.48a}$$

$$E_{0y} = (\xi + i\eta)\,P_{0y} \tag{1.48b}$$

$$E_{0z} = (\xi + i\eta)\,P_{0z} \tag{1.48c}$$

We shall return to a consideration of Eqs. (1.48).

If a molecule is acted on by an electric field whose x-component is:

$$E_x = a\,e^{i\omega t}$$

and it is also subjected to a velocity proportional resistance determined by g, the equation of the electronic motion in the x-direction becomes:

$$m\ddot{x} = -fx - g\dot{x} + a\,e\,e^{i\omega t} \tag{1.49a}$$

or:

$$\ddot{x} + \frac{g}{m}\,\dot{x} + \omega_0^2\,x = \frac{a\,e}{m}\,e^{i\omega t} \tag{1.49b}$$

where:

$$\omega_0^2 = 4\pi^2\,\nu_0^2 = \frac{f}{m}$$

Now a solution, $x = x_0\,e^{i\omega t}$ is assumed; the solution is substituted into Eq. (1.49b), and the result is solved for x_0 to give:

$$x = \frac{a\,e}{m\,(\omega_0^2 - \omega^2) + i\,\omega\,g}\,e^{i\omega t} \tag{1.50}$$

If there is no resistance to the electronic motion Eq. (1.49b) becomes:

$$\ddot{x} + \omega_0^2\,x = \frac{a\,e}{m}\,e^{i\omega t} \tag{1.51}$$

The solution of this equation is:

$$x = \frac{a\,e}{m\,(\omega_0^2 - \omega^2)}\,e^{i\omega t} + C_1\,e^{i\omega_0 t} + C_2\,e^{i\omega_0 t} \tag{1.52}$$

where the last two terms are the solution to the homogeneous equation:

$$\ddot{x} + \omega_0^2\,x = 0$$

while the first term is the particular solution.

A cursory glance at the Lorentz development to this point does not seem to indicate any connection between this development and its assumed objective, the Interruption broadening of spectral lines. It is at this point, however, that connection does become apparent.

1.7. Equivalence of molecular collisions and a damping force

If a molecule does not undergo a collision, its vibrating electron will follow the electric force associated with the incident radiation. If the frequency of the incident radiation is the same as the natural vibrational frequency of the electron, a classical resonance condition may be expected to prevail, and, but for a regulating factor, the increase of the vibrational amplitude without limit would be expected.

The collision of the molecule with another will serve to limit this amplitude increase, for Lorentz supposes that by virtue of the encounter the molecular vibration will be changed into a vibration of a different kind.

The electron will carry out this new vibration until another collision essentially stops this vibration and starts still another one, and so on. Thus, in this way, in analogy to a resistance proportional to the velocity, the amplitude of the vibrations will be limited.

In the absorbing gas then the state of vibration of each molecule will be interrupted over and over again by a large number of blows. Let A blows be distributed among N molecules per unit of time. The mean length of time between blows, that is, the mean time that the electronic vibration in the molecule proceeds undisturbed is:

$$\tau = N/A$$

Let $f(\vartheta)$ be the probability that a molecule will not be struck during a time interval ϑ. The chance that it will be struck in the next interval $\mathrm{d}\vartheta$ is $\mathrm{d}\vartheta/\tau$ where τ has been defined by the last equation. Thus, the chance that the molecule will not be struck for a time $\vartheta + \mathrm{d}\vartheta$ is:

$$f(\vartheta)\left(1-\frac{\mathrm{d}\vartheta}{\tau}\right) \tag{1.53}$$

This should, of course, be the same as $f(\vartheta + \mathrm{d}\vartheta)$ or:

$$f(\vartheta) + \frac{\partial f(\vartheta)}{\partial\vartheta}\mathrm{d}\vartheta \tag{1.54}$$

If Eqs. (1.53) and (1.54) are equated, the result is:

$$\frac{\partial}{\partial\vartheta} f(\vartheta) = -\frac{f(\vartheta)}{\tau} \tag{1.55}$$

Eq. (1.55) has the solution:

$$f(\vartheta) = B\,\mathrm{e}^{-\vartheta/\tau}$$

where B is a constant. This constant may be evaluated as follows:

$$\int_0^\infty f(\vartheta)\,\mathrm{d}\vartheta = B\int_0^\infty \mathrm{e}^{-\vartheta/\tau}\,\mathrm{d}\vartheta = 1$$

$$\tau B = 1 \leftrightarrow B = 1/\tau$$

So that:

$$f(\vartheta) = \frac{1}{\tau}\,\mathrm{e}^{-\vartheta/\tau} \tag{1.56}$$

From Eq. (1.56) it is apparent that the number of molecules for which the time since the last collision is between ϑ and $\vartheta + \mathrm{d}\vartheta$ is given by $N/\tau\ \mathrm{e}^{-\vartheta/\tau}\,\mathrm{d}\vartheta$.

Now reconsider Eq. (1.52), the equation of displacement for the vibrating electron. It is now possible to evaluate C_1 and C_2 for each molecule by finding x and $\dot{x}$ for time $t - \vartheta$, that is, the time immediately subsequent to the last blow. In order to do this, suppose that immediately after a collision all orientations of the displacement and velocity vectors are equally probable.* This follows from the assumptions which have been made earlier with regard to the collisions, that is, the termination of one vibration and the inception of a different one by the collision. Further, when this condition holds:

$$x = \dot{x} = 0 \tag{1.57}$$

and Eq. (1.57) provides the two needed conditions for the evaluation of C_1 and C_2:

$$\frac{a\,e}{m\,(\omega_0^2 - \omega^2)}\,\mathrm{e}^{i\,\omega\,(t-\vartheta)} + C_1\,\mathrm{e}^{i\,\omega_0\,(t-\vartheta)} + C_2\,\mathrm{e}^{-i\,\omega_0\,(t-\vartheta)} = 0 \tag{1.58a}$$

$$\frac{a\,e\,i\,\omega}{m\,(\omega_0^2 - \omega^2)}\,\mathrm{e}^{i\,\omega\,(t-\vartheta)} + C_1\,i\,\omega_0\,\mathrm{e}^{i\,\omega_0\,(t-\vartheta)} - C_2\,i\,\omega_0\,\mathrm{e}^{-i\,\omega_0\,(t-\vartheta)} = 0 \tag{1.58b}$$

whose solution is:

$$C_1 = -\frac{a\,e}{2m\,(\omega_0^2 - \omega^2)}\left(1 + \frac{\omega}{\omega_0}\right)\mathrm{e}^{i\,\omega\,(t-\vartheta) - i\,\omega_0\,(t-\vartheta)} \tag{1.59a}$$

$$C_2 = -\frac{a\,e}{2\,m\,(\omega_0^2 - \omega^2)}\left(1 - \frac{\omega}{\omega_0}\right)\mathrm{e}^{i\,\omega\,(t-\vartheta) + i\,\omega_0\,(t-\vartheta)} \tag{1.59b}$$

When Eqs. (1.59a) and (1.59b) are substituted into Eq. (1.52) there results:

$$x = \frac{a\,e}{m\,(\omega_0^2 - \omega^2)}\,\mathrm{e}^{i\,\omega\,t} \times \times \left\{1 - \frac{1}{2}\left(1 + \frac{\omega}{\omega_0}\right)\mathrm{e}^{i\,(\omega_0 - \omega)\,\vartheta} - \frac{1}{2}\left(1 - \frac{\omega}{\omega_0}\right)\mathrm{e}^{-i\,(\omega_0 + \omega)\,\vartheta}\right\} \tag{1.60}$$

* Some forty years later Van Vleck and Weisskopf disagreed with this statement to which they attributed responsibility for certain discrepancies in the Lorentz results. We shall consider these discrepancies in Chapter 6. These two authors felt that in the case of thermal equilibrium all x and $\dot{x}$ orientation should not be equally probable, but rather should be governed by a Maxwell–Boltzmann distribution.

We may now find the average value of x over all collision times $\langle x \rangle$, as follows:

$$\langle x \rangle = \frac{1}{\tau} \int_0^\infty x\, e^{-\vartheta/\tau}\, d\vartheta \tag{1.61}$$

The integrals which occur in Eq. (1.61) are of the form:

$$\frac{1}{\tau} \int_0^\infty e^{u\vartheta - \vartheta/\tau} d\vartheta = \frac{1}{1 - u\tau} \tag{1.62}$$

if u is a pure imaginary. If Eq. (1.62) is utilized, the integration in Eq. (1.61) may be carried out quite easily to obtain:

$$\langle x \rangle = \frac{a\, e}{m\left(\omega_0^2 + \frac{1}{\tau^2} - \omega^2\right) + 2\frac{i\, m\, \omega}{\tau}} e^{i\omega t} \tag{1.63}$$

Compare this to Eq. (1.50). It can be seen that collisions have an effect on the electronic motion which is the same as a resistance to motion of $g = 2\, m/\tau$ and an elastic force of $f_c = f + m/\tau^2$. This fact may be noted by comparing the terms in the denominators on the right sides of Eqs. (1.50) and (1.63).

We return to a consideration of the radiation which is to be absorbed by this vibrating electron.

1.8. Effect of collisions on the radiation absorption coefficient

For simplicity, it can be assumed that propagation takes place in the z-direction. This would mean that the field vectors would contain a term $e^{i\omega(t-qz)}$ instead of simply $e^{i\omega t}$. Lorentz further assumes that only the x-components of **E** and **D** are different from zero. Since this is the case, Eq. (1.37a) reduces to:

$$-\frac{\partial}{\partial z}[H_{0y}\, e^{i\omega(t-qz)}] = \frac{1}{c}\frac{\partial}{\partial t}[D_{0x}\, e^{i\omega(t-qz)}]$$

or:

$$q\, H_y = \frac{1}{c} D_x \tag{1.64a}$$

Eq. (1.37b) becomes:

$$\frac{\partial}{\partial z}[E_{0x}\, e^{i\omega(t-qz)}] = -\frac{1}{c}\frac{\partial}{\partial t}[H_{0y}\, e^{i\omega(t-qz)}]$$

or:

$$q\, E_x = \frac{1}{c} H_y \tag{1.64b}$$

Combining Eqs. (1.64a) and (1.64b) gives:

$$D_x = c^2 q^2 E_x \tag{1.65a}$$

$$P_x = (c^2 q^2 - 1)\, E_x \tag{1.65b}$$

When Eqs. (1.48a) and (1.65b) are compared, it becomes apparent that:

$$c^2 q^2 - 1 = \frac{1}{\xi + i\eta} \tag{1.66}$$

According to Eq. (1.66) q may be arbitrarily broken down into a real and an imaginary part. Let q be given by:

$$q = \frac{1 - i\chi}{\omega'} \tag{1.67}$$

Now the exponential factor contained in the field vectors becomes:

$$e^{i\omega\left(t - z\frac{1-i\chi}{\omega'}\right)} \tag{1.68a}$$

so that the factor producing attenuation of the beam is:

$$e^{-\frac{\omega\chi}{\omega'}z} \tag{1.68b}$$

It is apparent from Eq. (1.68b) that $\omega\chi/\omega'$ determines the absorption; specifically it is a linear absorption coefficient or in Lorentz terminology "index of absorption".

Eq. (1.67) may now be substituted into Eq. (1.66) and the real and imaginary portions on each side of the resulting equation may be equated to obtain the two equations below.

$$\frac{2c^2}{\omega'^2} = \sqrt{\frac{(\xi+1)^2 + \eta^2}{\xi^2 + \eta^2}} + \frac{\xi}{\xi^2 + \eta^2} + 1 \tag{1.69a}$$

$$2\frac{c^2\chi^2}{\omega'^2} = \sqrt{\frac{(\xi+1)^2 + \eta^2}{\xi^2 + \eta^2}} - \frac{\xi}{\xi^2 + \eta^2} - 1 \tag{1.69b}$$

From these two equations ω' and χ may be obtained.

If $\xi = 0$, that is, if the frequency of the incident radiation differs only slightly from the natural frequency of the electron, Eq. (1.69b) becomes:

$$2\frac{c^2\chi^2}{\omega'^2} = \sqrt{1 + \frac{1}{\eta^2}} - 1 \tag{1.69c}$$

Over a distance $z = \lambda = \frac{2\pi c}{\omega}$ in air the amplitude decreases in the ratio $1/\exp\left(-\frac{\chi}{\omega'}2\pi c\right)$ according to Eq. (1.68b). Since this distance is so small, the absorption, and hence the factor $\frac{2\pi c\chi}{\omega'}$ can be expected to be very small. It follows that $\frac{c^2\chi^2}{w'^2}$ will be even smaller. This in turn means that $\eta \gg 1$ which allows the application of the binomial expansion to the radical in Eq. (1.69b). We rewrite this radical slightly before applying the binomial expansion as follows:

$$\left[1 + \left(\frac{2\xi + 1}{\xi^2 + \eta^2}\right)\right]^{1/2} \approx 1 + \frac{1}{2}\frac{2\xi + 1}{\xi^2 + \eta^2} - \frac{1}{8}\frac{(2\xi + 1)^2}{(\xi^2 + \eta^2)^2} \tag{1.70}$$

Higher terms in the binomial expansion are dropped as small. The result of substituting Eq. (1.70) into Eq. (1.69b) is:

$$2\frac{c^2\chi^2}{\omega'^2} = \frac{4\eta^2 - 4\xi - 1}{8(\xi^2 + \eta^2)^2} \tag{1.71}$$

As long as ξ is small compared with η^2, the numerator in Eq. (1.71) can be replaced by η^2. On the other hand, as soon as ξ attains the order of magnitude of η, the denominator becomes so large that the right side of Eq. (1.71) is small enough to be neglected. Thus, the numerator may be taken as $4\eta^2$ to a good approximation in all cases. Hence:

$$\frac{c\chi}{\omega'} = \frac{\eta}{2(\xi^2 + \eta^2)}$$

This equation yields the linear absorption coefficient:

$$k = \frac{\omega\chi}{\omega'} = \frac{\omega}{c}\left(\frac{c\chi}{\omega'}\right) = \frac{\omega}{2c}\frac{\eta}{\xi^2 + \eta^2} \tag{1.72}$$

This is the line shape which we have been seeking. This may be placed in a more meaningful form.

First, assume the polarization of the gas to be small so that $\alpha = 0$. Then Eq. (1.47) may be rewritten as:

$$\xi = \frac{2\pi}{N e^2}(\nu_0 - \nu) \tag{1.73a}$$

$$\eta = \frac{\omega g}{N e^2} = \frac{4\pi m}{N\tau e^2}\nu \tag{1.73b}$$

Eq. (1.72) may then be rewritten as:

$$k = \frac{N m e^2}{c\tau}\frac{\nu^2}{(\nu_0 - \nu)^2 + \frac{4m^2}{\tau^2}\nu^2} \tag{1.74}$$

By equating the right side of Eq. (1.74) to one-half $k_{\max}$, the half-width of the line is found to be:

$$\delta_I = \frac{4m}{\tau}\nu \tag{1.75}$$

where ν is the frequency of the line center.

Also the integrated absorption coefficient may be evaluated as follows:

$$S = \int_{+\infty}^{-\infty} k(\nu_0)\, d\nu_0 = \frac{N m e^2}{c\tau}\nu^2 \int_{+\infty}^{-\infty} \frac{d\nu_0}{(\nu_0 - \nu)^2 + \frac{4m^2}{\tau^2}\nu^2}$$

$$= \frac{2\pi}{\delta_I}\frac{N m e^2}{c\tau}\nu^2 \tag{1.76}$$

So that:

$$\frac{N m e^2}{c \tau} = \frac{S \delta_I}{2 \pi \nu^2} \tag{1.77}$$

The substitution of Eqs. (1.75) and (1.76) into Eq. (1.74) yields the more familiar form of the equation for the Lorentz line:

$$k(\nu_0) = \frac{S}{\pi} \frac{(\delta_I/2)}{(\nu_0 - \nu)^2 + (\delta_I/2)^2} \tag{1.78}$$

Before considering various facets of this equation and its immediate predecessor, it will be well for us to retrace our steps by a few. The $k(\nu_0)$ which has been attained attenuates, according to Eq. (1.68b), not the intensity but the field vectors, and, in the final analysis, it is not with the field vectors that we are concerned. Although we will find that an equation identical with Eq. (1.77) results for the absorption coefficient which is responsible for the intensity attenuation, the difference should be clearly understood.

Poynting's vector* may be written as:

$$\mathbf{P} = \mathbf{E} \times \mathbf{H} \tag{1.79}$$

P has the units energy per unit of surface area per unit of time and is indeed associated with the quantity which we desire, intensity of incident radiation. Thus, since the factor as given by Eq. (1.68b) occurs in both the electric and magnetic vectors, it will occur to the square in Eq. (1.78). Hence, the intensity of the incident radiation will be attenuated by the factor:

$$e^{-2kz} = e^{-\varkappa z} \tag{1.80}$$

As a result, Eq. (1.76) may be rewritten as:

$$S_2 = \int_{-\infty}^{+\infty} \varkappa(\nu_0)\, d\nu_0 = 2 \int_{-\infty}^{+\infty} k(\nu_0)\, d\nu_0 = 2S \tag{1.81}$$

It is apparent then that the equation for $k(\nu_0)$ will be the same as Eq. (1.78) with S_2 substituted for S. Let us now reinterpret $k(\nu_0)$ in Eq. (1.78) and consider this symbol as denoting the intensity absorption coefficient.

1.9. An alternate manner of obtaining the Lorentz result

We have obtained Lorentz's results after the fashion in which Lorentz obtained them, but now let us go back and obtain them in a slightly different form which we shall consider again at a later point.

Let us again carry out the integration of Eq. (1.61) according to Eq. (1.62) in a straightforward manner to obtain:

$$x = \Re\left\{\frac{e a e^{i\omega t}}{m(\omega_0^2 - \omega^2)}\left[1 - \frac{1 + (\omega/\omega_0)}{2\tau[-i(\omega_0 - \omega) + 1/\tau]} - \frac{1 - (\omega/\omega_0)}{2\tau[i(\omega + \omega_0) + 1/\tau]}\right]\right\} \tag{1.82a}$$

$$= \Re\{a e^{i\omega t}(x' - i x'')\} \tag{1.82b}$$

* We may recall this vector as the measure of energy flow in the field and as having the direction of this flow.

In order to write down the desired result from Eq. (1.82b), it is necessary to digress for a moment.

We begin by considering a molecule of electric dipole moment μ which makes an angle ψ with an external electric field a cos ω t. In this case now, we assume that after collision a Maxwell–Boltzmann distribution governs the dipole orientation (and hence, the orientation of x). If the last collision occurred at time t_0, and the energy due to the dipole-field interaction is $\boldsymbol{\mu} \cdot \mathbf{a}$, we may obtain the mean polarization as:

$$\langle m \rangle = \frac{\mu \int_0^\pi \cos\psi \exp\left[\mu\, a \cos\psi \cos(\omega\, t_0)/kT\right] \sin\psi\, d\psi}{\int_0^\pi \exp\left[\mu\, a \cos(\omega\, t_0)/kT\right] \sin\psi\, d\psi}$$

$$\doteq \frac{\mu^2 a \cos(\omega\, t_0) \int_0^\pi \cos^2\psi \sin\psi\, d\psi}{kT \int_0^\pi \sin\psi\, d\psi} = \frac{\mu^2 a}{3\,kT}\, \Re\left\{e^{i\,\omega\,t}\right\} \tag{1.83}$$

when we have expanded the exponential in a MacLaurin series and only retained the term containing a. The polarization per cubic centimeter may be obtained by substituting Eq. (1.83) for x in Eq. (1.61) and multiplying by N, the number of molecules per cubic centimeter. The result is:

$$P = \frac{N a\, \mu^2}{3\,kT}\, \Re\left\{\frac{e^{i\omega t}}{1 + i\,\omega\,\tau}\right\} = a\, \Re\left\{(c - i\,b)\, e^{i\omega t}\right\} \tag{1.84}$$

The work done on the molecule by the radiation field will surely be given by the average value of $\mathbf{a} \cos\omega t \dfrac{dp}{dt}$. Finally, if we divide this work by the energy flow in the field, $\dfrac{c a^2}{8\pi}$, we should determine the energy taken out of the field by the molecule, or, simply, the absorption coefficient.

Now

$$a \cos(\omega t) \frac{dP}{dt} = E \cos\omega t \left[\frac{N a\, \mu^2}{3\,kT}\, \frac{\omega^2\,\tau \cos(\omega t) - \omega \sin\omega t}{1 + \omega^2\,\tau^2}\right]$$

where:

$$\frac{dP}{dt} = \frac{N a\, \mu^2}{3\,kT}\, \frac{\omega^2\,\tau \cos(\omega t) - \omega \sin\omega t}{1 + \omega^2\,\tau^2} \tag{1.85a}$$

and the average value is:

$$\left\langle E \cos(\omega t) \frac{dP}{dt} \right\rangle = \frac{N a^2 \mu^2}{6\,kT}\, \frac{\omega^2\,\tau}{1 + \omega^2\,\tau^2} \tag{1.85b}$$

As a result we obtain:

$$k = \frac{\omega}{c}\, \frac{4\pi N \mu^2}{3\,kT}\, \frac{\omega\,\tau}{1 + \omega^2 \tau^2} \tag{1.86}$$

Now we note that the same result could have been obtained if we had set:

$$k = \frac{4\pi b \omega}{c} \tag{1.87}$$

from Eq. (1.84). This tells us then that, from Eq. (1.82b)

$$k = \frac{4\pi N \omega e x''}{c} = \frac{2\pi N e^2}{m c} \left(\frac{\omega}{\omega_0}\right) \left[\frac{1/\tau}{(\omega - \omega_0)^2 + (1/\tau)^2} - \frac{1/\tau}{(\omega + \omega_0)^2 + (1/\tau)^2}\right] \tag{1.88}$$

One is not compelled to arrive at Eq. (1.88) in this manner, but it is slightly more simple than other possible methods. Let us return to a consideration of Eq. (1.78).

Eq. (1.78) is, according to our narrative, the first practical shape which was obtained for a spectral line broadened by the presence of other molecules. Its popularity continues to wax fair after the passage of a good many years, although its domain of applicability has been rather seriously restricted. A few remarks concerning this line shape may be in order.

First off, it is apparent that a symmetrical distribution of spectral intensity about the unperturbed line position is predicted by the equation. Such an intensity distribution does indeed hold true under certain physical situations which we shall consider in detail somewhat later.

The half-width is given by Eq. (1.75). Quite obviously the parameter governing this half-width is the mean inter-collision time. Lorentz himself found and reported in the same article that the mean time which one would obtain from kinetic theory showed no very reasonable relation to the mean time called for by Eq. (1.75) to explain the observed half-widths of the spectral lines which he investigated.*

Finally, the inverse square dependence of the intensity on the frequency, a characteristic of the simpler Interruption Theories, is to be remarked. This, like the symmetrical distribution itself, has been justified for some physical situations and annihilated for others, the result, of course, being a more nearly proper assignment of its region of applicability.

We shall see the Interruption Theories develop until any obvious connection between them and the work of Michelson and Lorentz is almost completely obscured. Still, these gentlemen's efforts remain at the base of the pyramid, and many years and the labor of many individuals have been required to attain the level of obfuscation of this early work which exists in today's more sophisticated broadening explanations.

1.10. The end of the early period

This phase of the development of spectral line shape began with a list of broadening factors, and we shall end it with a similar list, this one contributed

* Let us note here that τ will depend on the molecular diameters. We shall see that, in general, Interruption theories are repeatedly required to predict diameters much greater than kinetic theory diameters in order to yield broadening results comparable with experiment.

by Lord Rayleigh. Before doing so, however, let us mention another contribution which Rayleigh tucked away in the text of the same article.

"... Is there no distinction between encounters first of two sodium atoms and secondly of an atom say of nitrogen?"[172] Or one might simply ask if self-broadening — the broadening of the spectral line being caused by the same type of molecule as the emitter — is the same as foreign broadening — the broadening molecule is not the same type as the emitter. We shall see that this is a very pregnant question.

Rayleigh decided that the causes which underlie the broadening of spectral lines in emission "... may be considered under five heads, and it appears probable that the list is exhaustive:

"(1) The translatory motion of the radiating particles in the line of sight, operating in accordance with Doppler's principle.

"(2) A possible effect of the rotation of the particles.

"(3) Disturbance depending on collision with other particles either of the same of another kind.

"(4) Gradual dying down of the luminous vibrations as energy is radiated away.

"(5) Complications arising from the multiplicity of sources in the line of sight. Thus if the light from a flame be observed through a similar one, the increase of illumination near the edge of the spectrum line is not so great as towards the edge, in accordance with the principles laid down by Stewart and Kirchhoff, and the line is effectively widened."[172]

CHAPTER 2

INTERRUPTION BROADENING

OUR FIRST few words concerned themselves with how spectral lines are broadened, and we will recall that this broadening results from (1) the radiating away of energy, (2) the Doppler shift, and (3) the fields of neighboring molecules. We shall not concern ourselves with (1) for the reader may find the subject satisfactorily detailed in a number of sources. Lord Rayleigh's discussion of (2)* should be sufficient to our purposes here, so that we can turn our complete attention to a consideration of (3). In this and the next chapter we shall consider the two most general theories advanced to account for spectral line broadening by neighboring molecules. In the last four chapters these theories will find frequent application in our considerations of the specialized phenomena to which we devote ourselves there. The fact that we show both the Interruption and Statistical Theories as special cases of a more general theory should cause us no concern. There is a great deal of a phenomenological nature to be learned from these theories and their individual development, so that the mere fact that they in turn can be shown to be the offspring of a more advanced theory hardly negates their importance. Further, one still generally appeals to a particular one of the two when faced with the interpretation of an experiment or the prediction of a line contour.

The background for our Interruption considerations is, of course, to be found in the Michelson discussion of Chapter 1. Here we begin with Lenz's first advance beyond this initial phase which occurred in 1924, and trace the evolution of the theory through the generalized case which actually yields either this Interruption Theory or the Statistical Theory of the next chapter.

2.1. The Lenz appeal to correspondences

In 1924 Lenz[120] began what may be considered a post-quantum mechanical continuation of the work of Michelson and Lorentz on "pressure" broadening.* We begin, with Lenz, by considering two types of molecular interaction which lead to a broadening of the spectral line:

(1) An interaction of a type such that all energy is radiated, that is, none of the radiation energy is transformed into translational energy of the molecules.

(2) A collision of the second kind in which excitation energy is completely transformed into translational energy.

* See *supra*, Chap. 1.

In the quantum mechanical case of a radiating gas, a certain percentage of the atoms present undergo radiating transitions. For collisions of the same kind the remainder of the excited atoms lose their excitation energy not by a radiating transition, but by a collision in which the excitation energy is transformed into translational energy. On the other hand, a classical oscillator radiates a portion of its energy and loses the rest on collision. Lenz appeals to the Bohr correspondence principle to equate the overall amounts of radiated and heat-converted energy in the two cases. We may obtain the desired amount of heat transformed energy in the classical case by a suitable selection of molecular diameter. It is apparent that we could obtain this amount by adjusting the molecular diameter to yield the required collision frequency. Now we consider a collision to have occurred when two molecular centers have approached each other to within this adjusted molecular diameter. It may then be assumed that the wave train which is being emitted is cut off at the collision. Then, as in the Lorentz case, a broadening which is proportional to the density results.

These considerations were essentially based on type (2) interactions. Lenz then advanced the hypothesis that type (1) interactions would yield the same result. Let us now consider the type (1) interactions from the Lenz point of view.

The assumption is first made that outside a certain separation of the two molecules, the field of the broadener produces no perturbing effect on the upper level of the emitter. At this minimum separation, however, the energy level of the emitter is disturbed by a large amount. Let us call the occurrence of this closest approach an optical collision. We then take as the classical analog of this sudden state perturbation the sudden frequency perturbation of the classical oscillator. Then between two such collisions the oscillator will radiate undisturbed.

During the short time of the optical collision on the other hand — that is, during the time during which the broadener is separated from the emitter by the amount or less than the amount of this minimum separation — radiation of a very different frequency is emitted. We simply neglect this "undefined emission" during collision, and assume that radiation ceases at collision and commences again after collision. Thus, we obtain the same type situation as that with which we have already dealt. Again a wave trains of length dependent on the time between collisions is emitted. According to this theory, the force laws which govern the interaction between emitter and broadener are of no specific importance.

Let us then replace our radiating atom by a classical oscillator. An equivalent form of Eq. (1.6b) gives for the amplitude:

$$J(\nu) = \text{const} \int_{-\tau/2}^{\tau/2} e^{2\pi i(\nu_0 - \nu)t}\, dt = \text{const} \frac{\sin \pi(\nu_0 - \nu)\tau}{\pi(\nu_0 - \nu)} \tag{2.1}$$

The square of Eq. (2.1) would yield Eq. (1.12). If the square of this equation is averaged over all collision times, where the probability of a collision time between τ and $\tau + d\tau$ is given by $f(\tau)\, d\tau$ as in Eq. (1.56) the result is:

$$I(\nu) = \text{const} \int_0^\infty \frac{\sin^2 [\pi(\nu_0 - \nu)\, \tau']}{\pi^2 (\nu_0 - \nu)^2} d\tau' e^{-\tau'/\tau_0} = \text{const} \frac{(1/2\pi\tau)}{(\nu_0 - \nu)^2 + (1/2\pi\tau)^2} \tag{2.2}$$

Eq. (2.2) is, of course, equivalent to Eq. (1.78). The proportionality between the pressure and the half-width is apparent from Eq. (2.2) through the inverse proportionality of mean time between collisions, τ_0, and half-width.

2.2. The phase shift definition of a collision and half-width

Weisskopf first attacked the problem of Interruption Broadening by foreign gas atoms in 1932.[208] In this paper, he, too, considered the radiating molecule as replaced by a classical oscillator.

Let us assume that $\omega_0(t)$ is the natural frequency of this oscillator as a function of time. $\omega_0(t)$ is a constant during the so-called transit time — the time interval between two optical collisions. $\omega_0(t)$ changes, however, during what Weisskopf considered the very short time of the collision. Let us develop a slightly more general equation than we would need for this case.

The varying electric dipole moment* of the oscillator is for this case:

$$\mu(t) = A(t) \exp \left[i \int_0^t \omega_0(\tau)\, d\tau \right] \tag{2.3}$$

The Fourier integral of Eq. (2.3) is then:

$$\mu(t) = \int_{-\infty}^{+\infty} J(\omega)\, e^{i\omega t}\, d\omega \tag{2.4}$$

so that the amplitude is given by

$$J(\omega) = \int_{-\infty}^{+\infty} M(t)\, e^{-i\omega t}\, dt = \int_{-0}^{+\infty} A(t) \exp \left\{ i \left[\left[\int_0^t \omega_0(\tau)\, d\tau - \omega t \right] \right] \right\} dt \tag{2.5a}$$

Weisskopf gave this equation as:

$$J(\omega) = \text{const} \int \exp \left[i \int_0^t \omega_0(\tau)\, d\tau - i\,\omega\, t \right] dt \tag{2.5b}$$

The intensity distribution in the spectral line would then be given by the absolute square of Eq. (2.5b).†

An interaction between broadener and emitter exists such that the frequency of the oscillator is shifted by $\Delta(r)$ where r is the emitter–broadener separation. It then follows that the phase shift in the undisturbed frequency during the optical collision is given by:

$$\eta_0 = \int_0^t \Delta(r)\, dt \tag{2.6}$$

Here t, the upper limit of the integral, is the time of collision.

* In Chap. 1 we considered the vibration of the photoelectron (Lorentz theory). Essentially, this amounts to the same thing. When this electron vibrates, it causes the noted variation in the dipole moment.

† As mentioned in Chap. 1, this could then be normalized at any time to the intensity over the line.

When η_0 has attained some value or other, an optical collision is assumed to have taken place, that is, we consider one wave train as terminated, and another completely independent wave train as initiated. Weisskopf arbitrarily assumes a value of $\eta_0 \sim 1$ for the phase shift required for the definition of an optical collision.

In order to justify qualitatively this assumption, let ϱ be the separation at closest approach of the two molecules and v their relative velocity and write:

$$r = \sqrt{|\mathbf{r}|^2} = \sqrt{\langle v^2 \rangle t^2 + \varrho^2 + 2\,\boldsymbol{\varrho} \cdot \mathbf{v}\,t} = \sqrt{\langle v^2 \rangle t^2 + \varrho^2}$$

since $\boldsymbol{\varrho}$ is perpendicular to $\mathbf{v}$.

Thus, Eq. (2.6) becomes:

$$\eta_0 = \int_{-\infty}^{+\infty} \Delta\left(\sqrt{v^2 t^2 + \varrho^2}\right) \mathrm{d}t \tag{2.7}$$

Weisskopf used the mean relative velocity squared here $\langle v \rangle^2$ while Margenau and Watson[218] used the mean square relative velocity $\langle v^2 \rangle$. As is well known, these two quantities differ by a factor $3\pi/8$. The reader may with obvious justification make the choice *chacun à son gout.*

Now Weisskopf assumed an interaction $\Delta(r) = C/r^6$ where C is given by $\alpha_1 \alpha_2 V$. α_1 and α_2 are the polarizabilities of the emitter and the broadener, respectively. From the Stark effect of the Na *D*-lines he took $\alpha_1 = 9.5 \times 10^{-23}$, and he assumed $\alpha_2 = 10^{-24}$. Finally, V was taken as 4 volts. If we let $x = \dfrac{vt}{\varrho}$ we may rewrite Eq. (2.7) as:

$$\eta_0 = \int_{-\infty}^{+\infty} \frac{C\,\mathrm{d}t}{\varrho^6 \left(\dfrac{v^2 t^2}{\varrho^2} + 1\right)^3} = \frac{C}{\varrho^5 v} \int_{-\infty}^{+\infty} \frac{\mathrm{d}x}{(x^2+1)^3} \sim 1 \tag{2.8)*}$$

* For our future utilization let us write this as:

$$\eta = 2\pi\,\eta_0 = \frac{2\pi C}{v\,\varrho^{n-1}}\,C_n \tag{2.8a'}$$

where

$$C_n = \int_{-\pi/2}^{\pi/2} \cos^{n-2}\varphi\,\mathrm{d}\varphi = \sqrt{\pi}\,\frac{\Gamma\left(\dfrac{n-1}{2}\right)}{\Gamma\left(\dfrac{n}{2}\right)}$$

In these equations φ is the angle between the distance of closest approach ϱ and the emitter-perturber direction. (See, for example, Fig. (2.1)). From Eq. (2.8a′) we may write:

$$\varrho = \left(\frac{2\pi C c_n}{v\,\eta}\right)^{1/(n-1)} \tag{2.8b'}$$

so that γ becomes:

$$\gamma = \frac{2}{\tau} = 2\pi \left(\frac{2\pi C c_n}{v\,\eta}\right)^{2/(n-1)} v N$$

$$= C^{\frac{2}{(n-1)}}\, v^{\frac{n-3}{n-1}}\, N\,(2\pi)^{\frac{n+1}{n-1}}\, c_n^{\frac{2}{n-1}}\, \eta^{-\frac{2}{n-1}}$$

where the assumption that $\eta_0 = 1$ determines an optical collision has been introduced. From Eq. (2.8) we may obtain for ϱ:

$$\varrho = J\left(\frac{C}{v}\right)^{1/5} \tag{2.9}†$$

where

$$J = \left(\int_{-\infty}^{+\infty} \frac{dx}{(1+x^2)^3}\right)^{1/5} = \left(\frac{3\pi}{8}\right)^{1/5} \approx 1$$

Finally, for a temperature of 500 and for values of the constants as given above Weisskopf obtained an optical collision diameter $\varrho = 6.8$ Å for the Na D-lines for $\eta_0 = 1$ which he considered a "plausible value". Actually, since we cannot strictly specify a phase shift which rigidly determines an optical collision, $\eta_0 = 1$ is probably as good a value as any.

The shape of the broadened spectral line is still given by Eq. (2.2). Let us now consider the physical theory implicit in the development.

2.3. The physical phenomenon implied by the Lenz–Weisskopf result

For conceptual clarity, let us refer to Fig. 2.1. In this figure we have represented the emitter path by the arrow, and the perturbers by the small circles. At point "b" on the emitter path the radiation of a "new" wave train is initiated, since under the Michelson–Lenz–Weisskopf theory a terminating optical collision has just been undergone by the emitter. As the emitter travels along its path toward point "c" it is perturbed by the surrounding atoms according to some potential interaction "law" of the form Cr^{-n} and small phase changes occur. The "A" perturbers in Fig. 2.1, however, are too distant to cause an optical collision in the Weisskopf sense (phase change of unity) so we ignore them. The effects of ignoring this we shall discuss later. When we arrive at the point "c" the second "B" perturber is now within the range requisite for inducing a phase change of unity. At this point, then, we cut off the emitted wave train and perform a Fourier analysis of the radiation emitted between points "b" and "c". Thus, we can see that, among other possible omissions, we have neglected (a) the effect of distant collisions and (b) the effect of phase changes greater than unity. Let us consider Omission (b) and defer our consideration of Omission (a) until a more *à propos* time.

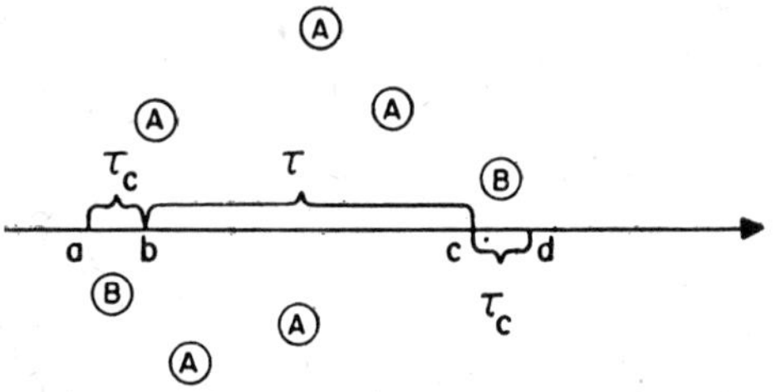

FIG. 2.1. A model of the physical conceptions inherent in the Weisskopf Interruption Theory

† In frequency units ν (as opposed to angular frequency units ω):

$$\varrho = \left(\frac{3\pi^2}{4}\frac{C}{v}\right)^{1/5} \tag{2.9'}$$

When we fail to consider phase changes greater than unity, as we implicitly do in the Weisskopf theory, we essentially neglect values of ω farther from line center than some value ω_l in the line wing. In essence, then, we can only expect legitimate application of the theory near line center.

It is well that we do so, for we shall see that the physical basis of the Interruption Theory is inapplicable to the so-called "line wing". On this basis, it is apparent the limit of applicability of the interruption distribution in the line wing will be given by $\Delta\omega = 2\pi C r^{-n}$ where $r = \varrho$. Then from Eq. (2.8a′) we obtain for the limit on $\Delta\omega$:

$$\Delta\omega \leq \left(\frac{v^n}{2\pi C c_n{}^n}\right)^{1/n-1} ; \quad \eta = 1 \tag{2.10}$$

From a slightly different viewpoint we shall later obtain about the same restriction.

Let us now return to the continued development of the Interruption Theory.

2.4. Line shift inclusion

Our line shape is given by Eq. (2.2), and a consideration of this equation shows that two of the important effects on a spectral line which arise out of the presence of foreign gas atoms are notably absent. These missing effects are (1) the shift of the line intensity maximum and (2) any asymmetry in the broadened line. An attempt to refine the theory so as to include these effects was first made by Lenz[121] in 1933. Although these efforts were not successful in the sense that the results of his work could be applied to a physical situation, they were certainly of importance as trail blazers. We shall see that Lenz introduced the now familiar replacement of the statistical time mean by the statistical mean, and that he may be considered the father of the "Correlation Function",* that currently popular interaction description. Before we consider Lenz's development, however, let us diverge into a few brief remarks about Interruption Theories.

The basis of the Interruption Theory concept is the idea of a sudden event, namely, the interaction of an emitter and a broadener, which has an instantaneous effect on the radiation being emitted. Basically then, if we are to expect this physical model to be reasonably descriptive, we must look to those situations where the broadeners are moving very rapidly or where, for some other reason, the time of collision is short enough as to be considered instantaneous. At this point we need go no further into what we shall discover the restrictions on the true Interruption Theories to be, but the requirement on the time of collision is important, since it is by his treatment of this time that Lenz hoped to introduce shift and asymmetry. Thus, although Lenz used the Fourier tool, his theory was not basically an Interruption one. The fact that it was necessary

* Certainly not to be compared with the Parameter of Pristine Prodigality, a term coined by Rev. Deuteronomy Webster who held forth in Boston from 1646 to 1649.

for him to introduce restrictions which effectively rendered it Interruption is hardly germane to this point.

Let us suppose that τ_0 is the mean time between collisions and, further, that 2τ is the time of collision. We take $f_0(\mathrm{d})$ to be the frequency change per unit time during the time of collision 2τ where ϱ is the optical collision diameter. As did Weisskopf, Lenz defined this diameter as the molecular separation at which the phase of "the classically substituted oscillators" has changed by one. Then the change of phase during collision is:

$$\eta = 2\tau f_0(\mathrm{d}) \tag{2.11}$$

The oscillator frequency is given by:

$$\omega_0(t) = \omega_0 + f(t) \tag{2.12}$$

We may express the amplitude as:

$$x = x_0 \, \mathrm{e}^{i\omega_0 t + i\eta(t)} \tag{2.13}$$

where:

$$\eta(t) = \int^{t} f(t)\,\mathrm{d}t$$

The amplitude in a Fourier expansion of Eq. (2.65) we may find from an analogy to Eq. (2.5b) as:

$$J(\omega) = \int_{-\infty}^{+\infty} \mathrm{e}^{i(\omega_0 - \omega)t + i\eta(t)}\,\mathrm{d}t \tag{2.14}$$

From Eq. (2.14) the intensity corresponding to the frequency ω is:

$$I(\omega) = |J(\omega)|^2 = \int_{-\infty}^{+\infty} \int_{-\infty}^{+\infty} \mathrm{e}^{i(\omega_0 - \omega)(t' - t) + i[\eta(t') - \eta(t)]}\,\mathrm{d}t\,\mathrm{d}t' \tag{2.15}$$

The variable is next changed to $T = t' - t$ where still $t = t$. Thus, Eq. (2.15) becomes:

$$I(\omega) = \int_{-\infty}^{+\infty} \mathrm{d}T\,\mathrm{e}^{i(\omega_0 - \omega)T} \int_{-\infty}^{+\infty} \mathrm{e}^{i\Delta\vartheta(t, T)}\,\mathrm{d}t \tag{2.16a}$$

where

$$\Delta\eta(t, T) = \eta(t + T) - \eta(t) \tag{2.16b}$$

In Eq. (2.16a) Lenz introduced the "Correlation Function", a name to be applied much later to the time integral over the interaction function. Actually this second integral in Eq. (2.16a) does provide the entire description of the broadening interaction and thus constitutes the crux of the whole consideration.

Now let $\mathbf{r}$ be the position vector of the oscillator-emitter and let $\mathbf{r}_k$ be the position vector of the k-th disturber. Lenz considered all disturbers the same, and the frequency disturbance due to the k-th disturber as $\varphi(\mathbf{r} - \mathbf{r}_k)$, simply some function of their separation. Then he let the perturbations due to various broadeners be additive so that:

$$f(t) = \sum_k \varphi(\mathbf{r} - \mathbf{r}_k) \tag{2.17}$$

We shall encounter this "binary" interaction approximation repeatedly, and, of course, the reason for its inclusion is the practical impossibility of its exclusion. Obviously, complete correctness would require one big interaction between the emitter and all the broadeners to include interactions between the broadeners themselves. The difficulties are obvious. Fortunately the additive, binary interaction is, in general, a rather good approximation.

The rectilinear velocities of the emitter and the k-th disturber are given by $\mathbf{v}$ and $\mathbf{v}_k$ respectively. The position vectors at time $t = 0$ are taken as $\mathbf{r}^a$ and $\mathbf{r}_k^a$. Finally, Lenz assumed that close or central collisions are not of importance compared to those in the neighborhood of the optical collision diameter. By the above definitions:

$$\varphi(\mathbf{r} - \mathbf{r}_k) = \varphi[(\mathbf{r}^a - \mathbf{r}_k^a) + (\mathbf{v} - \mathbf{v}_k)\, t] \tag{2.18}$$

so that, by Eqs. (2.16b), (2.17), and (2.18):

$$\Delta\eta\,(0, T) = \sum \Delta\eta_k\,(0, T) = \sum \int_0^T \varphi[\mathbf{r}^a - \mathbf{r}_k^a + (\mathbf{v} - \mathbf{v}_k)\, t]\, \mathrm{d}t \tag{2.19}$$

Lenz assumed that the second integral in Eq. (2.16a) is the statistical time mean value of $\mathrm{e}^{i\Delta\eta(t, T)}$. This assumption provides a device for the evaluation of the integral.

Let us digress for a moment from the matter at hand in order to justify Lenz's next step. If we are considering the behavior of a collection of N_1 particles, we may first conveniently set up a $2N_1$ dimensional phase space, the coordinates of which are the $2N_1$ position coordinates and conjugate momenta of the N_1 particles. Any point in this phase space will now represent a certain momentum and spatial configuration for the N_1-particle system, and this point is generally referred to as a representative point. The behavior of this representative point conveniently tells us the behavior of our system, and Boltzmann's impressively titled "ergodic hypothesis" tells us what we may expect this behavior to be. The ergodic hypothesis* states that the representative point will successively pass through all points in phase space compatible with the total energy of the system. We cannot prove from this hypothesis, but we can certainly infer from it, that the statistical time mean value of a quantity will be the same as the statistical mean value of the quantity. Thus, the second integral in Eq. (2.16a) which Lenz has assumed to be the time mean of the integrand may be evaluated by finding the mean of the integrand.

If we take u_k, v_k, w_k as our components of $\mathbf{v}_k$ and x_k^a, y_k^a, z_k^a as our components of $\mathbf{r}_k^a$, the volume element of the phase space of our broadeners will be:

$$\mathrm{d}V = \prod_k \mathrm{d}x_k^a\, \mathrm{d}y_k^a\, \mathrm{d}z_k^a\, \mathrm{d}u_k\, \mathrm{d}v_k\, \mathrm{d}w_k$$

Now let us assume a Maxwell–Boltzmann distribution, and we then obtain:

$$\int_{-\infty}^{+\infty} \mathrm{e}^{i\Delta\eta}\, \mathrm{d}t = \langle \mathrm{e}^{i\Delta\eta} \rangle = \frac{\int \mathrm{e}^{i\Delta\eta(0, T)}\, \mathrm{e}^{E/kT}\, \mathrm{d}V}{\int \mathrm{e}^{E/kT}\, \mathrm{d}V} \tag{2.20}$$

* To be quite technically correct, we should probably introduce the so-called quasi-ergodic hypothesis, but the ergodic one will be sufficient to our purposes.

We have made the assumption that all broadeners are the same. On this basis, Lenz wrote Eq. (2.20) as the product of N_1 identical integrals. Although Lenz does not mention it, we should always keep in mind that all such transformations to integral products † are not wholly justified by the assumption of identical particles. Previous occupancy of certain portions of space by other broadeners would still limit the integration for any one broadener to the unoccupied portions of space. Thus, in cases of this kind, an additional assumption must be made to the effect that these previously occupied parts of space need not be excluded. ‡

We let:

$$E = \sum \frac{m_k}{2} (\mathbf{v}_k \cdot \mathbf{v}_k) \qquad (2.21a)^{**}$$

$$|\mathbf{v}| = c; \quad |\mathbf{v}_k| = c_k; \quad c_0^2 = \frac{2kT}{m_k}; \quad m_i = m_k = m_0 \qquad (2.21b)$$

Utilizing Eqs. (2.21) the k-th identical integral becomes, after integration of the denominator in Eq. (2.20):

$$\frac{1}{\pi^{3/2} V c_0^3} \int_{-\infty}^{+\infty} e^{-c_k^2/c_0^2 + i\Delta\eta_k(0,T)} \, dx_k \, dy_k \, dz_k \, du_k \, dv_k \, dw_k \qquad (2.22)$$

Next it is assumed that $\Delta\eta_k(0, T)$ is very small except in the immediate neighborhood of the emitter. Now let:

$$\int e^{i\Delta\eta_k(0,T)} \, dx_k \, dy_k \, dz_k = V + \Upsilon_k \qquad (2.23)$$

where V is the volume in the configuration space of the k-th emitter, and Υ_k is the integral of $e^{i\Delta\eta_k(0,T)}$ over the spheres of effective $\Delta\eta_k(0, T)$. Utilizing this in Eq. (4.22), we obtain:

$$1 + \frac{1}{\pi^{3/2} V c_0^3} \int_{-\infty}^{+\infty} \Upsilon_k e^{-c_k^2/c_0^2} \, du_k \, dv_k \, dw_k = 1 + \frac{N\Upsilon}{N_1} \qquad (2.24a)$$

where:

$$\Upsilon = \frac{1}{\pi^{3/2} c_0^3} \int_{-\infty}^{+\infty} [e^{i\Delta\eta_k(0,T)} - 1] \, e^{-c_k^2/c_0^2} \, dx_k \cdots dw_k \qquad (2.24b)$$

Thus, when Eq. (2.20) is considered as the product of N_1 identical integrals of the form Eq. (2.24), the result is:

$$\int_{-\infty}^{+\infty} e^{i\Delta\eta} \, dt = \int_{-\infty}^{+\infty} e^{N\Upsilon - \frac{mc^2}{m_0 c_0^2}} \, du \, dv \, dw \qquad (2.25)$$

In Eq. (2.25) we have finally integrated over the velocity space of the emitter after having introduced the Boltzmann factor $e^{-mc/m_0 c_0^2}$ for the emitter. The

† Notice that the fact that $\Delta\eta$ is a sum over the various broadeners means the integral is actually a product of integrals, each concerned with an individual broadener. If all broadeners are the same the result follows.

‡ This is also implied by Eq. (2.17), that is, by neglecting interactions between broadeners.

** Kinetic energy is here taken as total energy.

approximation: $1 + N\Upsilon/N_1 = e^{N\Upsilon/N_1}$ has been used, $1 + N\Upsilon/N_1$ being the first two terms in the power series expansion of $e^{N\Upsilon/N_1}$. In addition m is the mass of the emitter.

The integral of Eq. (2.25) may only be evaluated by graphical methods. Lenz considered certain special cases.

The limits on Eq. (2.19) may as well be $-1/2\,T$ and $+1/2\,T$. Then:

$$\Delta\eta_k(0, -T) = -\Delta\eta_k(0, T) \tag{2.26}$$

from Eqs. (2.19) and (2.23). Now, in order to carry out the integration in Eq. (2.23), let the x-axis of our spatial coordinate system correspond to $\mathbf{v} - \mathbf{v}_k$, and define quantities as follows:

$$\varrho^2 = (y - y_k)^2 + (z - z_k)^2 \tag{2.27a}$$

$$\varrho\beta = x^a - x^a_k + |\mathbf{v} - \mathbf{v}_k|\,\alpha \tag{2.27b}$$

$$\varrho\xi = x^a - x_k^a \tag{2.27c}$$

$$\varrho n = |\mathbf{v} - \mathbf{v}_k|\,T \tag{2.27d}$$

Thus, $r^2 = \varrho^2(1 + \beta^2)$ where $\mathbf{r}$ is the emitter-broadener separation, β is the cotangent of the angle between $\mathbf{r}$ and the x-axis, while ξ is the tangent of this angle.

We assume an interaction potential function of the form $\varphi = -ar^{-p}$. Then by virtue of Eqs. (2.27), Eq. (2.19) becomes:

$$\left.\begin{aligned}\Delta\eta_k(0, T) &= -\int_{-T/2}^{T/2} ar^{-p}\,\mathrm{d}\alpha = -a\int_{\xi-n/2}^{\xi+n/2} [\varrho^2(1+\beta^2)]^{-p/2}\left[\frac{\varrho\,\mathrm{d}\beta}{|\mathbf{v} - \mathbf{v}_k|}\right] \\ &= -g\int_{\xi-n/2}^{\xi+n/2} \frac{\mathrm{d}\beta}{(1+\beta^2)^{p/2}}\end{aligned}\right\} \tag{2.28a}$$

where

$$g = \frac{1}{\varrho^{p-1}}\,\frac{a}{|\mathbf{v} - \mathbf{v}_k|} \tag{2.28b}$$

On the basis of the high p normally present the simplification may be made:

$$\psi(\zeta) = \int_{-\infty}^{\zeta} \frac{\mathrm{d}\beta}{(1+\beta^2)^{p/2}} \tag{2.29}$$

Now Eq. (2.29) is very small for ζ less than around -1, a constant for ζ greater than around $+1$, and nearly rectilinear for intermediate ζ. We then take for $\psi(\zeta)$:

$$\begin{aligned} &0 \text{ for } \zeta < -\varkappa \\ &\chi(\zeta + \varkappa) \text{ for } -\varkappa \leq \zeta s + \varkappa \\ &2\varkappa\mu \text{ for } \zeta > \varkappa \end{aligned} \tag{2.30}$$

Lenz found by graphical integration that for $6 \leq p \leq 10$, $\varkappa$ has a value of 0.75 and $\varkappa\mu$ ranges between 0.57 and 0.44.

We may take $dx_k\, dy_k\, dz_k = 2\pi\varrho^2\, d\varrho\, d\xi$ so that from Eq. (2.23):

$$\Upsilon_k = -\int_0^{\infty} 2\pi\varrho^2\, d\varrho \int_{-\infty}^{+\infty} [e^{i\Delta\eta_k(0,T)} - 1]\, d\xi \tag{2.31}$$

In integrating over ξ we need only consider $n > 0$ according to Eq. (2.26). Also, due to Eqs. (2.29) and (2.30) the integral reduces to one in the two regions $\xi + n/2 \geq -\varkappa$ and $\xi - n/2 \leq +\varkappa$. Now integration over ξ gives for the two cases shown:

$$\text{for } n \geq 2\varkappa\text{:}\quad 2\varkappa + n + (2\varkappa - n)\, e^{-2ig\chi\varkappa} + \frac{2}{ig\chi}(e^{-2ig\chi\varkappa} - 1) \tag{2.32a}$$

$$\text{for } n \leq 2\varkappa\text{:}\quad 2\varkappa + n - (2\varkappa - n)\, e^{-ig\chi\eta} + \frac{2}{ig\chi}(e^{-2ig\chi\eta} - 1) \tag{2.32b}$$

In attempting to carry out the remaining integral, Lenz divided the integrand into two parts corresponding to Eqs. (2.32a) and (2.32b).* Thus:

$$\Upsilon_k = \Upsilon_k^{(1)} + \Upsilon_k^{(2)} \tag{2.33}$$

Let ϱ be the optical collision diameter. From Eqs. (2.28b), (2.21b), and (2.30) Lenz expressed ϱ as:

$$\varrho_0 = 2\varkappa\left(\frac{a}{c_0}\right)^{1/p-1} \tag{2.34}†$$

which is arrived at by assuming a phase change of 1 (or 2π) to have occurred at ϱ separation. Now let:

$$\Lambda = \frac{c_0 T}{\varrho_0}; \qquad \sigma = \frac{|\mathbf{v} - \mathbf{v}_k|}{c_0} \tag{2.35}$$

At the boundary between the regions of Eqs. (2.32), from Eqs. (2.27d) and (2.35) $2\varkappa\varrho_{\text{lim}} = \sigma\Lambda\varrho$. Thus, we get for the value of Eq. (2.28a) at the boundary:

$$\gamma_0 = 2\varkappa\chi g_{\text{lim}} = \frac{2\varkappa\chi}{\sigma^p \Lambda^{p-1}} \tag{2.36}$$

Due to the high powers present in the denominator of Eq. (2.36) γ_0 is either $\ll 1$ or $\gg 1$ except in a small transition region.

From Eqs. (2.27d), (2.28b), and (2.35):

$$n = \sigma\Lambda\frac{\varrho_0}{\varrho}; \qquad g = \frac{1}{\sigma}\left(\frac{\varrho_0}{2\varkappa\varrho}\right)^{p-1} \tag{2.37a}$$

and let:

$$\gamma_1 = 2\varkappa\chi g = \frac{2\varkappa\chi}{\sigma}\left(\frac{\varrho_0}{2\varkappa\varrho}\right)^{p-1}; \qquad \gamma_2 = \chi g n = 2\varkappa\chi\Lambda\left(\frac{\varrho_0}{2\varkappa\varrho}\right)^{p} \tag{2.37b}$$

Further:

$$2\pi\left(\frac{2\varkappa}{\varrho_0}\right)^3 \varrho^2\, d\varrho = -\frac{2\pi}{p-1}\left(\frac{2\varkappa\chi}{\sigma\gamma_1}\right)^{3/p-1}\frac{d\gamma_1}{\gamma_1} = -\frac{2\pi}{p}\left(\frac{2\varkappa\chi\Lambda}{\gamma_2}\right)\frac{d\gamma_2}{\gamma_2}. \tag{2.37c}$$

* η goes from ∞ to 0 as ϱ goes from 0 to ∞.

† It would appear that we should have $(2\varkappa)^{1/p-1}$ here instead of $2\varkappa$. We simply use the Lenz result above, however.

We may thus introduce Eqs. (2.37) into Eqs. (2.32) and then introduce the result into Eq. (2.31) to obtain:

$$\frac{\Upsilon_k}{\varrho_0^3} = \varphi_1(\gamma_0)\,\sigma^{-3/(p-1)} + \psi_1(\gamma_0)\,\Lambda\sigma^{1-2/(p-1)} \tag{2.38a}$$

$$\frac{\Upsilon_k}{\varrho_0^3} = \varphi_2(\gamma_0)\,\Lambda^{3/p} + \psi_2(\gamma_0)\,\sigma\Lambda^{1+2/p} \tag{2.39a}$$

$$\varphi_1(\gamma_0) = -\frac{2\pi}{p-1}\frac{(2\varkappa\chi)^{3/(p-1)}}{(2\varkappa)^2}\int_{\gamma_0}^{+\infty}(1-e^{-i\gamma_1})\frac{d\gamma_1}{\gamma_1^{1+3/(p-1)}} \tag{2.38b}$$

$$\psi_1(\gamma_0) = -\frac{2\pi}{p-1}\frac{(2\varkappa\chi)^{2/(p-1)}}{(2\varkappa)^2}\int_{\gamma_0}^{+\infty}(1-e^{-i\gamma_1})\frac{d\gamma_1}{\gamma_1^{1+2/(p-1)}} \tag{2.38c}$$

$$\varphi_2(\gamma_0) = -\frac{2\pi}{p}\frac{(2\varkappa\chi)^{3/p}}{(2\varkappa)^2}\int_0^{\gamma_0}(1-e^{-i\gamma_2})\frac{d\gamma_2}{\gamma^{1+3/p}} \tag{2.38b}$$

$$\psi_2(\gamma_0) = -\frac{2\pi}{p}\frac{(2\varkappa\chi)^{2/p}}{(2\varkappa)^2}\int_0^{\gamma_0}\left[(1+e^{-i\gamma_2})+\frac{2}{i\gamma_2}(e^{-i\gamma_2}-1)\right]\frac{d\gamma_2}{\gamma_2^{1+2/p}} \tag{2.38c}$$

An inspection of the above integrals indicates that they are properly convergent. If we now let $z = \sigma\Lambda^{1-1/p}$ then, since $z\varkappa\chi \sim 1$ (see Eq. (2.30)) $\Upsilon_0 \sim z^{-p}$ according to Eq. (2.36). If σ, Λ, and, as a consequence, z take values from 0 to ∞ then γ_0 is obviously $\ll 1$ or $\gg 1$ except right around $z = 1$. In the case $\gamma_0 \ll 1$, it is apparent that we may neglect Eq. (2.39a) compared to Eq. (2.38a), and in the case $\gamma_0 \gg 1$ the reverse is true. Thus, there results:

$$\frac{\Upsilon_k^{(1)}}{\varrho_0^3} = A_1\,\sigma^{-3/(p-1)} + B_1\,\Lambda\sigma^{1-2/(p-1)} \quad \text{for } \sigma\Lambda^{1-1/p} > 1 \tag{2.39a}$$

$$\frac{\Upsilon_k^{(2)}}{\varrho_0^3} = A_2\,\Lambda^{3/p} + B_2\,\sigma\Lambda^{1+2/p} \quad \text{for } \sigma\Lambda^{1-1/p} < 1 \tag{2.39b}$$

where:

$$A_1 = \varphi_1(0);\quad B_1 = \psi_1(0);\quad A_2 = \varphi_2(\infty);\quad B_2 = \psi_2(\infty) \tag{2.39c}$$

Eqs. (2.39a) and (2.39b) are not valid near the indicated limits, and in addition "cumbersome expressions" must be introduced for the transition region. These expressions are neglected as "keine Rolle spielen".

This essentially completes the evaluation of Υ_k. Now it will be necessary to integrate over T in Eq. (2.16a) and over velocity space in the transformed version of the second integral in Eq. (2.16a) in order to obtain the intensity distribution. Integration over T infers integration over Λ in the equations for Υ_k.*

Lenz established a polar coordinate origin at the endpoint of $\mathbf{v}$. We let the angle between $\mathbf{v}$ and $\mathbf{v}-\mathbf{v}_k$ be φ and take this angle as our polar angle. Then

* See *supra*, Eq. (2.35).

we take $\mathbf{v} - \mathbf{v}_k$ as the radial coordinate and ϑ as the azimuthal coordinate. Our volume element in the velocity space of the k-th emitter becomes:

$$du_k\, dv_k\, dw_k = -(|\mathbf{v} - \mathbf{v}_k|)^2\, d(|\mathbf{v} - \mathbf{v}_k|)\, d\vartheta\, d(\cos\varphi) = -2\pi c_0\, \sigma^2\, d\sigma\, d(\cos\varphi) \tag{2.41}$$

where, since our integral will possess azimuthal symmetry, we have integrated over $d\vartheta$, and where Eq. (2.35) has been used for the substitution for $\mathbf{v} - \mathbf{v}_k$. Now let us apply the law of cosines to the vector triangle consisting of $\mathbf{v}$, $\mathbf{v}_k$, and $\mathbf{v} - \mathbf{v}_k$ where the absolute values of the three vectors are given by Eqs. (2.21b) and (2.35).

$$c_k^2 = c^2 + c_0^2\, \sigma^2 - 2cc_0\, \sigma \cos\varphi \tag{2.42}$$

or

$$\left(\frac{c_k^2}{c_0^2}\right) = \sigma^2 + \sigma_0^2 - 2\sigma\sigma_0 \cos\varphi; \quad \sigma_0 = \frac{c}{c_0} \tag{2.42}$$

We next substitute Eqs. (2.41) and (2.42) into Eq. (2.24a) and integrate over $\cos\varphi$ from -1 to $+1$ to obtain:

$$\Upsilon = \frac{1}{\sqrt{\pi}\,\sigma_0} \int_0^\infty [e^{-(\sigma-\sigma_0)^2} - e^{-(\sigma+\sigma_0)^2}]\, \Upsilon_k(\sigma)\, \sigma\, d\sigma \tag{2.43}$$

Now the following function is formed:

$$F_j(\sigma_0) = \frac{1}{\sqrt{\pi}\,\sigma_0} \int^\infty [e^{-(\sigma-\sigma_0)^2} - e^{-(\sigma+\sigma_0)^2}]\, \sigma^{j+1}\, d\sigma \tag{2.44}$$

When Eqs. (2.40) are utilized in Eq. (2.43) integrals of the form of Eq. (2.44) will result, and the relevant values of j will be: $j = -\dfrac{3}{p-1},\ 0,\ 1 - \dfrac{2}{p-1},\ 1.$ Since we may expect large p values, j is always between -1 and 1 in value. In order to find $F_j(0)$, apply l'Hospital's rule to Eq. (2.44) to obtain:

$$\left.\begin{aligned} F_j(\sigma_0) &= \lim_{\sigma_0 \to 0} \left\{ \frac{1}{\sqrt{\pi}} \int^\infty [2(\sigma-\sigma_0)\, e^{-(\sigma-\sigma_0)^2} + 2(\sigma+\sigma_0)\, e^{-(\sigma+\sigma_0)^2}]\, \sigma^{j+1}\, d\sigma \right\} \\ &= \frac{4}{\sqrt{\pi}} \int_0^\infty e^{-\sigma^2}\, \sigma^{j+2}\, d\sigma = \frac{2}{\sqrt{\pi}}\, \Gamma\left(1 + \frac{j+1}{2}\right) \end{aligned}\right\} \tag{2.45a}$$

Eq. (2.45) holds then for $\sigma_0 = 0$ which in turn means that $c = 0$ and we consider the disturbers at rest. Now for the p values which may arise $F_j(0)$ has values between 1.13 and 1. The other limiting case of $F_j(\sigma_0)$ arises for $\sigma_0 \gg 1$. In this case, the second term in Eq. (2.44) drops out as small. The integral only has appreciable value near $\sigma = \sigma_0$ so that:

$$F_j(\sigma_0) = \sigma_0^j; \quad \sigma_0 \gg 1 \tag{2.45b}$$

where $\sigma_0 = 4$ is large enough.

Lenz allowed Eq. (2.45a) to suffice, and, further, on the assumption of low gas densities, he neglected the region of Eq. (2.45b) and simply used Eq. (2.39a).

We shall not show this, but it might be mentioned here that at pressures of more than one atmosphere Eq. (2.45b) must be utilized. Now from Eqs. (2.39b) and (2.39c) we may obtain:

$$N\Upsilon = N\varrho_0^3[A_1 F_{j_1}(\sigma_0) + \Lambda B_1 F_{j_2}(\sigma_0)] \tag{2.46a}$$

$$j_1 = -\frac{3}{p-1}; \quad j_2 = 1 - \frac{2}{p-1} \tag{2.46b}$$

The Doppler effect is next introduced by replacing ω, the natural frequency of our substituted oscillator, by $\omega_c = \omega_0(1 - u/c_2)$, the frequency shifted by an emitter velocity of u in the x-direction.*

Eqs. (2.46a) and (2.25) yield for Eq. (2.16a) the following:

$$I(\omega) = \Re\left\{\int_0^\infty e^{i\frac{\Lambda}{l} + N\Upsilon - \frac{m}{m_0}\sigma_0^2} du\, dv\, dw\, d\Lambda\right\} \tag{2.47}$$

The real part has been taken here, since by Eq. (2.26) we may take the real part, and then take as lower integral limit zero instead of minus infinity. Also in Eq. (2.47) $1/l = (\omega_c - \omega)\varrho_0/c_0 = (\omega_c - \omega)T_0$, m_0 is the mass of a broadener, and m is the mass of the emitter. Integration has been taken over Λ which replaces T according to Eq. (2.35). Integration of Eq. (2.47) may be carried out over Λ if Eq. (2.46a) is kept in mind. The result is:

$$I(\omega) = \Re\left\{\int_0^\infty e^{-\frac{m}{m_0}\sigma_0^2} du\, dv\, dw\, G(\sigma)\right\} \tag{2.48a}$$

where:

$$G(\sigma_0) = e^{N\varrho_0^3 A_1 F_2 j_1(\sigma_0)}/i(\omega_c - \omega)T_0 + N\varrho_0^3 B_1 F(\sigma_0) \tag{2.48b}$$

The real part of B_1 is negative. Lenz considered the special case of mercury where $\sigma_0 < 1$ so that in Eq. (2.48a) $J(\omega) \doteq \Re[G(0)]$. In addition, for the case $m_0/m \sim 1$, the integrand of Eq. (2.48a) may be regarded as a "Zackenfunction" whose only appreciable value occurs for $\sigma_0 \sim m_0/m$. Lenz took $\sigma_{0_{max}} = 2$ and disregarded the Doppler effect. The result is:

$$I(\omega) = \Re\left[G\left(\sqrt{\frac{m_0}{m}}\right)\right] \tag{2.49a}$$

Let us develop the exponential in Eq. (2.49a) in a series and take only the first two terms to obtain:

$$I(\omega) = \Re\left\{\frac{1 + N\varrho_0^3 F_{j_1}[\Re(A_1) + i\Im(A_1)]}{i(\omega_0 - \omega)T + N\varrho_0^3 F_{j_2}[\Re(B_1) + \Im(B_1)]}\right\}$$

$$= \frac{1 + \varepsilon(\omega_0 - \Delta\omega_j - \omega)T_0}{(\omega_0 + \Delta\omega_0 - \omega)^2 + (\delta/2)^2} \tag{2.49a}$$

$$T_0\Delta\omega_0 = N\varrho_0^3 F_{j_2}(\sigma_m)\Im[B_1]; \qquad \frac{T_0}{2}\delta = N\varrho_0^3 F_{j_2}(\sigma_m)\Re[B_1] \tag{2.49b}$$

$$\varepsilon = \frac{\Im[A_1]\ F_{j_1}(\sigma_m)}{\Re[B_1]\ F_{j_2}(\sigma_m)}; \qquad T_0 = \frac{\varrho_0}{c_0}; \qquad \sigma_m = \sqrt{\frac{m_0}{m}} \tag{2.49c}$$

* Cf. *supra*, Sec. 1.5.

We can see by a consideration of Eq. (2.49a) that this very restricted line shape equation does yield a shift as given by $\Delta\omega_0$ of the intensity maximum and an asymmetrical line, a measure of whose asymmetry is given by ε. As in other dispersion type line shapes δ is the half-width of the broadened spectral line. Finally, T_0 is the time that a collision lasts, and it can be seen that as this goes to zero the line loses its asymmetrical shape and the shift disappears.

2.5. A specific evaluation of the Lenz half-width

The half-width may well be evaluated more precisely for future comparison. From Eqs. (2.39c) and (2.38c) we may write:

$$B_1 = \psi_1(0) = -\frac{2\pi}{p-1}\frac{(2\varkappa\chi)^{2/(p-1)}}{(2\varkappa)^2}\int_0^\infty (1-\mathrm{e}^{-i\gamma_1})\frac{\mathrm{d}\gamma_1}{\gamma^{1+2/(p-1)}}$$

$$= -\frac{2\pi}{5}\frac{(2\varkappa\chi)^{2/5}}{(2\varkappa)^2}\int_0^\infty (1-\mathrm{e}^{-i\gamma_1})\frac{\mathrm{d}\gamma_1}{{\gamma_1}^{7/5}}.$$

We shall only sketch the method developed by Jensen[98] for the evaluation of an integral of this type. The integral is first transformed as follows:

$$\int_0^\infty (1-\mathrm{e}^{i\gamma_1})\frac{\mathrm{d}\gamma_1}{{\gamma_1}^{1+2/(p-1)}} = \frac{i}{2/(p-1)}\varphi\left(\frac{2}{p-1}\right)$$

where:

$$\varphi\left(\frac{2}{p-1}\right) = (-i)^{1-\frac{2}{p-1}}\int_0^\infty \mathrm{e}^{-ix}\,x^{\frac{2}{p-1}}\,\mathrm{d}x = \lim_{\varepsilon\to 0}\lim_{a\to\infty}\int_\varepsilon^a \mathrm{e}^{-ix}\,x^{-\frac{2}{p-1}}\,\mathrm{d}x$$

On the substitution $w = ix$, we obtain for φ:

$$\varphi = (-i)^{1-2/(p-1)}\lim_{\varepsilon\to 0}\lim_{a\to\infty}\int_{w=i\varepsilon}^{w=ai}\mathrm{e}^{-w}\,w^{-2/(p-1)}\,\mathrm{d}w$$

We consider φ in the complex plane where φ would be an integration along the axis of imaginaries. By using Cauchy's Integral Theorem we may equate this to the sum of integrals (a) along a circular path about the origin from $i\varepsilon$ to ε, (b) along the axis of reals from ε to a, and (c) along a circular path from a to ia. After carrying out these operations we finally obtain:

$$\varphi\left(\frac{2}{p-1}\right) = (-i)^{1-2/(p-1)}\,\Gamma\left(1-\frac{2}{p-1}\right) = \frac{\pi\left(\frac{2}{p-1}\right)}{\sin\left[\pi\left(\frac{2}{p-1}\right)\right]}\,\frac{(-i)^{1-2/(p-1)}}{\Gamma\left(1+\frac{2}{p-1}\right)}$$

so that

$$\int_0^\infty (1-\mathrm{e}^{i\gamma_1})\frac{\mathrm{d}\gamma_1}{{\gamma_1}^{1+2/(p+1)}} = \frac{\pi}{\sin\left[\pi\left(\frac{2}{p-1}\right)\right]}\,\frac{\exp\left[i\,\frac{\pi}{2}\left(\frac{2}{p-1}\right)\right]}{\Gamma\left(1+\frac{2}{p-1}\right)}$$

$$= \frac{\pi}{\sin\left(\frac{2}{5}\pi\right)}\,\frac{\mathrm{e}^{i\pi/5}}{\Gamma\left(\frac{5}{7}\right)}$$

and

$$\Re(B_1) = -3.7845\frac{(2\varkappa\chi)^{2/5}}{(2\varkappa)^2}$$

On substitution from Eqs. (2.49c) and (2.34), Eq. (2.49c) becomes:

$$\delta = 7.977\, N(c_0)^{3/5}\,(a)^{2/5}\,F_{j_2} \tag{2.50}$$

where we have utilized the value $\varkappa\chi = 0.57$ since $p = 6$.*

Although Lenz has introduced into the Interruption Theory a consideration of the time of collision, thus obtaining a line which is shifted and asymmetrical, he has obtained an equation for this line which is only valid in a very restricted domain. As has been mentioned, the considered equation is only valid for pressures up to one atmosphere and for frequencies near the line center. Thus, the results are restricted in precisely the fashion in which Interruption results in general are restricted, that is, to low pressures and the line core. Most important, however, is the singling out of the collision function Eq. (2.25) and its treatment by the replacement of the time average by a statistical average.

We have remarked above that the inclusion of the collision time means the negation of the Interruption concept, for the idea of a sudden, short duration collision is incompatible with a consideration of the time of this collision. However, distant collisions, that is to say, collisions causing phase shifts of less than unity could certainly be considered as sudden. Thus, their inclusion which we now consider may be treated as pure Interruption.

2.6. Line shift without collision time

In analogy to Eq. (2.15) we begin with the equation:

$$\begin{aligned} I(\nu) &= \left|\int e^{2\pi i(\nu-\nu_0)t + i\Delta(t)}\,dt\right|^2 \\ &= \iint\limits_{-\infty}^{+\infty} e^{2\pi i(\nu-\nu_0)(t''-t') + i[\Delta(t'')-\Delta(t')]}\,dt\,dt' \qquad (2.51) \\ &= \int\limits_{-\infty}^{+\infty} e^{2\pi i(\nu-\nu_0)t}\,dt \int\limits_{-\infty}^{+\infty} e^{i[\Delta(t+t')-\Delta(t')]}\,dt' \end{aligned}$$

As had Lenz, Lindholm[125] considered the second integral as the statistical time mean value of the integrand and set this time mean equal to $\exp(-A(t) + iB(t))$ so that:

$$I(\nu) = \int\limits_0^\infty e^{-A(t)}\cos[2\pi(\nu-\nu_0)\,t + B(t)]\,dt \tag{2.52a}$$

Next it is assumed that but three different phase changes occur on collision.† These we shall designate as η_a, η_b, η_c. This in turn would mean that three diffe-

* See *supra*, Eq. (2.30) and subsequent.

† As we shall see, we could initially assume some other number and later extend the number of allowed phase changes as desired.

rent "differential collision cross-sections", σ_i, occur, where:

$$\sigma_i = \int |J(\eta_i)|^2 \sin\vartheta \, d\vartheta \tag{2.53}$$

The total cross-section is $\sigma = \sigma_a + \sigma_b + \sigma_c$, and the mean time between collisions is τ. The probability of $n + m + l$ collisions during the time t is then:

$$\frac{1}{(n+m+l)!} \left(\frac{t}{\tau}\right)^{n+m+l} e^{-t/\tau} \tag{2.54 a}$$

and the probability that, of these $(n + m + l)$ collisions, n will be of type a corresponding to a phase shift η_a, m will be type b, and l will be type c is:

$$\frac{(n+m+l)!}{n!\,m!\,l!} \left(\frac{\sigma_a}{\sigma}\right)^n \left(\frac{\sigma_b}{\sigma}\right)^m \left(\frac{\sigma_c}{\sigma}\right)^l \tag{2.54 b}$$

Thus, the probability that n a-collisions, m b-collisions, and l c-collisions will occur in time t is:

$$\left(\frac{\sigma_a}{\sigma}\right)^n \left(\frac{\sigma_b}{\sigma}\right)^m \left(\frac{\sigma_c}{\sigma}\right)^l \frac{!}{n!\,m!\,l!} \left(\frac{t}{\tau}\right)^{n+m+l} e^{-t/\tau} \tag{2.55}$$

As in Lenz' considerations:

$$\Delta(t+t') - \Delta(t') = \sum_k [\Delta_k(t+t') - \Delta_k(t')] = n\eta_a + m\eta_b + l\eta_c \tag{2.56}$$

In this case then the mean value of $e^{i[\Delta(t+t')-\Delta(t')]}$ which Lindholm assumed equal to the second integral in Eq. (2.51) is given by:

$$\left\langle e^{i[\Delta(t+t')-\Delta(t')]}\right\rangle = \sum_{n=0}^{\infty} \sum_{m=0}^{\infty} \sum_{l=0}^{\infty} \left(\frac{\sigma_a}{\sigma}\right)^n \left(\frac{\sigma_b}{\sigma}\right)^m \left(\frac{\sigma_c}{\sigma}\right)^l \frac{1}{n!\,m!\,l!} \left(\frac{t}{\tau}\right)^{n+m+l} e^{-t/\tau} \left\{e^{i[n\eta_a + m\eta_b + l\eta_c]}\right\}. \tag{2.57 a}$$

No normalizing denominator is needed in Eq. (2.57a), since the factor multiplying the exponential is already a normalized probability. Since: $\sum_{n=0}^{\infty} \left(\frac{\sigma_a t e^{i\eta_a}}{\sigma\tau}\right)^n \frac{1}{n!} = \exp\left[\frac{\sigma_a t}{\sigma\tau} e^{i\eta_a}\right]$, etc., Eq. (2.57a) becomes

$$\left\langle e^{i[\Delta(t+t')-\Delta(t')]}\right\rangle = e^{\frac{t}{\tau\sigma}[\sigma_a e^{i\eta_a} + \sigma_b e^{i\eta_b} + \sigma_c e^{i\eta_c}]} e^{-t/\tau} \tag{2.57 b}$$

Utilizing Eq. (2.57b) in Eq. (2.51) we obtain:

$$\begin{aligned} I(\nu) &= \Re\left\{\int e^{2\pi i(\nu-\nu_0)t} e^{\frac{t}{\tau\sigma}[\sum_i \sigma_i(\cos\eta_i + i\sin\eta_i) - \sigma]} dt\right\} \\ &= \int e^{-\frac{t}{\tau\sigma}\sum_i [\sigma - \sigma_i\cos\eta_i]} \cos\left[\frac{t}{\tau\sigma}\sum_i \sigma_i \sin\eta_i + 2\pi(\nu-\nu_0)t\right] dt \end{aligned} \tag{2.52 b}$$

where we have taken the real part of exp $(i \sin\eta_i)$ since the intensity is a real quantity. Thus, from Eqs. (2.52) there results:

$$A = t\sum_i \frac{\sigma_i}{\sigma\tau}[\sigma - \sigma_i\cos\eta_i] = \alpha t \tag{2.58 a)*}$$

$$B = t\sum_i \frac{\sigma_i}{\sigma\tau}\sin\eta_i = \beta t \tag{2.58 b}$$

* Lindholm gives $[1 - \sigma_i \cos\eta_i]$.

and Eq. (2.52) becomes:

$$\left.\begin{aligned} I(\nu) &= \int_0^\infty e^{-\alpha t} \cos\{[2\pi(\nu-\nu_0)+\beta]\, t\}\, dt \\ &= \frac{\text{const}}{\left[(\nu-\nu_0)+\dfrac{\beta}{2\pi}\right]^2+\left[\dfrac{\alpha}{2\pi}\right]^2} \end{aligned}\right\} \quad (2.59)$$

In obtaining Eq. (2.59) the collision time has been neglected in that we simply took the various phase changes as having occurred. A comparison of Eq. (2.59) with Eq. (2.49a), however, shows that this neglect of collision time has only effected a change from Lenz' result insofar as the line asymmetry is concerned. Eq. (2.59), of course, yields no line asymmetry. This difference between the two equations arises principally from the approximations which Lenz was obliged to utilize.

It might also be noted that we restricted ourselves to three phase changes η_a, η_b, and η_c, in setting up Eq. (2.55), but there is no reason to restrict Eqs. (2.58) for the shift and half-width in this manner. We note again the collision integral, Eq. (2.57a).

2.7. Reason for Weisskopf theory failure to yield shift†

Let us digress for a moment in order to consider an interesting aspect of the situation which has been mentioned by Burkhardt[19] and Unsold[221]. To begin with, let us rewrite Eqs. (2.52) as:

$$\alpha = v\, N \sum_i \sigma_i \left(1 - \frac{\sigma_i}{\sigma} \cos \eta_i\right) = v\, N \sum_i 2\sigma_i \sin \frac{2\eta_i}{2} \qquad (2.60\text{a})$$

$$\beta = v\, N \sum_i \sigma_i \sin \eta_i. \qquad (2.60\text{b})$$

From Eqs. (2.59) and (2.60a) it is apparent that the damping constant γ may be written as:

$$\gamma = 2\alpha \doteq 2v\, N \sum_i 2\sigma_i \sin^2 \frac{\eta_i}{2} \qquad (2.61)$$

If, as usual, the distance of closest approach is taken as ϱ, then the cross section σ_i for a phase change η_i can be taken as 2π where we now suppose there to be a continuum of cross sections instead of the discrete set of Eqs. (2.60). In addition, the assumption of this continuum replaces the summations in Eqs. (2.60) by integrations. These changes in Eqs. (2.60) together with the

† Cf. *supra*, Sec. 2.3.

substitution for η_i of Eq. (2.8a′) into these equations yield:

$$\left.\begin{aligned}\gamma &\doteq 2v\,N \int_0^\infty 4\pi\,\varrho\,d\varrho\,\sin^2 \frac{2\pi\,C\,c_n}{v\varrho^{n-1}}\\ &= C^{\frac{2}{n-1}}\,v^{\frac{n-3}{n-1}}\,N\,8\pi^{\frac{n+1}{n-1}}\,c_n^{\frac{2}{n-1}} \int_0^\infty \sin^2 \frac{1}{x^{n-1}}\,x\,dx\end{aligned}\right\}\quad (2.62\text{a})$$

$$\left.\begin{aligned}\beta &= \omega_0 - \omega_m = vN \int_0^\infty 2\pi\,\varrho\,d\varrho\,\sin \frac{2\pi\,C\,c_n}{v\varrho^{n-1}}\\ &= C^{\frac{2}{n-1}}\,v^{\frac{n-3}{n-1}}\,N\,2\pi^{\frac{n+1}{n-1}}\,c_n^{\frac{2}{1-n}} \int_0^\infty \sin \frac{2}{x^{n-1}}\,x\,dx\end{aligned}\right\}\quad (2.62\text{b})$$

where now:

$$\frac{\eta}{2} = \frac{\pi\,C c_n}{v\varrho^{n-1}} = \frac{1}{x^{n-1}}$$

Now let us consider the case of $n = 4$ as illustrating a rather interesting circumstance relating to the Weisskopf theory. (We note that in order to obtain agreement between Eqs. (2.9) and (2.62a) we must needs choose $\eta_0 = 0.64$.)

We may recall the failure of the Weisskopf theory to yield a line shift under any conditions.* What we propose to demonstrate is that the neglect on the part of the Weisskopf theory of collisions resulting in phase changes of less than unity, while accounting for most of the broadening, is responsible for the absence of a line shift.

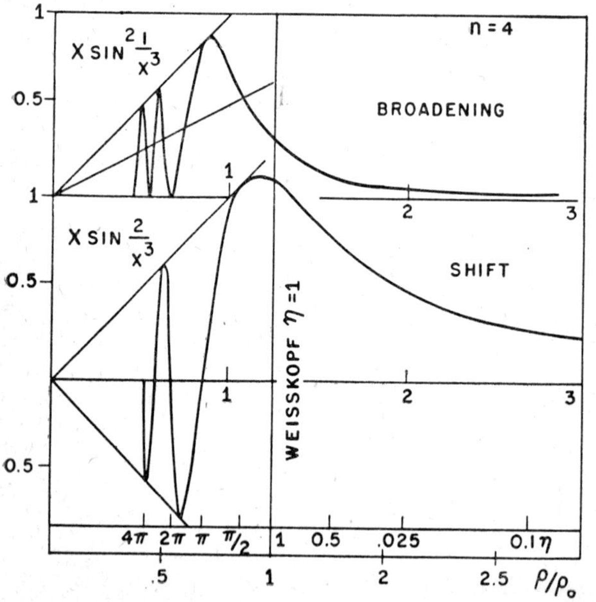

FIG. 2.2. The broadening and shift contributions with collision phase shift for $n = 4$.
(After Unsold[221] and Lindholm[233])

To begin with, let us plot x $\sin^2(1/x^3)$ (from Eq. (2.62a)) and x $\sin(2/x^3)$ (from Eq. (2.62b)) against ϱ/ϱ_0. The result appears in Fig. 2.2, and let us consider this figure. The vertical line in about the center of the figure is the Weisskopf collision radius. Now the area

* This is not true for his more detailed theory for which see *infra*, Chap. 5. This theory does not appear to have been applied, however.

under the upper curve gives us the integral in Eq. (2.62a) and is hence a measure of the broadening. In like manner, the area under the lower curve is a measure of the shift. The Weisskopf theory only considers those areas to the left of the ordinate specifying the Weisskopf collision radius. It thus becomes quite apparent that with the adoption of this radius we obtain almost all the broadening effects (insofar as this theory is concerned) while we obtain almost none of the shift effects. In addition, these curves serve to illustrate the relative importance of near and distant collisions in broadening and shift.

If the η_i are very small then, since $\sin \eta_i \doteq \eta_i$:

$$\omega - \omega_m = \beta \doteq vN\Sigma \sigma_i \eta_i \tag{2.63}$$

In Fig. 2.3 is illustrated a wave train during whose emission strong collisions ($\eta > 1$), which lead primarily to broadening, and weak collisions ($\eta < 1$), which lead primarily to shift, have been undergone by the emitter. Let us suppose the wave train to be essentially rendered completely inchoherent at a strong collision insofar as Fourier analysis is concerned, and let us determine the mean frequency between two strong collisions separated by a time interval T. In order to do this, it would appear quite reasonable to add to the number of vibrations of the natural frequency, $\omega_0 T$, the number of additional vibrations due to the weak collisions $TvN \sum_{\eta_i < 1} \sigma_i \eta_i$ to obtain:

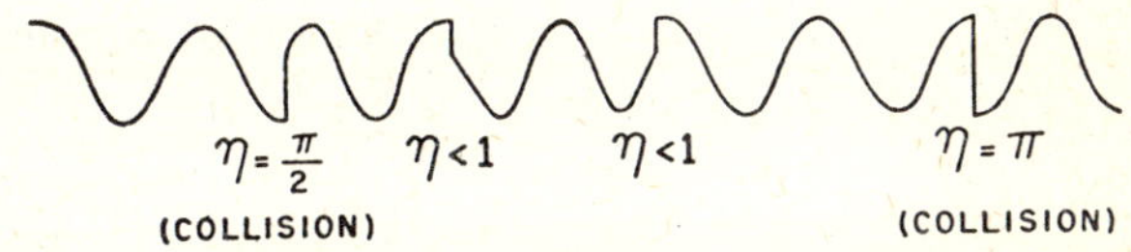

FIG. 2.3. An illustration of the phase shifts induced by collision.
(After Unsold[203] and Lindholm[233])

$$\omega_m = \frac{1}{T}\left\{\omega_0 T + TvN \sum_{\eta_i < 1} \sigma_i \eta_i\right\} \tag{2.64}$$

The most cursory glance serves to show the agreement between Eqs. (2.63) and (2.64).

Finally, a consideration of Table 2.1 yields values for shift and width from Eqs. (2.62) according to this theory. The case $n = 2$, in addition to yielding a symmetrical line yields:

$$\gamma_2 = 8\pi^2 \frac{C^2 N}{v}\left\{0.923 - \ln \eta_m + \frac{\eta_m{}^2}{24} - \cdots\right\} \tag{2.65}$$

the reason for its exclusion from the table being self evident. In addition, the table gives the η_0 necessary for agreement with the Weisskopf theory.

TABLE 2.1

n_n	γ_n	η_{0n}	β_n	γ_n/β_n
3	$4\pi^3\, CN$	0.64	—	—
4	$3.88\, C^{2/3} v^{1/3}\, N$	0.64	$33.4\, C^{2/3} v^{1/3}\, N$	1.16
6	$17.0\, C^{2/5} v^{3/5}\, N$	0.61	$6.16\, C^{2/5} v^{3/5}\, N$	2.80

2.8. Lindholm general theory to include collision time[126]

At this point we diverge from true Interruption concepts in the mainstream of this development, the difference from the Lenz situation being that now we shall return to them only as limiting cases. Therefore a few remarks may be in order.

We have dwelt at sufficient length on the collision integral so that the reader should be reasonably well aware of its existence. Lenz initiated its serious considerations as we have remarked, and, in so doing, strayed from the Interruption idea. Lindholm likewise devoted his attention to it, but the returned to the path. In his later work he was to stray after Lenz, and no subsequent author showed any desire to return. We will next consider this work of Lindholm in which the Interruption Theory appears only as a limiting case. Later Foley was to devote his attention to this same integral after having applied quantum mechanics to its derivation, but he was destined to give it the name by which it is familiar today. Following after him was Anderson with further study of the by now familiar integral. With these works we shall deal as we encounter them. In our immediate consideration, however, we describe Lindholm's method of inclusion not only of the small phase shifts of distant collisions but also of the large phase shifts of extremely close collisions.

Let the collision time be s_i, and the phase change per unit of time during the collision be k_i. Then, if we assume the phase to change linearly during the collision, the total phase change on collision will be $\eta_i = k_i s_i$. Let us consider Fig. 2.4. x and y measure times back to t' and to t'' respectively as shown. The five

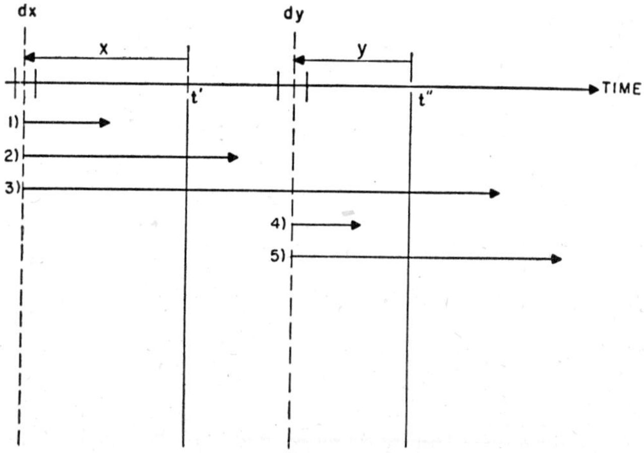

FIG. 2.4. Various possible collisions classifiable as to duration. (After Lindholm[126])

arrows below the time axis represent five different collisions whose duration is represented by the respective arrow lengths. Only those collisions which occur at least partly during $t'' - t'$ are to be included.

In analogy to Eq. (2.55) the probability that n_{xi} collisions will begin during the time interval dx is:

$$\left(\frac{\sigma_i}{\sigma}\frac{dx}{\tau}\right)^{n_{xi}} \frac{1}{n_{xi}!\,e^{dx/\tau}} \tag{2.66a}$$

It can be seen directly from our definitions that these collisions will contribute to $e^{i[\Delta(t'')-\Delta(t')]}$ the factor

$$e^{in_{x_i}k_i\varphi(s_i-x)} \tag{2.67a}$$

where:

$$\left.\begin{aligned} \varphi(s_i-x) &= 0 && \text{for } s_i-x<0 \\ \varphi(s_i-x) &= s_i-x && \text{for } 0<s_i-x<t''-t' \\ \varphi(s_i-x) &= t && \text{for } t''-t'<s_i-x \end{aligned}\right\} \tag{2.67b}$$

Eq. (2.67b) is merely a restatement of our restriction of the collision such that at least a portion of it occurs within $t''-t'$.

In analogy with Eq. (2.66a) we get for the n_{yi} probability:

$$\left(\frac{\sigma_i}{\sigma}\frac{dy}{\tau}\right)^{n_{yi}} \frac{1}{n_{yi}!\,e^{dy/\tau}} \tag{2.66b}$$

and in analogy to Eq. (2.67a) we obtain:

$$e^{in_{y_i}k_i\Psi(s_i-y)} \tag{2.67c}$$

where:

$$\left.\begin{aligned} \psi(s_i-y) &= s_i && \text{for } s_i-y<0 \\ \psi(s_i-y) &= y && \text{for } 0<s_i-y \end{aligned}\right\} \tag{2.67d}$$

From Eqs. (2.61a) and (2.67c) and in analogy with Eq. (2.57a), we now obtain the mean value of $e^{i[\Delta(t'')-\Delta(t')]}$ as:

$$\langle e^{i[\Delta(t'')-\Delta(t')]}\rangle = \sum_{n_{x_1 1}=0}^{\infty}\cdots\sum_{n_{x_j 1}=0}^{\infty}\;\sum_{n_{x_2 1}=0}^{\infty}\cdots\sum_{n_{y_j i}=0}^{\infty}\;\prod_{i}\prod_{dx_l}\prod_{dy_j}\frac{1}{n_{xli}!\,e^{dxl/\tau}}\times$$

$$\times\left(\frac{\sigma_i}{\sigma}\frac{dx_l}{\tau}\right)^{n_{xli}}\frac{1}{n_{y_j i}!\,e^{\frac{dy_j}{\tau}}}\left(\frac{\sigma_i}{\sigma}\frac{dy_j}{\tau}\right)^{n_{y_j i}} e^{i[n_{xli}k_i\varphi(s_i-x)+n_{y_j i}k_i\psi(s_i-y)]} \tag{2.68}$$

In Eq. (2.68) the x_l represent the possible dx time intervals during which included collisions may originate. The y_i represent the possible dy time intervals during which included collisions may originate. Thus, n_{xl1} is the number of collisions of the type one originating with some probability during the dx_l time interval. Again we can see that the summations in Eq. (2.68) are the MacLaurin series expansion for the exponential. We, of course, have such an exponential for each dx_l (or dy_j) interval, and they are all to be multiplied together. Let us consider the dx_l intervals. We have a long series of exponentials corresponding to the various dx_l intervals which are to be multiplied together. This is the same as adding the exponents of these exponentials, and, since we may define dx_l as small as we like, we may change this sum of exponents to an integral over x. The same holds true for y, and, when we have summed over i in the exponent, we obtain as the contribution to our mean value:

$$\left.\begin{aligned} \exp \sum_i \Bigg[\int_0^\infty \left(\frac{\sigma_i \mathrm{d}x}{\sigma\tau} \, \mathrm{e}^{i k_i \varphi(s_i - x)} - \frac{\mathrm{d}x}{\tau} \right) \\ + \int_0^t \left(\frac{\sigma_i \mathrm{d}y}{\sigma\tau} \, \mathrm{e}^{i k_i \psi(s_i - y)} - \frac{\mathrm{d}y}{\tau} \right) \Bigg] \end{aligned} \right\} \tag{2.69a}$$

In the second integral of Eq. (2.69a) we do not extend the upper limit beyond t since this would then duplicate the values given by the x-integral. Since $\sum_i \sigma_i = \sigma$ the following relation holds:

$$\sum_i \int \frac{\mathrm{d}x}{\tau} = \sum_i \int \frac{\sigma_i}{\sigma} \frac{\mathrm{d}x}{\tau}$$

Finally, let us sum over all collision times s_i, and neglect the summation over i which may be introduced at any subsequent point. Our mean value is then:

$$\left.\begin{aligned} \langle \mathrm{e}^{i[\Delta(t'') - \Delta(t')]} \rangle = \exp \sum_{s_i=0}^{\infty} \Bigg[\int_0^\infty \left(\frac{\sigma_i \mathrm{d}x}{\sigma\tau} \, \mathrm{e}^{i k_i \varphi(s_i - x)} - \frac{\sigma_i \mathrm{d}x}{\sigma\tau} \right) \\ + \int_0^t \left(\frac{\sigma_i \mathrm{d}y}{\sigma\tau} \, \mathrm{e}^{i k_i \Psi(s_i - y)} - \frac{\sigma \mathrm{d}y}{\sigma\tau} \right) \Bigg] \end{aligned} \right\} \tag{2.69b}$$

From Eqs. (2.52a) and (2.69a) we now obtain:

$$\left.\begin{aligned} A = \sum_{s_i=0}^{\infty} \Bigg[\int_0^\infty \left(\frac{\sigma_i \mathrm{d}x}{\sigma\tau} \cos\,[k_i\, \varphi(s_i - x)] - \frac{\sigma_i \mathrm{d}x}{\sigma\tau} \right) \\ + \int_0^t \left(\frac{\sigma_i \mathrm{d}y}{\sigma\tau} \cos\,[k_i\, \psi(s_i - y)] - \frac{\sigma_i \mathrm{d}y}{\sigma\tau} \right) \Bigg] \end{aligned} \right\} \tag{2.70a}$$

$$B = \sum_{s_i=0}^{\infty} \left\{ \int_0^\infty \frac{\sigma_i \mathrm{d}x}{\sigma\tau} \sin\,[k_i\, \varphi(s_i - x)] + \int_0^t \frac{\sigma_i \mathrm{d}y}{\sigma\tau} \sin\,[k_i\, \psi(s_i - y)] \right\} \tag{2.71a}$$

Our function $\varphi(s_i - x)$ has been defined by Eq. (2.67b). As an example this equation yields for the first integration Eq. (2.70a):

$$\begin{aligned} \sum_{s_i=0}^{\infty} \int_0^\infty \left(\frac{\sigma_i \mathrm{d}x}{\sigma\tau} \cos\,[k_i\, \varphi(s_i - x)] - \frac{\sigma_i \mathrm{d}x}{\sigma\tau} \right) &= \sum_{s_i=0}^{t} \int_t^\infty \left(\frac{\sigma_i \mathrm{d}x}{\sigma\tau} \cos\,[k_i \cdot 0] - \sigma_i \frac{\mathrm{d}x}{\sigma\tau} \right) \\ + \sum_{s_i=0}^{t} \int_0^t \left(\frac{\sigma_i \mathrm{d}x}{\sigma\tau} \cos\,[k_i (s_i - x)] - \frac{\sigma_i \mathrm{d}x}{\sigma\tau} \right) &+ \sum_{s_i=t}^{\infty} \int_{s_i}^\infty \left(\frac{\sigma_i \mathrm{d}x}{\sigma\tau} \cos\,[k_i \cdot 0] - \sigma_i \frac{\mathrm{d}x}{\sigma\tau} \right) \\ + \sum_{s_i=t}^{\infty} \int_{s_i-t}^{s_i} \left(\frac{\sigma_i \mathrm{d}x}{\sigma\tau} \cos\,[k_i (s_i - x)] - \frac{\sigma_i \mathrm{d}x}{\sigma\tau} \right) & \\ &+ \sum_{s_i=t}^{\infty} \int_0^{s_i-t} \left(\frac{\sigma_i \mathrm{d}x}{\sigma\tau} \cos\,[k_i (s_i - t)] - \sigma_i \frac{\mathrm{d}x}{\sigma\tau} \right) \end{aligned}$$

In this manner we obtain Eqs. (2.70b) and (2.71b)

$$\left.\begin{aligned} A &= \sum_{s_i=0}^{t} \frac{\sigma_i}{\sigma\tau}\left[(1-\cos\eta_i)(t-s_i)-\sin\eta_i\frac{2}{k_i}+2s_i\right] \\ &+\sum_{s_i=t}^{\infty}\frac{\sigma_i}{\sigma\tau}\left[(1-\cos k_i t)(s_i-t)-\sin k_i t\frac{2}{k_i}+2t\right] \end{aligned}\right\} \quad (2.70\text{b})$$

$$\left.\begin{aligned} B &= \sum_{s_i=0}^{t} \frac{\sigma_i}{\sigma\tau}\left[(1-\cos\eta_i)\frac{2}{k_i}+\sin\eta_i(t-s_i)\right] \\ &+\sum_{s_i=t}^{\infty}\frac{\sigma_i}{\sigma\tau}\left[(1-\cos k_i t)\frac{2}{k_i}+\sin k_i t(s_i-t)\right] \end{aligned}\right\} \quad (2.71\text{b})$$

The direct substitution of Eqs. (2.70b) and (2.71b) into Eq. (2.52a) would now give us an expression, if a rather cumbersome one, for the intensity distribution, but before doing this, let us digress for the purpose of simplification.

2.9. Specification of and approximations to the general theory

First we shall assume, with Lindholm, that all perturbers possess the same velocity relative to the emitter and let this velocity be the relative mean velocity $\langle v\rangle$.* Next, it is assumed that the perturber paths are straight lines. The frequency perturbation is taken as the van der Waals' interaction $\Delta\nu = -bR^{-6}$ where R is the broadener-emitter separation. If the distance of closest approach is ϱ, and the phase change as before is η_i, we obtain from Eq. [2.7):

$$\eta_i = 2\pi\int_{-\infty}^{+\infty}\Delta\nu\,\mathrm{d}t = 2\pi b\int_{-\infty}^{+\infty}\frac{\mathrm{d}t}{(\langle v\rangle^2 t^2+\varrho^2)^3} = \frac{3\pi^2 b}{4\langle v\rangle\varrho^5} \quad (2.72\text{a})$$

The "length in space of the collision" we now take as proportional to ϱ and as given by $2\varkappa\varrho$. Let:

$$x = \varrho\left[\frac{8\langle v\rangle}{3\pi^2 b}\right]^{1/5}; \qquad y = \frac{\langle v\rangle t}{2\varkappa}\left[\frac{8\langle v\rangle}{3\pi^2 b}\right]^{1/5} \quad (2.73)$$

so that Eq. (2.72) becomes:

$$\eta_i = \frac{2}{x^5} \quad (2.72\text{b})$$

In addition:

$$s_i = \frac{2\varkappa\varrho}{\langle v\rangle} = \frac{2\varkappa}{\langle v\rangle}\left[\frac{3\pi^2 b}{8\langle v\rangle}\right]^{1/5} x \quad (2.74\text{a})$$

$$k_i = \frac{\eta_i}{s_i} = \frac{\langle v\rangle}{\varkappa}\left[\frac{8\langle v\rangle}{3\pi^2 b}\right]^{1/5}\frac{1}{x^6} \quad (2.74\text{b})$$

$$t = \frac{2\varkappa}{\langle v\rangle}\left[\frac{3\pi^2 b}{8\langle v\rangle}\right]^{1/5} y \quad (2.74\text{c})$$

$$\sigma_i = 2\pi\varrho\,\mathrm{d}\varrho = 2\pi\left[\frac{3\pi^2 b}{8\langle v\rangle}\right]^{2/5} x\,\mathrm{d}x \quad (2.74\text{d})$$

* We may recall that in the Lenz derivation not a constant velocity but a Maxwell-Boltzmann distribution of the u_k, v_x, and w_k was assumed. We shall see that this apparently has little effect.

where Eqs. (2.74) follow from Eq. (2.73) and the definitions of the various quantities. Also let:

$$l = N\,4\pi\varkappa \left[\frac{3\pi^2 b}{8\langle v\rangle}\right]^{3/5}; \qquad k = (\nu - \nu_0)\frac{4\pi\varkappa}{\langle v\rangle}\left[\frac{3\pi^2 b}{8\langle v\rangle}\right]^{1/5} \tag{2.74e}$$

where N is the number of perturbers per unit of volume.

Now, if, in a straightforward if somewhat laborious manner, we substitute the results of Eqs. (2.70b), (2.71b), (2.73), and (2.74) into Eq. (2.52a) we obtain:

$$I(\nu) = \int_0^\infty e^{-l\Phi(y)} \cos\left[ky + l\,\Psi(y)\right] dy \tag{2.75}$$

where:

$$\left.\begin{aligned}\Phi(y) = &\int_0^y \left[\left(1 - \cos\frac{2}{x^5}\right)(y - x) - \sin\frac{2}{x^5}\,x^6 + 2x\right] x\,dx\\ &+ \int_y^\infty \left[\left(1 - \cos\frac{2y}{x^6}\right)(x - y) - \sin\frac{2y}{x^5}\,x^6 + 2y\right] x\,dx\end{aligned}\right\} \tag{2.76a}$$

$$\left.\begin{aligned}\Psi(y) = &\int_0^y \left[\left(1 - \cos\frac{2}{x^5}\right)x^6 + \sin\frac{2}{x^5}(y - x)\right] x\,dx\\ &+ \int_y^\infty \left[\left(1 - \cos\frac{2y}{x^6}\right)x^6 + \sin\frac{2y}{x^6}(x - y)\right] x\,dx\end{aligned}\right\} \tag{2.77a}$$

The fact that $\Phi(y)$ and $\Psi(y)$ may not be evaluated directly should hardly come as a surprise. Let us consider $\Phi(y)$. We may transform this function into a sum of the form $\int_0^\infty + \int_0^y$. The integral with limits zero and infinity may be evaluated exactly after Jensen[98]. The other integral may be expanded in a series, and it can then be shown that for $y > 1$ this integral may be neglected. The integrals in $\Phi(y)$ may also be written as $\int_0^\infty + \int_0^y$. Again the integral with limits zero and infinity may be directly evaluated, and it is found that, when the latter integral is expanded in a series, this remaining integral is small for $y < 0.5$. The integrals were numerically evaluated by Lindholm for $0.5 < y < 1$, and he concluded that the result for $y < 0.5$ could be extended up to approximately $y = 1$. In this way Eqs. (2.76) were found to be:

$$\Phi(y) = 0.795\,y + 0.165 \quad \text{for } y > 1 \tag{2.76b}$$

$$\Phi(y) = 0.369\,y^{4/3} + 0.591\sqrt{y} \quad \text{for } y < 1 \tag{2.76c}$$

$$\Psi(y) = 0.577\,y + 0.227 \quad \text{for } y > 1 \tag{2.77b}$$

$$\Psi(y) = 0.213\,y^{4/3} + 0.591\sqrt{y} \quad \text{for } y < 1 \tag{2.77c}$$

When Eqs. (2.76b), (2.76c), (2.77b), and (2.77c) are substituted into Eq. (27.5) the result is:

$$I(\nu) = I_0 + I_\infty \tag{2.78a}$$

where:

$$I_0 = \int_0^1 e^{-l[0.369\, y^{4/3} + 0.591\, y \sqrt{y}\,]} \cos\left[ky + l\left(0.213\, y^{4/3} + 0.591 \sqrt{y}\right)\right] dy \tag{2.78b}$$

$$\left.\begin{aligned} I_\infty &= e^{-l[0.165]} \int_1^\infty e^{-l[0.795y]} \cos\left[y(k + 0.577\, l) + 0.227\, l\right] dy \\ &= e^{-0.960l}\, \frac{0.795l \cos(0.804l + k) - (k + 0.579l)\sin(-0.804l + k)}{(0.795l)^2 + (k + 0.577l)^2} \end{aligned}\right\} \tag{2.78c}$$

Numerical calculation is required for Eq. (2.78b), and in a general form this is as far as Lindholm carried the calculation. Let us now consider some of the special cases of Eq. (2.78a).

For very high pressures l, as given by Eq. (2.74e), is obviously very large. It is apparent from Eq. (2.78c) that we may neglect I_∞ under these conditions. In Eq. (2.78b) the integrand, due to the behavior of the exponential with large l, will be very small unless y is very small. When y is small $y^{4/3} \ll y^{1/2}$ so that we may write:

$$\left.\begin{aligned} I(\nu) &= I_0 = \int_0^\infty e^{-l[0.591\sqrt{y}\,]} \cos\left[ky + l\left(0.591 \sqrt{y}\right)\right] dy \\ &= \sqrt{\frac{\pi}{2}}\, \frac{0.591l}{(-k)^{3/2}} \exp\left[-\frac{(0.591l)^2}{2(-k)}\right] = c_1 N \langle v \rangle^{6/5} (\Delta\nu)^{-3/2} e^{-N^2C_2/\Delta\nu} \end{aligned}\right\} \tag{2.79}$$

for $k < 0$. $I(\nu) = 0$ for $k > 0$ which means that the line is displaced to the red, and the intensity falls to zero at the frequency of the undisplaced line. We shall see in Chapter 3 that this is precisely the Statistical Theory result for the case of van der Waals' forces, $n = 6$. Therefore this reduces to the Statistical theory for very high pressures. It also reduces to the Statistical theory for very large k which, we may remark from Eq. (2.74e), may occur for frequencies differing a great deal from that of the line center. We should keep in mind this reduction to the Statistical Theory for (1) high pressures and (2) frequencies in the line wing.

Lindholm made use of the results of the Statistical Theory in his evaluation of $\varkappa$. If we let $\varkappa = 16/3\pi = 1.7$ instead of 0.75 as given after Eq. (2.30) we obtain exact agreement with this Theory. Since the Statistical Theory had yielded such excellent agreement with experiment at high pressure Lindholm chose 1.7 as the value for $\varkappa$ on the basis of the resulting agreement. He justified his disagreement with Lenz in the following fashion:

First, let us recall that the time duration of the collision is essentially $2\varkappa$ and consider Fig. 2.5. Curve (a) in this figure is the actual curve for the phase integral $2\pi b \int_{-\infty}^{t} \times \frac{dt}{(v^2t^2+\varrho^2)^3}$. Curve (b) is the Lenz approximate curve where $\varkappa = 0.75$, and curve (c) is the Lindholm curve where $\varkappa = 1.7$. Lindholm felt that the larger $\varkappa$ value is more justified in that it includes the slow frequency shifts at either end of the true curve (a).

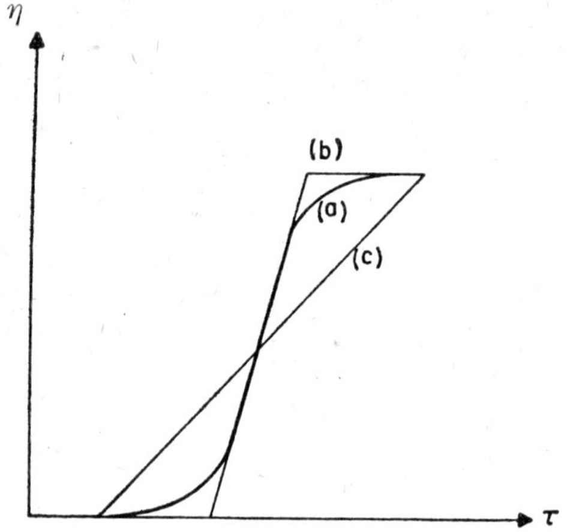

FIG. 2.5. The Lenz and Lindholm approximations of the course of the phase change. (After Lindholm[126])

Now let us consider the special case of Eq. (2.78a) corresponding to Lenz' calculation for low pressure and spectral positions near line center. Lindholm took N and $(\nu - \nu_0)$ small enough so that Eq. (2.74e) goes to zero, and as a result Eq. (2.78b) becomes $I_0 = 1$. Now let us expand Eq. (2.78c) in a series and keep only second order terms in l and k. The result is:

$$\left.\begin{aligned} I_\nu &= I_0 + I_\infty \\ &= 1 + (1 - 0.960\,l)\,\frac{[(0.795\,l) - (k + 0.577\,l)\,(k + 0.804\,l)]}{(0.795\,l)^2 + (k + 0.577\,l)^2} \\ &= 0.795\,\frac{1 - (k + 0.577l)\,\dfrac{.227}{.795} - 0.165\,l}{(0.795l)^2 + (0.577l + k)^2} \end{aligned}\right\} \quad (2.80\text{a})$$

Of particular importance here is the illustration of the reduction to the Interruption Theory for (1) low pressure (small N) and (2) frequencies near line center (small $\nu - \nu_0$)).

A comparison of Eq. (2.49a) and (2.80a) is in order. It can be seen that, from the above equation and Eq. (2.74a), the line shift and half-width are given by:

$$\Delta = \frac{.577}{\dfrac{2\pi\varkappa}{\langle v\rangle}\left[\dfrac{3\pi^2 b}{8\langle v\rangle}\right]^{1/5}} = \frac{.577\,l}{c} = .097\,N\,\langle v\rangle^{3/5}\,b^{2/5} \qquad (2.81\text{a})$$

$$\delta/2 = \frac{.795l}{c} = 2.68\,N\,\langle v\rangle^{3/5}\,b^{2/5} \qquad (2.81\text{b})$$

The linear dependence on the pressure and the dependence on the temperature through $\langle v\rangle$ of both the shift and the half-width may be noted from Eqs. (2.81).

We may compare this with Eq. (2.50). Before doing so, however, it will be necessary to rewrite Eq. (2.50) slightly. The constant b differs from the constant a in the latter equation by virtue of the fact that Lenz used units of angular velocity ω while Lindholm utilized units of frequency ν.

Thus Eq. (2.50) must be multiplied by the factor $(1/2\pi)\,(2\pi)^{2/5}$, and one obtains 2.65 instead of 7.977. The difference between these two equations arises chiefly from the fact that Lindholm used only the mean relative velocity, while Lenz used a distribution of velocities. A reasonable basic agreement exists between the two theory results, however.

Let us turn our attention to the behavior of Eqs. (2.78b) and (2.78c) in the wings of the line. We consider Eq. (2.75). Using Eqs. (2.76b), (2.76c), (2.77b), and (2.77c) this equation may be written as:

$$\left.\begin{aligned} I(\nu) = & \int_0^\infty e^{-l[0.369\, y^{4/3} + 0.591 \sqrt{y}\,]} \\ & \cos\left[ky + l\left(0.213\, y^{4/3} + 0.591 \sqrt{y}\right)\right] dy \\ & - \int_1^\infty e^{-l[0.369\, y^{4/3} + 0.591 \sqrt{y}\,]} \\ & \cos\left[ky + l\left(0.213\, y^{4/3} + 0.591 \sqrt{y}\right)\right] dy \\ & + \int_1^\infty e^{-l[0.795 y + 0.165]} \cos\left[ky + l(0.577\, y + 0.277)\, dy\right. \end{aligned}\right\} \quad (2.82)$$

For k very large the cosine function, and hence the integrand will fluctuate very rapidly for large y resulting in the usual cancellation effect. Thus, we may restrict ourselves to small y. As a result of this restriction, the first integral in Eq. (2.82) makes the greatest contribution to $I(\nu)$. In evaluating this integral — a rather lengthy operation — Lindholm utilized the substitution $ky = z^2$ which allowed him to expand certain portions of the integrand in power series to arrive at integrals of the form:

$$\int_0^\infty e^{-sz + iz^2} z^\alpha \, dz = \frac{1}{2} \sum_{n=0}^{\infty} (-1)^n \, (i)^{\frac{\alpha+1+n}{2}} \, \Gamma\left(\frac{\alpha+1+n}{2}\right) \frac{s^n}{n!}$$

He thus obtained the following two equations for the red and violet wings of the line respectively:

$$I(\nu) = \frac{0.741\, l}{(-k)^{3/2}} + \frac{0.254\, l}{(-k)^{7/3}} - \frac{0.129\, l^3}{(-k)^{5/2}} + \frac{0.532\, l^2}{(-k)^{17/6}} \quad (2.83\text{a})$$

$$I(\nu) = \frac{0.507\, l}{k^{7/3}} - \frac{0.614\, l^2}{k^{17/6}} + \frac{0.320\, l^3}{k^{10/3}} + \cdots \quad (2.83\text{b})$$

Lindholm obtained more general results than Lenz had obtained primarily because he assumed the same relative velocity for all perturbers, a not very serious approximation. Thus, his method of including the time of collision in the Fourier analysis yielded, for the case of van der Waals' forces, Eqs. (2.78c) which can be evaluated for specific vases.

Further he demonstrated clearly the limiting positions occupied by the Statistical and Interruption Theories.

2.10. Adiabaticity in line broadening

One is accustomed to consider an adiabatic process as one in which the entropy of the system — or the degree of disorder thereof — remains unchanged. Now if the probability that a molecule is in State X has the value unity before and after the occurrence of some phenomenon which affects the molecule "system", the degree of disorder of this system remains unchanged and hence, the occurrence of the phenomenon constitutes an adiabatic process. On the otherhand, if the probability for the state X is unity at the initiation of the process and changes by virtue of the occurrence of this process to 0.6 for the state X, 0.3 for state Y, and 0.1 for state Z, the degree of system disorder has changed, and, in consequence, a diabatic* process has occurred. In other words if, due to the molecular interaction which occurs during the collision and is responsible for the broadening, the molecule changes state, the encounter has not been an adiabatic one. We will recall the manner — and we shall detail it in Chapter 6 — in which the expansion coefficients in time dependent perturbation theory are a measure of the probability, after a given time, that the system which was originally in a state X may now be found in some other state Y. Therefore for a specific consideration of the adiabaticity or otherwise of a broadening encounter, we would do well to look to these coefficients. Let us first derive an expression for them which will be explanatory in this regard.

Out time-dependent Schrödinger equation is, of course:

$$i\hbar\dot{\Psi} = H(t)\,\Psi \tag{2.84}$$

At any given time "t" this Hamiltonian will have eigenfunctions and eigenvalues as given by:

$$H(t)\,\varphi_n(t) = E_n(t)\,\varphi_n(t) \tag{2.85}$$

Since the solutions of Eq. (2.85) form a complete, orthonormal set, we may expand the solution to Eq. (2.84) in terms of them:

$$\Psi = \sum a_n(t)\,\varphi_n(t)\exp\left[-\frac{i}{\hbar}\int_0^t E_n(t')\,dt'\right] \tag{2.86}$$

We substitute Eq. (2.86) into Eq. (2.84), multiply through on the left by $\bar{\varphi}_k$, and integrate over all space to obtain the following equation for the state growth coefficients:

$$\dot{a}_k = -\sum a_n\, e^{-\frac{i}{\hbar}\int_0^t [E_k - E_n]\,dt'} \int \bar{\varphi}_k \frac{\partial\varphi_n}{\partial t}\,dt \tag{2.87}$$

* In certain circles "non-adiabatic" is considered as an "atrocious, pleonastic synonym", and the author quit using the term as soon as he discovered how bad it was.

Eq. (2.85) may be differentiated with respect to time,

$$\frac{\partial H}{\partial t}\varphi_n + H\frac{\partial \varphi_n}{\partial t} = \frac{\partial E_n}{\partial t}\varphi_n + E_n\frac{\partial \varphi_n}{\partial t}$$

multiplied through on the left by $\bar{\varphi}_k$ and integrated over all space to obtain:

$$\int \bar{\varphi}_k \frac{\partial H}{\partial t}\varphi_n \,\mathrm{d}t + \int \bar{\varphi}_k H \frac{\partial \varphi_n}{\partial t}\,\mathrm{d}t = E_n \int \bar{\varphi}_k \frac{\partial \varphi_n}{\partial t}\,\mathrm{d}t \qquad (2.88)$$

We next find an alternate expression for the second integral on the left of Eq. (2.88). Due to the Hermitian quality of the Hamiltonian we may write:

$$\int \bar{\varphi}_k H \frac{\partial \varphi_n}{\partial t}\,\mathrm{d}\tau = \int (H\bar{\varphi}_n)\frac{\partial \varphi_n}{\partial t}\,\mathrm{d}\tau = E_k \int \bar{\varphi}_k \frac{\partial \varphi_n}{\partial t}\,\mathrm{d}\tau$$

This equality enables us to re-write Eq. (2.88) as:

$$\int \bar{\varphi}_k \frac{\partial \varphi_n}{\partial t}\,\mathrm{d}t = -\frac{\int \bar{\varphi}_k \left(\frac{\partial H}{\partial t}\right)\varphi_n \,\mathrm{d}t}{E_k - E_n}$$

This equality may be substituted into Eq. (2.87) and the latter integrated with the result:

$$a_k = \int_0^t \frac{\left(\frac{\partial H}{\partial t'}\right)_{kn}}{(E_k - E_n)} \mathrm{e}^{-\frac{i}{\hbar}\int_0^t (E_k - E_n)\,\mathrm{d}t''}\,\mathrm{d}t' \qquad (2.89)$$

Eq. (2.89) is a measure then of the adiabaticity of a collision inducing an interaction representable by H', since it tells us — or more exactly its absolute square tells us — the rate at which another states, "k", will be occupied by our system originally in the state "m". It is obvious that for an adiabatic process, the absolute value of Eq. (2.89) must be much less than one. Many authors* have discussed adiabaticity, but our discussion will most closely parallel that of Margenau and Lewis[27].

Before paralleling these gentlemen, however, let us remark briefly on the qualitative matter of level separation as pointed up by the resonance denominator of Eq. (2.89). When a collision occurs the level of the emitter is perturbed or distorted either up or down by the interaction of the broadener. Now it is apparent that the closer the distorted emitter level lies to another of that system's levels, the more probable is a collision induced transition to the latter. With which, of course, adiabaticity is destroyed. It is apparent from these considerations that the widely separated levels of visible atomic spectra will probably not be subject to diabatic pressures while the more and more closely spaced molecular levels giving rise to infrared and microwave spectra would seem more prone to such influences. All of which is rather qualitative but somewhat informative.

* More than one.

We shall now consider two cases: (1) Diabaticity would cause transitions among the sub-levels of a degenerate state. For example, transitions either before or after the magnetic degeneracy is removed by the Stark Effect between these magnetic levels. (2) Diabaticity causes transitions between non-degenerate levels.

For a state which is twofold degenerate and whose degeneracy is removed in first order by a perturbation H' the denominator in Eq. (2.89) is given by $2|H_{12}'|$. If the state happens to be more than twofold degenerate then this equality will not hold precisely; however, it will be an order-of-magnitude relation which will be sufficient unto our purposes here. Eq. (2.89) now becomes:

$$\left| \int_{-\infty}^{+\infty} \frac{\frac{\mathrm{d}}{\mathrm{d}t'}|H_{12}'(t')|}{|H_{12}'(t')|} \exp\left[-\frac{i}{\hbar}\int_{-\infty}^{t'} 2|H_{12}'(t'')|\,\mathrm{d}t''\right] \mathrm{d}t' \right| \ll 1 \tag{2.90}$$

An inspection of this equation should be sufficient to evoke those conditions which will satisfy the inequality represented by it. The imaginary character of the exponential means, of course, an oscillating function which will integrate to zero over a sufficiently long period of time if nothing too drastic happens to its multiplicative factor. These shall constitute the adiabaticity requirements for the present situation:

(1) The perturbation or collision time is much greater than the period of oscillation, $\hbar/2|H_{12}'|$. This assures many oscillations of the exponential function.

(2) The multiplicative factor $\mathrm{d}/\mathrm{d}t'\ |H_{12}'(t')|/|H_{12}'(t')| = \mathrm{d}/\mathrm{d}t' \ln|H_{12}'(t')|$ is a slowly varying function of the time during an average period of oscillation. This keeps any drastic behavior of this function from interfering with the cancellation effect of the oscillating exponential.

Let us remark that our exponential is, of course, the phase shift. Thus, our requirement (1) requires large phase shifts for adiabaticity. This in turn means collisions of closest approach less than the optical collision radius. Condition (2) can be shown to reduce to Condition (1), and condition (1) basically requires long encounters and this would appear to infer low velocities.

In the degenerate case those slow collisions falling within the optical collision radius are adiabatic while those falling without are diabatic.

We turn our attention to the non-degenerate case. Since, in actuality, the degeneracy was removed in the "degenerate" case, the real difference between that situation and this one is the large difference between level separation in that and the present case. In the present case, the key approximation is:

$$E_k - E_n \doteq E_k^{(0)} - E_n^{(0)}$$

that is to say, the level separation under the influence of the broadening perturbation is about the same as the unperturbed separation. This allows us to write our adiabaticity criterion as:

$$\frac{1}{|E_k^{(0)} - E_n^{(0)}|} \left| \int_{+\infty}^{-\infty} \frac{\mathrm{d}}{\mathrm{d}t'} H_{kn}'(t') \exp\left[-\frac{i}{\hbar}(E_k^{(0)} - E_n^{(0)})\,t'\right] \mathrm{d}t' \right| \ll 1 \tag{2.91}$$

Let us first consider the case of slow collisions, that is, of collisions for which the period of oscillation, $\hbar/|E_k^{(0)} - E_n^{(0)}|$ is much less than the time of collision, τ_p. We may integrate Eq. (2.91) by parts and make use of the fact that the perturbing potential disappears at both infinities to obtain:

$$\frac{1}{\hbar}\left|\int_{-\infty}^{+\infty} H_n(t') \exp\left[-\frac{i}{\hbar}(E_k^{(0)} - E_n^{(0)})\, t'\right] dt'\right| \ll 1 \qquad (2.92)$$

We again have the situation wherein the inequality holds if H_{kn}' is but slowly varying over the oscillation period. Here as in the degenerate case, this can be demonstrated for the classical path assumption. As a result of this all slow collisions are adiabatic.

In the case of fast collisions wherein the collision time is much less than the oscillation period we can take the exponential in Eq. (2.91) as unity. The equation may then be integrated with result:

$$|H_{ku}'(\infty) - H_{kn}'(-\infty)|/|E_k^{(0)} - E_n^{(0)}| = 0 \ll 1$$

so that the adiabatic criterion is seen to be satisfied for fast collisions. Let us therefore repeat ourselves:

In the degenerate case those slow collisions falling within the optical radius are adiabatic.

In the non-degenerate case all fast and all slow collisions are adiabatic.

2.11. Spitzer's rotational adiabaticity

In our considerations of the last section we have gone into what might be called the practical problems involved in an adiabatic collision — or otherise. What we consider now is a somewhat different phenomenon, although one can certainly relate it closely to adiabatic considerations, and its interest is principally pedagogical.

Let us suppose that our emitting molecule exists initially in an environment wherein there are no fields to establish a unique direction in space. As a consequence the magnetic quantum number associated with a given state has no physical significance. Now let us suppose that a broadening atom approaches, and that this broadener is possessed of an electric field, a field which will probably be — and we so assume it — radial. Now an electric field exists in the neighborhood of our emitter, and the z-axis of this emitter may be expected to be aligned along it. Of course, various orientations of the emitter correspond to various m-values and, hence, to various states. At some initial time the field of the broadener may be supposed switched on. Now whatever the emitter orientation is at this time, there will correspond an m-state. But from this temporal point on, the movement — for we do suppose there to be a movement — of the broadener will result in a rotation of the emitter z-axis since it is defined as along the radial field. Suppose the field to be of strength sufficient to swing the molecule around so that it maintains the same orientation with respect to its own z-axis. The m-value will remain unchanged so one may say that the collision is adiabatic.

On the other hand suppose the collision to be a distant one and the field of the perturber to be practically too weak to affect the emitter. Then surely we cannot expect the rotation of the emitting molecule, and, from the point of view ennunciated here, the collision is diabatic. And yet, we may still claim adiabaticity on the basis of the considerations of the last section, and adiabatic theory may be quite legitimately applied.

What appears to evolve here is a two class adiabaticity system, the first of which has been described in the previous section and to which one should usually appeal. The second class of spatial adiabaticity we may generally ignore, but of its existence we should certainly be aware.

Spitzer, utilizing Schwinger's computations[187] for the matrix elements, $\vartheta^{-1} \int \bar{\psi}_s \frac{\partial \psi_r}{\partial t} \, d\tau$, in a rotating coordinate system, carried out a calculation which led him to a minimum value for the phase shift on collision which would allow the reasonable utilization of the spatially adiabatic approximation.

The adiabatic equations

$$H(t)\,\psi_s(t) = E_s(t)\,\psi_s(t)$$

are assumed to have the eigenvalues

$$E_s(t) = \frac{q_s}{[r(t)]^n}$$

so that the adiabatic phase shift is given by:

$$\eta = \int_{-\infty}^{+\infty} \frac{2\pi E_s(t)}{h} \, dt = \frac{q_s}{h} \int_{-\infty}^{+\infty} \frac{dt}{(R^2 + v^2 t^2)^{n/2}}$$

Spitzer showed that the adiabatic approximation is valid if

$$\eta > c_n K_{sr} \left| \frac{\Lambda_s}{\Lambda_s - \Lambda_r} \right| \tag{2.93a}$$

and invalid if

$$\eta < c_n \left| \frac{\Lambda_s}{\Lambda_s - \Lambda_r} \right| \tag{2.93b}$$

where:

$$c_n = \frac{\sqrt{\pi}\,\Gamma[\frac{1}{2}(n-1)]}{\Gamma(\frac{1}{2}n)}$$

$$K_{sr} = \dot{\vartheta}^{-1} \int \bar{\psi}_s \frac{\partial \psi r}{\partial t} \, d\tau$$

$$q_s = \Lambda_s q_s; \quad |\Lambda_s - \Lambda_r| = 1 \text{ (lowest value)}$$

Eqs. (2.93) then give the Spitzer criteria for the utilization of the adiabatic hypothesis when the rotating electronic states are utilized. We might remark that when the rotation is ignored and adiabaticity is assumed, the two incorrect approximations tend to compensate each other.

Finally, Spitzer obtained also a limit for the applicability of the Interruption analysis

$$|\nu - \nu_0| < \frac{P}{2\pi}\left(\frac{\hbar\, v^n}{c_n\, q_s}\right)^{1/n-1} \tag{2.94}$$

which restricts the validity in the line wings. The symbol P is the only previously undefined one in Eq. (2.94) and this a rather arbitrary allowable error that is inserted.

Let us note that if we let $P = 1$, $q_s/\hbar = 2\pi C$, Eq. (2.94) very nearly ($C_n^n \neq C_n$) corresponds to Eq. (2.10), the earlier obtained wing limit for the Interruption Theory.

2.12. The correlation function

Our interruption considerations began with a particular viewpoint, and then the emphasis gradually shifted to an associated but really quite different one. To begin with, the broadening effects of collisions were treated by supposing the collisions to be sudden drastic events which interrupted the otherwise smooth flow of emitted radiation, the manner of this interruption being of less importance than the fact of it. Now then, after such an assumption was made Fourier analysis was applied to the description of this phenomenon. We discovered, however, that the effects of near and distant collisions were not included in such a consideration, and, in including these effects, we maintained the Fourier method but moved away from the Interruption concept. Therefore, our classification of even more advanced stages of this developing theory as Interruption Theories may be somewhat presumptuous, but we shall continue to appeal to genealogical arguments in this regard. In what has preceded, our considerations have been completely classical ones. In what is to follow our considerations will be basically quantum ones, and, if we wish, we may find in them the quantum justification which may at times have appeared lacking in what has gone before.

Here, as earlier, we are going to assume a classical path, a subject which we shall discuss at some length later on. In this classical path we are supposing that the location of the particles under consideration can be specified so that, for example, separations between them may be precisely determinable. Such classical paths are *a priori* assumptions in our previous classical theories, but they pose the necessity for justification when quantum theories are under consideration. Having made this determination with respect to translational motion and molecular location, let us write down the Hamiltonian for our physical system [46]:

$$H = H_1(p_l, q_l) + V\big(p_l, q_l, Q_i(t)\big) + S(p_l, q_l, p_s, q_s) + H(p_s, q_s) \tag{2.95}$$

where H_1 is the unperturbed Hamiltonian of the molecules, V that of the molecular interactions responsible for our broadening, S that of the molecule-field interaction, and H_4 is the Hamiltonian of the unperturbed field.

We take the function $\Psi_{a_n}{}^{(0)}$ to represent the unperturbed system molecule plus radiation field wherein the molecule is in an upper state, and there are no photons in the field. On the other hand the function $\Psi_{b_s}^{(0)}$ represents the system in a state such that the molecule is in a lower state, and there is a photon of frequency ω_s in the field. Now we shall expand the wave function for our system perturbed by the broadening interactions in terms of these states. The coefficients in this expansion will be time dependent. Further, they will tell us the probability for the appearance of a particular frequency in the broadened spectral line after a sufficiently long period of time is allowed to pass so that some sort of transition must have been made. Let us detail this.

Our assumed solution will be of the form:

$$\psi = \sum_n a_n(t)\, \Psi_{a_n}^{(0)} + \sum_s b_s(t)\, \Psi_{b_s}^{(0)} \tag{2.96}$$

We will recall that the zeroth-order wave functions for our system molecule plus field is a product of the molecular wave function and the harmonic oscillator function for the virtual oscillators in the field. Also, let us remark that V is independent of the field coordinates, so that there can be no change in the number of photons in the field as far as the operator V is concerned. On the other hand S is a function of these field coordinates, so that there must be some change in the number of photons in the field for the elements of this operator to exist. Thus, V_{kj} vanishes if the number of photons changes, and S_{kj} vanishes if such is not the case. Let us substitute Eq. (2.96) into the Schrödinger equation to obtain:

$$\left.\begin{aligned} H\psi &= i\hbar \sum \dot{a}_n \Psi_{a_n}^{(0)} + i\hbar \sum \dot{b}_s \Psi_{b_s}^{(0)} \\ &+ \sum a_n E_n \Psi_{a_n}^{(0)} + \sum b_s (E_s + h\nu_s)\, \Psi_{b_s(0)} \end{aligned}\right\} \tag{2.97}$$

where the temporal differentiation has already been carried out. Let us now suppose that no optical transition takes place. In accordance with our discussion above then, we obtain:

$$\dot{a}_n = -\frac{i}{\hbar} V_{mn}(t)\, a_m(t) \exp\left[-i(E_m - E_n)\, t\right] \tag{2.98a}$$

We multiply through on the left by $\Psi_{b_f}^{(0)}$ and integrate over all space, remembering our discussion of V and S above, to obtain:

$$\dot{b}_s = S_{if}\, b_s \exp\left[-\frac{i}{\hbar}(E_i - E_f - \hbar\omega_s)\, t\right] - \frac{i}{\hbar} V_{nf}(t)\, a_n(t) \exp\left[-\frac{i}{\hbar}(E_n - E_f)\, t\right] \tag{2.98b}$$

where, of course, $(H_1)_{nn} = E_n$, $(H_4)_{ss} = \hbar\omega_s$, and so on.

In Eqs. (2.98) the V_{kj} are the matrix elements of the molecular interactions and the S_{kj} are the matrix elements of the molecule-field interaction. Now under the adiabatic collision assumption Eq. (2.98a) yields:

$$a_n(t) = \frac{V_{mn}(t) a_m(t)}{E_m - E_n} \exp\left[-\frac{i}{\hbar}(E_m - E_n)\, t\right] \tag{2.99}$$

We shall substitute for V_{nf} in Eq. (2.98b) from Eq. (2.99) to obtain:

$$\dot{b}_s = S_{if} a_i \exp\left[-\frac{i}{\hbar}(E_i - E_f - \hbar\omega_s)\,t\right] - \frac{i}{\hbar} b_s\, P_f(t) \tag{2.100}$$

where $P_f(t) = E_n - E_f$.

We now choose a_i as:

$$a_i = \mathrm{e}^{-\gamma t}\,\mathrm{e}^{-\frac{i}{\hbar}\int\limits_0^t P_i(t)\,\mathrm{d}t} \tag{2.101}$$

and we may now solve Eq. (2.100) for the probability amplitude for the emission of a quantum of frequency ω_s:

$$b_s(T) = \exp\left[-\frac{i}{\hbar}\int\limits_0^T P_f\,\mathrm{d}t\right]\int\limits_0^T \mathrm{d}t\, S_{if}\exp\left[-\gamma T - \frac{i}{\hbar}\int^t (P_i(t) - P_f(t))\,\mathrm{d}t + i\omega t\right] \tag{2.102}$$

where $\omega = \omega_{if} - \omega_s$, and we have supposed $b_i(0) = 0$.

If the natural line width factor γ is neglected and the time of collision T is extended to infinity, we essentially obtain the Weisskopf expression for the Fourier amplitude.

As we have noted on several occasions, the Fourier amplitude may be written as:

$$b(\omega) = \sqrt{N}\int\limits_0^\infty \mathrm{d}t\exp\left[-i\int\limits_0^t P(x)\,\mathrm{d}x + i\omega t\right] \tag{2.103}$$

where $P(t) = P_i(t) - P_f(t)$ from which:

$$I(\omega) = N\int\limits_0^\infty \mathrm{d}t_1\int\limits_0^\infty \mathrm{d}t_2\exp\left[-i\int\limits_{t_2}^{t_1} P(x)\,\mathrm{d}x + i\omega(t_1 - t_2)\right] \tag{2.104a}$$

which becomes, when we let $t_1 - t_2 = \tau$ and $t_2 = t_0$:

$$\left.\begin{aligned} I(\omega) &= N\int\limits_{-\infty}^{+\infty}\mathrm{d}\tau\int\limits_0^\infty \mathrm{d}t_0\exp\left[-i\int\limits_{t_0}^{t_0+\tau} P(x)\,\mathrm{d}x + i\omega\tau\right] \\ &= \Re\left\{N\int\limits_0^\infty \mathrm{d}\tau\int\limits_0^\infty \mathrm{d}t\exp\left[-i\int\limits_{t_0}^{t_0+\tau} P(x)\,\mathrm{d}x + i\omega\tau\right]\right\}\end{aligned}\right\} \tag{2.104b}$$

again we encounter Eq. (2.25):

$$C(\tau) = \int\limits_0^\infty \mathrm{d}t_0\exp\left[-i\int\limits_{t_0}^{t_0+\tau} P(x)\,\mathrm{d}x\right] \tag{2.105}$$

With Eq. (2.105) we encounter the so-called "Correlation Function" which has become rather fashionable of late. Its appeal is obvious, since, its Fourier transform being the intensity distribution virtually by definition, one concerns oneself primarily with this function. Margenau and Lewis[217] have provided a rather lucid qualitative interpretation.

It is certainly apparent that this Correlation Function will be large when τ is zero, that is, $C(0)$ is large. Now suppose our exponential changes rapidly from its value for τ zero as we increase the value of τ. We will then have a function somewhat similar to a delta function as our Correlation Function, and with bright alertness we will be able to say that our integrated exponent quickly loses correlation with its previous values. Matters are all of a "more or less" nature here, but, at any rate, this delta function for our Correlation Function will mean that the intensity will be more or less independent of the frequency. Were the delta function exactly a delta function, the peak would occur for τ zero, and intensity would be completely independent of frequency. This would mean a broad — though not very meaningful — intensity distribution. As we tend toward this delta function, however, we could except our intensity distribution to be broad. Thus, rapid loss of correlation results in wide intensity distributions. The reverse situation follows rather naturally. As we gain correlation among the values of our integrated exponential our Correlation Function broadens and our intensity distribution narrows. Indeed the half width of the Correlation Function is nearly the reciprocal of the half width of the intensity distribution. One may deal with this function in many ways; Foley began by assuming:

$$\int_{t_0}^{t_0+\tau} P(x)\,\mathrm{d}x = \sum_{i=1}^{n} \eta_i \tag{2.106}$$

under the assumptions that (1) the time interval involved is long in comparison to the duration of the collision (2) which is vanishingly small. These then result in the approximation of the sum of an integral number of collision phase shifts as given by Eq. (2.106). The fact that this approximation is applicable to the case of low densities and high velocities arises out of these assumptions.

Let us first refer to Eq. (1.56) for the probability that the molecule will undergo a collision during τ where as always τ_0 is the mean time between collisions. Note that $\tau/\tau_0 = A$ the number of collisions during time τ.* As in Eq. (2.54a) the probability of n collisions in time τ is given by $\left(\frac{\tau}{\tau_0}\right)^n \frac{1}{n!} \mathrm{e}^{-\tau/\tau_0}$. From Eqs. (2.105) and (2.106) the average value of the correlation function for this case of n collisions is:

$$\int \mathrm{d}t_0 \exp\left[-i \sum_{i=1}^{n} \eta_i\right] = \prod_{i=1}^{n} \int_{-\infty}^{+\infty} \mathrm{d}\eta\, p(\eta)\,(\cos\eta - i\sin\eta) = (A - iB)^n \tag{2.107}$$

the $p(\eta)\,\mathrm{d}\eta$ being the distribution in phase shifts. Thus, A and B are the average values ($A = \int \mathrm{d}\eta\, p(\eta) \cos\eta$, etc.) of $\cos\eta$ and $\sin\eta$ respectively.

Now Foley established a cut-off distance ϱ_0, time τ_0, and phase shift η_0 for the intermolecular forces which may later be extended. We have already averaged over n collisions, and Foley completed the calculation and obtained $C(\tau)$ by

* Jablonski[96] objected to Foley's utilization of $1/\tau_0\, \mathrm{e}^{-\tau/\tau_0}$ for the distribution of collisions. He felt that a Maxwell-Boltzmann distribution was called for, but, as Foley[41] pointed out, the random distribution for the broadening case seems to have precedent if nothing else as justification. This would not appear to be entirely conclusive.

averaging over the various possible values of n as follows:†

$$C(\tau) = \sum_{n=0}^{\infty} \frac{e^{-\tau/\tau_0}}{n!} \left(\frac{\tau}{\tau_0}\right)^n (A - iB)^n = \exp\left[\frac{A - iB}{\tau_0}\tau - \frac{\tau}{\tau_0}\right] \tag{2.108}$$

since the series is the well known one for the exponential.

The direct substitution of Eq. (2.108) into Eq. (2.104b) yields:

$$\left.\begin{aligned} I(\omega) &= \Re\left\{N \int_0^{\infty} d\tau \exp\left[\frac{A-1}{\tau_0}\tau - i\frac{B-\omega}{\tau_0}\tau\right]\right\} \\ &= \frac{(1-A)/\pi\tau_0}{[(1-A)/\tau_0]^2 + [B/\tau_0 - \omega]^2} \end{aligned}\right\} \tag{2.109}$$

It is quite apparent from Eq. (2.109) that we have obtained a line center shift of $\frac{B}{\tau_0}$. The half-width is, of course, $\frac{2(1-A)}{\tau_0}$. Since τ_0 appears in both shift and width results, it follows that a linear density proportionality exists. The constants occurring in Eq. (2.109) may be evaluated in a quite straightforward manner.

We commence by supposing interactions of the form $V = \beta h \gamma^{-p}$ so that the phase shift η is given by:

$$\eta = \int_{-\infty}^{+\infty} \frac{V_i - V_f}{h} dt = \int_{-\infty}^{+\infty} \frac{\beta_i - \beta_f}{(\varrho^2 + v^2 t^2)^{p/2}} dt = \frac{\gamma}{v\varrho^{p-1}} \tag{2.110}$$

where v is, as usual, the relative velocity and ϱ the distance of closest approach.

Eq. (1.13) tells us that:

$$\frac{1}{\tau_0} = \frac{v}{l} = N\pi\varrho^2 v \tag{2.111a}$$

where v is the mean relative velocity and l is the mean free path. Now, however, let us introduce a normalized distribution of velocities $f(v)\,dv$ and a distribution of interaction force constants g (not a continuous distribution). The reason for

† Mizushima[159] raised the objection that "... His derivation of ... Eq. (4.109) ... is inadequate in that he replaced the sum of averages by the average of sums ..."[159] A little consideration serves to clarify Mizushima's contention. On the left side of Eq. (2.107) there is obviously a sum over η. Now the individual terms of this sum are not reproduced by assigning various values to n, but, traced back through Eq. (2.107), various sums result in Eq. (2.108). Thus, the average in Eq. (2.108) is one over sums, as Mizushima noted. Mizushima carried through this calculation in what appears to be the more proper order to obtain:

$$I(\nu) = \frac{c}{\pi} \frac{\delta}{(\nu_0 - \nu_0 - \Delta)^2 + \delta^2} \tag{2.108a'}$$

where:

$$\delta = \int F(s)\,(1 - \cos\eta_s)\, ds/2\pi \tag{2.108b'}$$

$$\Delta = \int F(s) \sin\eta_s\, ds/2\pi \tag{2.108c'}$$

and $F(s)\,ds$ is the number of collisions with collision parameter between s and $s + ds$ per unit time.

the latter distribution should become apparent when we recall the dependence of this interaction constant on the molecular states involved. When these two distributions are applied and averaged over, Eq. (2.111a) becomes:

$$\frac{1}{\tau_0} = N \sum_k \int_0^\infty dv\, f(v)\, g_k\, \pi\, \varrho_{vk}^2\, v \tag{2.111b}$$

which is surely the expression for the number of collisions which produce a phase shift greater than our cut-off phase shift η_0. Eq. (2.111b) may be rewritten according to Eq. 2.110) as:

$$\left.\begin{aligned} \frac{1}{\tau_0} &= N \sum_k \int_0^\infty dv\, f(v)\, g_k\, \pi \frac{\gamma_k^{2/(p_k-1)}\, v^{(p_k-3)/(p_k-1)}}{\eta^{2/(p_k-1)}} \\ &= N\pi \sum_k g_k \langle v^{(p_k-3)/(p_k-1)} \rangle \frac{\gamma^{2/(_k-1)}}{\eta^{2/(p_k-1)}} \end{aligned}\right\} \tag{2.111c}$$

The line half-width has been determined as $\frac{2(1-A)}{\tau_0}$ or:

$$\left.\begin{aligned} &= \frac{2}{\tau_0}\left[1 - \int_{-\infty}^{+\infty} \cos\eta\, p(\eta)\, d\eta\right] = \frac{2}{\tau_0}\left[1 + \int_{-\infty}^{+\infty}\left[2\sin^2\frac{\eta}{2} - 1\right] p(\eta)\, d\eta\right] \\ &= \frac{4}{\tau_0}\int_{-\infty}^{+\infty} \sin^2\frac{\eta}{2}\, p(\eta)\, d\eta \end{aligned}\right\} \tag{2.112}$$

Since:

$$\int_{-\infty}^{+\infty} p(\eta)\, d\eta = 1$$

Foley took $1/\tau_0\, p(\eta)\, d\eta$ as:

$$\frac{1}{\tau_0}\, p(\eta)\, d\eta = \frac{D\, \gamma_k^{2/(p_k-1)}\, g_k\, f(v)\, v^{(p_k-3)/(p_k-1)}}{\eta^{(p_k+1)(p_k-1)}\,(p_k-1)}\, d\eta \tag{2.113}$$

where D is a normalization factor.

When Eq. (2.113) is substituted into Eq. (2.112) the result is:

$$\delta = 8\pi N \sum_k g_k \frac{\gamma^{2/p_k-1}}{(p_k-1)} \langle v^{(p_k-3)/(p_k-1)} \rangle \int_0^\infty \frac{d\eta\, \sin^2\eta/2}{|\eta|^{(p_k+1)/(p_k-1)}} \tag{2.114}$$

Integration may be carried out from zero to infinity (and the result doubled) in the case of all phase shifts positive. If this is not the complete integration must, of course, be carried out.

Eq. (2.113) may be again used to obtain the phase shift as:

$$\left.\begin{aligned} \frac{B}{\tau_0} &= \frac{1}{\tau_0}\int \sin\eta\, p(\eta)\, d\eta \\ &= 2\pi N \sum_k g_k \frac{\gamma^{2/(p_k-1)}}{(p_k-1)} \langle v^{p_k-3/p_k-1} \rangle \int_0^\infty \frac{d\eta\, \sin\eta}{|\eta|^{(p_k+1)/(p_k-1)}} \end{aligned}\right\} \tag{2.115}$$

where again only positive phase shifts have been assumed.

The integrals in Eqs. (2.114) and (2.115) are amenable to evaluation in terms of gamma functions. For foreign gas broadening in molecular spectra to which we are restricting ourselves here, an interaction of the form β/r^p yields a shift to width ratio of:

$$\frac{\delta}{\Delta} = \frac{1}{2}\cot\left\{\frac{p-3}{2(p-1)}\pi\right\} \tag{2.116}$$

Anderson[5] was later able to make use of the Correlation Function in a manner yielding interesting results rather rapidly. He did restrict himself to additive forces due to different broadeners and began with the following expression for the frequency perturbation due to these broadeners:

$$\Delta\omega = \sum_{i=1}^{n} \Delta\omega(R_i) \tag{2.117}$$

We may now write our Correlation Function as:

$$C(\tau) = \left\langle \exp\left[i\int_0^\tau \nu(t)\,\mathrm{d}t\right]\right\rangle = \left\langle \prod_{i=1}^{n} \exp\left[i\int_0^\tau \nu(R_i)\,\mathrm{d}t\right]\right\rangle = \left\langle \exp\left[i\int_0^\tau \nu(R)\,\mathrm{d}t\right]\right\rangle^n \tag{2.118}$$

The average of the product has been replaced by the product of the averages* since we suppose there to be no interactions between broadeners, and, further, all broadeners are like. If we take as a distance of closest approach along the supposed straight line trajectory of our broadeners, ϱ, and x_0 as the position of the particle at the start of the time interval, the emitter-broadener separation is surely:

$$R^2 = (x_0 + \langle v\rangle\, t)^2 + \varrho^2$$

The averaging indicated in Eq. (2.118) is carried out over ϱ and x_0 with the result:

$$\left.\begin{aligned} C(\tau) &= \frac{1}{V^n}\left\{\int_0^\infty 2\pi\varrho\,\mathrm{d}\varrho \int_{-\infty}^{+\infty} \mathrm{d}x_0 \exp\left[i\int_0^\tau \nu(R)\,\mathrm{d}t\right]\right\}^n \\ &= \left[1 - \frac{2\pi N}{n}\int_0^\infty \varrho\,\mathrm{d}\varrho \int_{-\infty}^{+\infty} \mathrm{d}x_0 \left(1 - \exp\left[i\int \nu(R)\,\mathrm{d}t\right]\right)\right]^n \\ &= \exp\left[-N\,V'(\tau)\right] \end{aligned}\right\} \tag{2.119}$$

since:

$$\lim_{n\to\infty}\left[1 - \frac{A}{n}\right]^n = \exp\left[-A\right]$$

and where:

$$V'(\tau) = 2\pi\int_0^\infty \varrho\,\mathrm{d}\varrho \int_{-\infty}^{+\infty} \mathrm{d}x_0 \left\{1 - \exp\left[i\int_0^\tau \nu\left(\sqrt{(x_1 + \langle v\rangle\, t)^2 + \varrho^2}\right)\mathrm{d}t\right]\right\} \tag{2.120}$$

* For precedent see Eq. (7.25) and for discussion see *infra*, Chap. 4.

Anderson was able to find a limit in the high pressure (large separation from line center) case and in the low pressure (frequencies near line center) case. Numerical evaluations were required for intermediate situations, however. For the low pressure case or for those frequencies near line center N is comparatively small — or, we might consider $\langle v \rangle$ as large — the x_0 may be dropped from the exponential and the range of integration of x_0 is restricted to $\langle v \rangle$. The limits on the temporal integration are extended and we obtain:

$$\left.\begin{aligned} C(\tau) &= \langle v \rangle \tau \int_0^\infty \varrho \, \mathrm{d}\varrho \left\{1 - \exp\left[i \int_{-\infty}^{+\infty} \nu\left(\sqrt{\varrho^2 + \langle v^2 \rangle t^2}\right) \mathrm{d}t\right]\right\} \\ &= (\sigma_r + \sigma_i) \langle v \rangle \tau = \sigma \langle v \rangle \tau \end{aligned}\right\} \tag{2.121}$$

which results in an intensity expression:

$$I(\omega) = \frac{N \langle v \rangle \sigma_r}{(\omega - \omega_0 - N \langle v \rangle \sigma_i)^2 + (N \langle v \rangle \sigma_s)^2}. \tag{2.122}$$

It is rather apparent that we have obtained an Interruption type distribution. In finding this distribution as a limiting case of a more general distribution we supposed that (1) the perturber density was low and/or (2) the perturber velocity was high and/or (3) the frequencies of interest to us were near the center of the spectral line as opposed to being in the line "wing" or at large distances from the line center.

Since these approximations were used in obtaining the Interruption form, we can suppose the use of the form or the theory to be justified when these conditions are satisfied. Which, indeed, it is.

The limit which exists in the other extreme — low velocities, frequencies in the line wing, and high densities — is of Statistical form, a form which we intend to consider in detail in the next chapter. We will also defer our consideration of this reduction until then.

We shall largely complete our Interruption considerations by turning our attention to the generalized quantum derivation of the spectral line shape. In this case, however, we introduce first a particular case of this general situation, primarily because it offers such direct and delightful authentication of the Lorentz line shape. Before even turning to our special case, it would behoove us to make a few brief remarks on the density matrix.

2.13. The density matrix

Let us again consider the representative point in phase space which we mentioned preceding Eq. (2.20). We specify the density in phase space by Υ, and we suppose this density to be a measure of the probability that the representative point is in the volume element corresponding to a certain configuration and momentum state per unit phase volume. Since the representative point is of necessity somewhere in phase space:

$$\int \ldots \int \Upsilon \, \mathrm{d}q_1 \ldots \mathrm{d}q_{3N'} \, \mathrm{d}p_1 \ldots \mathrm{d}p_{3N'} = 1 \tag{2.123a}$$

and the mean value of a quantity F is defined as:

$$\langle F\rangle = \int \ldots \int \cdot F \Upsilon \, dq_1 \ldots dp_{3N'} \tag{2.123b}$$

in accordance with the definition of a mean value under the aegis of any other distribution function.

For convenience, let us now take a collection of N such representative points in the phase space, representing a system of systems. Here Υ will be the actual system density, and we could replace one by B in Eq. (2.123a). Corresponding to this density in phase space, we may obtain and utilize a "density matrix" as the quantum mechanical analog.

We may define the components of this density matrix as:

$$\Upsilon_{nm} = \frac{1}{N'} \sum_{i=1}^{m} \bar{a}_{im}(t)\, a_{in}(t) = \langle \bar{a}_m a_n \rangle \tag{2.124}$$

Now, as usual, the average value of some observable F over one of the systems is:

$$\left.\begin{aligned} \langle F\rangle &= \int \bar{\Psi} F \Psi \, d\tau = \sum_{m,n} \int \bar{a}_m \bar{\psi}_m F a_n \psi_n \, d\tau \\ &\sum_{m,n} F_{mn} \bar{a}_m a_n \end{aligned}\right\} \tag{2.125a}$$

so that the average value of F over all systems is:

$$\langle F\rangle = \sum_{m,n} F_{mn} \langle \bar{a}_m a_n \rangle = \sum_{m,n} F_{mn} \Upsilon_{nm} \tag{2.125b}$$

which, according to the rules of matrix multiplication is:

$$\langle F\rangle = \sum_n (\Upsilon F)_{nn} = Tr \, ||\Upsilon F|| \tag{2.126}$$

In Eq. (2.126) "Tr" is the trace or diagonal sum of the product matrix $||\Upsilon F||$. Thus, the density matrix provides an averaging medium which we shall utilize subsequently.

We may obtain another useful relation by recalling the state growth equation which we have noted on several occasions:

$$\frac{\partial a_n}{\partial t} = -\frac{i}{\hbar} \sum_k H_{nk} a_k \tag{2.127}$$

From Eq. (2.127) we may infer that:

$$\begin{aligned} \frac{\partial \Upsilon_{nm}}{\partial t} &= \frac{\partial}{\partial t} \langle \bar{a}_m a_n \rangle = \left\langle \bar{a}_m \frac{\partial a_n}{\partial t} \right\rangle + \left\langle \frac{\partial \bar{a}_m}{\partial t} a_n \right\rangle \\ &= -\frac{i}{\hbar} \sum_k \left(H_{nk} \langle \bar{a}_m a_k \rangle - \bar{H}_{mk} \langle \bar{a}_k a_n \rangle \right) \\ &= -\frac{i}{\hbar} \sum_k (H_{nk} \Upsilon_{km} - \Upsilon_{nk} H_{km}) = -\frac{i}{\hbar} [(H\Upsilon)_{nm} - (\Upsilon H)_{nm}] \end{aligned}$$

or in matrix form:

$$i\hbar \frac{\partial}{\partial t} ||\Upsilon|| = ||H|| \, ||\Upsilon|| - ||\Upsilon|| \, ||H|| \tag{2.128}$$

2.14. Another quantum justification of the Lorentz equation

To begin with, the absorption coefficient for the absorption by a gas may be represented as:

$$\varkappa = 4\pi \frac{\omega}{c} N [\Im(\chi)] \tag{2.129a}$$

where ω is the frequency of the absorbed radiation and χ is either the electric or magnetic susceptibility. We may recall that the induced dipole moment is related to the susceptibility by

$$\mu(t) = \Re[\chi \boldsymbol{E} e^{-i\omega t}] \tag{2.129b}$$

where we now suppose our electric field to be:

$$\boldsymbol{E}(t) = \boldsymbol{E} \cos \omega t = \Re[\boldsymbol{E} e^{-i\omega t}] \tag{2.129c}$$

Now our general procedure here will be to (a) find the average value $\mu(t)$ using Eq. (2.126), (b) ascertain χ from Eq. (2.129b) by a utilization of the results of (a), and (c) find $\varkappa$ by an insertion of the results of (b) in Eq. (2.129), an apparently straightforward sequence of operations.

We must begin by gaining some knowledge of Υ, and the physical conditions of the problem will furnish us this information. To begin with, we recall the Van Vleck–Weisskopf distribution of $\mu(t)$ — through the amplitude distribution — as an application of the Maxwell–Boltzmann energy distribution. This leads us to our averaging density matrix at time $t_0 = t - \vartheta$ where t_0 is the time of collision:

$$||\Upsilon_0(t_0)|| = e^{\sigma/kT} ||e^{-H(t_0)/kT}|| = C ||e^{-H(t_0)/kT}|| \tag{2.130a}$$

and the normalizing condition:

$$Tr ||\Upsilon|| = 1$$

tells us that:

$$1/C = e^{-\sigma/kT} = Tr ||e^{-H(t_0)/kT}|| \tag{2.130b}$$

Further:

$$||H(t)|| = ||H_0|| - ||\mu|| \cdot \boldsymbol{E}(t) = ||H_0|| + ||V|| \cos \omega t \tag{2.131}$$

where now H_0 is the Hamiltonian of the isolated molecule and the dipole moment operator is given by $||\mu||$.

We are now in a position to utilize Eq. (2.126) for finding the average value of $||\mu(t)||$ as:

$$\langle \mu(t) \rangle = Tr ||\mu \langle \Upsilon(t) \rangle|| \tag{2.132}$$

The density matrix is to be averaged over the random distribution $1/\tau \, e^{-\vartheta/\tau} d\vartheta$ of times $t_0 = t - \vartheta$ since the last collision, an obviously necessary procedure. Insofar as $||\Upsilon||$ is concerned, it certainly may be presumed dependent on the time of the last collision t_0 such that $||\Upsilon|| = ||\Upsilon(t, t_0)||$. In addition $||\Upsilon(t, t_0)||$ must satisfy Eq. (2.128) and Eq. (2.130) through:

$$||\Upsilon(t_0, t_0)|| = ||\Upsilon_0(t_0)|| \tag{2.133}$$

Firstly then

$$||\langle\Upsilon(t)\rangle|| = \frac{1}{\tau}\int_0^\infty ||\Upsilon(t, t-\vartheta)||\, e^{-\vartheta/\tau}\, d\vartheta \tag{2.134}$$

and since:

$$\left[\frac{\partial}{\partial t}||\Upsilon(t, t_0)||\right]_{t_0=t-\vartheta} = \frac{\partial}{\partial t}||\Upsilon(t, t-\vartheta)|| + \frac{\partial}{\partial\vartheta}||\Upsilon(t, t-\vartheta)||$$

we may obtain from Eq. (2.134):

$$\left.\begin{aligned}
\frac{\partial}{\partial t}||\langle\Upsilon(t)\rangle|| &= \int_0^\infty \left[\frac{\partial}{\partial t}||\Upsilon(t, t_0)||\right]_{t_0=t-\vartheta} \frac{e^{-\vartheta/\tau}}{\tau}\, d\vartheta \\
&\quad - \int_0^\infty \frac{\partial}{\partial\vartheta}||\Upsilon(t, t-\vartheta)|| \frac{e^{-\vartheta/\tau}}{\tau}\, d\vartheta \\
&= -\frac{i}{\hbar}\int_0^\infty [||H(t)||\,||\Upsilon(t, t-\vartheta)|| \\
&\quad - ||\Upsilon(t, t-\vartheta)||\,||H(t)||]\frac{e^{-\vartheta/\tau}}{\tau}\, d\vartheta \\
&\quad - ||\Upsilon(t, t-\vartheta)||\frac{e^{-\vartheta/\tau}}{\tau}\Big|_0^\infty - \frac{1}{\tau^2}\int_0^\infty ||\Upsilon(t, t-\vartheta)||\, e^{-\vartheta/\tau}\, d\vartheta \\
&= -\frac{i}{\hbar}[||H(t)||\,||\langle\Upsilon(t)\rangle|| - ||\langle\Upsilon(t)\rangle||\,||H(t)||] \\
&\quad - \frac{1}{\tau}[||\langle\Upsilon(t)\rangle|| - ||\Upsilon_0(t)||]
\end{aligned}\right\} \tag{2.135}$$

by Eq. (2.128) and an integration by parts.

Next the transformation:

$$||D(t)|| = ||\langle\Upsilon(t)\rangle|| - ||\Upsilon_0(t)|| \tag{2.136}$$

may be introduced into Eq. (2.135). $||D(t)||$ is essentially a measure of the variation of the density matrix $||\Upsilon(t)||$ from a density matrix describing a condition of instantaneous thermal equilibrium.

Under this transformation Eq. (2.135) becomes:

$$\left.\begin{aligned}
\frac{\partial}{\partial t}||D(t)|| &= -\frac{i}{\hbar}[||H(t)||\,||D(t)|| - ||D(t)||\,||H(t)||] \\
&\quad - \frac{1}{\tau}||D(t)|| - \frac{\partial}{\partial t}||\Upsilon_0(t)||
\end{aligned}\right\} \tag{2.137}$$

Let us write out a typical matrix element of $(\partial/\partial t)\,||D(t)||$ from Eq. (2.137):

$$\frac{\partial}{\partial t}D_{mn} = -\frac{i}{\hbar}\sum_k [H_{mk}D_{kn} - D_{mk}H_{kn}] - \frac{i}{\tau}D_{mn} - \frac{\partial}{\partial t}(\Upsilon_0(t))_{mn} \tag{2.138a}$$

We may surely assume that $||H_0||$ has been diagonalized so that Eq. (2.138a) becomes:

$$\begin{aligned}\frac{\partial}{\partial t} D_{mn} &= -\frac{i}{\hbar} \sum_k \\ &\quad \times [H_{mk}{}^0 D_{km} \delta_{mk} - D_{mk} H_{kn}{}^0 \delta_{kn} + (V_{mk} D_{kn} - D_{mk} V_{kn}) \cos \omega t] \\ &\quad - \frac{1}{\tau} D_{mn} - \frac{\partial}{\partial t} (\Upsilon_0(t))_{mn} \\ &= -\frac{i}{\hbar} (H_{mm}{}^0 - H_{nn}{}^0) D_{mn} + \cdots\end{aligned}$$

$$\left.\begin{aligned}\frac{\partial}{\partial t} D_{nm} &= -\frac{i}{\hbar} (E_m{}^0 - E_n{}^0) D_{mn} - \frac{i}{\hbar} \sum_k [V_{mk} D_{kn} - D_{mk} V_{kn}] \\ &\quad - \frac{1}{\tau} D_{mn} - \frac{\partial}{\partial t} (\Upsilon(t))_{mn}\end{aligned}\right\} \quad (2.138\text{b})$$

From Eq. (2.138b) we may obtain simply by rearrangement:

$$\left.\begin{aligned}\left(\frac{\partial}{\partial t} + i\omega_{mn} + \frac{1}{\tau}\right) D_{mn}(t) &= -\frac{\partial}{\partial t} (\Upsilon_0(t))_{mn} \\ - \frac{i}{\hbar} \sum_k (V_{mk} D_{kn} - D_{mk} V_{kn}) \cos \omega t\end{aligned}\right\} \quad (2.139)$$

This is the apropos point in the development for the introduction of necessary restriction to a weak incident radiation field. This approximation allows us to drop the summation on the right of Eq. (2.139) due to the resultant smallness of the V_{nm}. In addition, this assumption means that the density matrix $||\Upsilon_0(t)||$ should not be very different from the density matrix for an isolated molecule at temperature T. Now this latter density matrix is given by:

$$||\Upsilon^{(0)}|| = C^{(0)} ||\mathrm{e}^{-H_0/kT}||; \qquad 1/C^{(0)} = Tr\,||\mathrm{e}^{-H_0/kT}|| \qquad (2.140\text{a})$$

and, since we have presupposed $||H_0||$ diagonal:

$$\Upsilon_{mn}{}^{(0)} = \Upsilon_m{}^{(0)} \delta_{mn}; \quad \varrho_m{}^{(0)} = \frac{\mathrm{e}^{-E_m/kT}}{Tr\,||\mathrm{e}^{-E_n/kT}||} \qquad (2.140\text{b})$$

Karplus and Schwinger[100] demonstrated the manner in which an operator of the form $||\mathrm{e}^{A+B}|| = ||\mathrm{e}^{H(t)/kT}||$ could be expanded in a series. The weakness of the field and the consequent smallness of the V_{mn} were then their basis for cutting off this expansion after the second term with the result:

$$(\mathrm{e}^{-H(t)/kT})_{mn} = \mathrm{e}^{-E_m/kT} \delta_{mn} + \frac{\mathrm{e}^{-E_m/kT} - \mathrm{e}^{-E_n/kT}}{E_m - E_n} V_{mn} \cos \omega t \qquad (2.141\text{a})$$

$$1/C = Tr\,||\mathrm{e}^{-H(t)/kT}|| + Tr\,||\mathrm{e}^{-H_0/kT}|| + \frac{1}{kT} Tr\,||\mathrm{e}^{-H_0/kT} \boldsymbol{\mu} \cdot \boldsymbol{E}|| \cos \omega t \quad (2.141\text{b})$$

Finally, the utilization of Eqs. (2.140) and (2.141) in Eq. (2.130a) yields:

$$(\Upsilon_0(t))_{mn} = \Upsilon_m^{(0)}\,\delta_{mn} + (\Upsilon_m^{(0)} - \Upsilon_n^{(0)})\,\frac{V_{mn}}{\hbar\,\omega_{mn}}\cos\omega t \qquad (2.142)$$

The substitution of Eq. (2.142) into Eq. (2.139), recalling that the sum has been dropped in this latter equation, now results in:

$$\left(\frac{\partial}{\partial t} + i\omega_{mn} + \frac{1}{\tau}\right) D_{mn}(t) = \omega(\Upsilon_m^{(0)} - \Upsilon_n^{(0)})\,\frac{V_{mn}}{\hbar\,\omega_{mn}}\cos\omega \qquad (2.143)$$

An integrating factor may be utilized after the normal fashion to obtain the steady state solution to this first order equation as:

$$\left.\begin{aligned} D_{mn}(t) &= \frac{\omega}{\omega - \omega_{mn} + i/\tau}(\Upsilon_n^{(0)} - \Upsilon_m^{(0)})\,\frac{V_{mn}}{2\,\hbar\,\omega_{mn}}\,e^{-i\omega t} \\ &+ \frac{\omega}{\omega - \omega_{mn} - i/\tau}(\Upsilon_n^{(0)} - \Upsilon_m^{(0)})\,\frac{V_{mn}}{2\,\hbar\,\omega_{mn}}\,e^{i\omega t} \end{aligned}\right\} \qquad (2.144)$$

To Eq. (2.144) we add $\|\Upsilon_0(t)\|$ as indicated by Eq. (2.136) to obtain:

$$\left.\begin{aligned} \langle\Upsilon_{mn}(t)\rangle &= \Upsilon_m^{(0)}\,\delta_{mn} \\ &+ \left(\frac{\omega}{\omega - \omega_{mn} + i/\tau} - 1\right)(\Upsilon_n^{(0)} - \Upsilon_m^{(0)})\,\frac{V_{mn}}{2\,\hbar\,\omega_{mn}}\,e^{-i\omega t} \\ &+ \left(\frac{\omega}{\omega - \omega_{mn} - i/\tau} - 1\right)(\Upsilon_n^{(0)} - \Upsilon_m^{(0)})\,\frac{V_{mn}}{2\,\hbar\,\omega_{mn}}\,e^{i\omega t} \end{aligned}\right\} \qquad (2.145)$$

The simple substitution of Eq. (2.145) into Eq. (2.132) yields:

$$\left.\begin{aligned} \boldsymbol{\mu}(t) &= \sum_{m,n} \boldsymbol{\mu}_{nm}\,\boldsymbol{\mu}_{mn}\cdot\boldsymbol{E}\,e^{-i\omega t}\left(1 - \frac{\omega}{\omega - \omega_{mn} + i/\tau}\right)\frac{\Upsilon_n^{(0)} - \Upsilon_m^{(0)}}{2\,\hbar\,\omega_{mn}} \\ &+ \sum_{m,n} \boldsymbol{\mu}_{mn}\,\boldsymbol{\mu}_{nm}\cdot\boldsymbol{E}\,e^{i\omega t}\left(1 - \frac{\omega}{\omega - \omega_{mn} - i/\tau}\right)\frac{\Upsilon_n^{(0)} - \Upsilon_m^{(0)}}{2\,\hbar\,\omega_{mn}} \end{aligned}\right\} \qquad (2.146)$$

Now, due to the spherical symmetry of the isolated molecule, $\boldsymbol{\mu}_{nm}\boldsymbol{\mu}_{mn}\cdot\boldsymbol{E}$ may be replaced by $\frac{1}{3}\boldsymbol{\mu}_{nm}\cdot\boldsymbol{\mu}_{mn}\boldsymbol{E} = \frac{1}{3}|\boldsymbol{\mu}_{nm}|^2\boldsymbol{E}$. When the substitution has been made in Eq. (2.146) we may use the resulting equation and Eq. (2.146b) to obtain:

$$\left.\begin{aligned} \chi &= \sum_{m,n} \tfrac{1}{3}|\boldsymbol{\mu}_{mn}|^2\left(1 - \frac{\omega}{\omega - \omega_{mn} + i/\tau}\right)\frac{\Upsilon_n^{(0)} - \Upsilon_m^{(0)}}{h\,\omega_{mn}} + \cdots \\ &= \sum_{m,n} \tfrac{1}{6}|\boldsymbol{\mu}_{mn}|^2\left(2 - \frac{\omega}{\omega - \omega_{mn} + i/\tau} + \frac{\omega}{\omega + \omega_{mn} + i/\tau}\right)\frac{\Upsilon_n^{(0)} - \Upsilon_m^{(0)}}{\hbar\,\omega_{mn}} \end{aligned}\right\} \qquad (2.147)$$

from which:

$$\left.\begin{aligned} \Im[\chi] &= \sum_{m,n}\frac{1}{6\hbar}|\boldsymbol{\mu}_{mn}|^2\,\frac{\omega}{\omega_{mn}} \\ &\left[\frac{1/\tau}{(\omega - \omega_{mn})^2 + 1/\tau^2} + \frac{1/\tau}{(\omega + \omega_{mn})^2 + 1/\tau^2}\right](\Upsilon_n^{(0)} - \Upsilon_m^{(0)}) \end{aligned}\right\} \qquad (2.148)$$

Eqs. (2.148) and (2.129a) then may be combined to give:

$$\left.\begin{aligned}\alpha &= \frac{2\pi}{3}\frac{\omega^2}{c}\frac{N}{kT}\sum_{m,n}|\mu_{mn}|^2\\&\left[\frac{1/\tau}{(\omega-\omega_{mn})^2+1/\tau^2}+\frac{1/\tau}{(\omega+\omega_{mn})^2+1/\tau^2}\right]\frac{1-e^{-\hbar\omega_{mn}/kT}}{\hbar\,\omega_{mn}/kT}\Upsilon_n^{(0)}\end{aligned}\right\}\quad(2.149)$$

The Lorentz absorption coefficient equation has thus been afforded still another quantum justification subject, of course, to the approximations introduced.

2.15. The quantum intensity distribution

We have studied several approximations to what the "true" — if this word is meaningful — intensity distribution should be. Some of them have been rather difficult of obtention, others almost ludicrously otherwise. We now wish to develop the "true" intensity distribution.

Anderson[3], Bloom and Margenau[15], and Margenau and Lewis[217] have provided us with derivations of this quantum intensity distribution. We may obtain the result directly by a rather intuitive approach. In such an approach we first recall the familiar expression for the classical intensity distribution of the radiation from on oscillating dipole:

$$I(\omega) = \frac{2\,\omega^4}{3\pi c^3}\left|\int_{-\infty}^{+\infty}\mu(t)\,e^{-i\omega t}\,dt\right|^2 \quad (2.150)$$

where, of course, the Fourier transform for the electric dipole moment has already been given in Eq. (2.3) and subsequent. When one supposed that operators replace the physical observables of Eq. (2.150), then the quantum expectation value for the intensity may be written as:

$$I(\omega) = \frac{2\,\omega^4}{3\,\pi c^3}\,Tr\left\{\Upsilon\int_{-\infty}^{+\infty}\mu(t)\,e^{-i\omega t}\,dt\int_{-\infty}^{+\infty}\mu(t')\,e^{i\omega t'}\,dt'\right\} \quad (2.151)$$

where the Υ is the density matrix and the $\mu(t)$, of course, are matrices. Objections can and have been raised to such a simple obtention of Eq. (2.151), so let us briefly consider one of the more rigorous Bloom and Margenau derivations.

Our first basis is established in terms of the eigenstates of the isolated, unperturbed molecule:

$$H_0\,\varphi_j = E_j\,\varphi_j \quad (2.152)$$

To this we add two other basis sets. The first describes the molecular system when perturbed by both collisions, V, and the field, S,

$$i\hbar\frac{\partial\Psi}{\partial t} = (H_0 + V + S)\,\Psi \quad (2.153)$$

while the second concerns itself only with collisions:

$$i\hbar\frac{\partial\Phi_i}{\partial t} = (H_0 + V)\,\Phi_i \quad (2.154)$$

and:

$$\Phi_i = \sum_j U_{ij}\varphi_j \tag{2.155}$$

where the boundary conditions,

$$\Phi_i(t_0) = \varphi_i \Rightarrow U(t_0) = 1 \tag{2.156}$$

follow logically and U is a Time Development Operator (TDO). This TDO may be considered as shaping the behavior of the time independent basis with time so as to yield the proper time dependent behavior. For example, letting $H(t) = H_0 + V$ we can take:

$$\Phi_i \doteq \exp\left[-\frac{i}{\hbar}\int_0^t H(t')\,dt'\right]\varphi_i; \quad U = \exp\left[-\frac{i}{\hbar}\int_0^t H(t')\,dt'\right] \tag{2.157}$$

and substitution into Eq. (2.154) shows that this would satisfy the equation. Further we require:

$$i\hbar\dot{U} = (H_0 + V)\,U$$

and inspection shows, for example, that Eq. (2.157) indeed satisfies this. With this diversion on TDO's let us return to the main development.

The Ψ_i may in turn be expressed in terms of the basis Φ_j:

$$\Psi = \Sigma\, a_j \Phi_j$$

where now

$$i\hbar\dot{a}_i = \sum_j S_{ji}\,a_j$$

and we remark the two step procedure which has been used to avoid the assumption of adiabaticity. We could have simply obtained the expansion of Ψ in the basis set φ_i directly by appealing to Eq. (2.153) immediately. When this is done adiabatic approximations to the actual situation are obtained because S alone is responsible for the transition, the operator V being restricted from such activity. In this basis to basis procedure, however, an expression for the collision smeared levels is first obtained, and, subsequently, the radiation (S) induced transition between a pair of such levels is considered. The smearing of these levels is partially brought about by adiabatic effects.

The eigenvalues of any operator may be expressed as:

$$\left.\begin{aligned} P_{ij} &= \int \bar{\Phi}_i P \Phi_j\,d\tau = \sum_{kl} U_{ki}\,U_{lj}\int \varphi_k P \varphi_l\,d\tau \\ &= (U^\dagger P^s U)_{ij} = (P^H)_{ij} \end{aligned}\right\} \tag{2.159}$$

and Eq. (2.158) becomes:

$$i\hbar\dot{a} - S^H a \tag{2.160}$$

of interated solution:

$$a(t) = \left[1 + \sum_{s=1}^{\infty} I^{(s)}(t)\right] a(t_0) \tag{2.161a}$$

where:

$$I^{(s)}(t) = (i\hbar)^{-s}\int_{t_0}^{t} dt_s \int_{t_0}^{t_s} dt_{s-1}\int_{t_0}^{t_{s-i}} \cdots \int_{ts}^{tr} dt_1\, S^H(t_s)\cdots S^H(t_1) \tag{2.161b}$$

As we are well aware this means that our transition probability is given by:

$$\left.\begin{aligned} |a_k(t)|^2 = |a_k|^2 &+ \sum_i \sum_{s=1}^{\infty} (a_k \bar{a}_i \bar{I}_{ki}^{(s)} + a_i \bar{a}_k I_{ki}^{(s)}) \\ &+ \sum_{il} a_i \bar{a}_l \sum_{rs} I_{ki}^{(s)} \bar{I}_{kl}^{(r)} \end{aligned}\right\} \quad (2.162)$$

From this one can obtain the increment in level population which occurs after an infinite time and hence obtain the spectral intensity distribution which can be equated with Eq. (2.151) and which we hardly need carry further. We should remark that in the final expression terms associated with both absorption and induced emission occur.

Our intention here has been to indicate the general fashion in which the rigorous proof of Eq. (2.151) was developed, and we have gone far enough to do this. Perhaps we should note that in the actual evaluation of Eq. (2.162) only the integrals $I_{ki}^{(1)} \bar{I}_{ki}^{(1)}$ were evaluated leading to an obvious degree of approximation in the final result.

2.16. Another adiabatic theory

One more theory awaits our consideration, this being an adiabatic one due to Mizushima[159]. Commencing with the Hamiltonian

$$H_1 + S + V + H_2$$

Which is analogous to that of Eq. (2.95). Mizushima eliminates the diabaticity from the state growth equation:

$$i \frac{\partial}{\partial t} a_j = \sum_i \left(H - i \frac{\partial}{\partial t} \right)_{ji} a_i$$

by eliminating the partial time derivative on the right hand side of this equation. Subsequently, a series of unitary transformations yields, for the author, the "generalized Fourier integral formula". By assuming **T** — the matrix diagonalizing $H_1 + H_2 + V$ — Mizushima obtained an expression which when integrated yields the Weisskopf result.

We have already noted in connection with Foley's theory* that Mizushima disagreed with the Foley procedure and finally obtained instead of the Foley result:

$$I(\nu) = \frac{c}{\pi} \frac{\delta}{(\nu - \nu_0 - \Delta)^2 + \delta^2} \quad (2.108\text{a}')$$

$$\delta = \int F(\varrho)\,(1 - \cos \eta_\varrho)\, d\varrho / 2\pi \quad (2.108\text{b}')$$

$$\Delta = \int F(\varrho) \sin \eta_\varrho\, d\varrho / 2\pi \quad (2.108\text{c}')$$

$$\eta - \int \frac{\Delta p}{h}\, dt \quad (2.160)$$

* See *supra*, Sec. 2.12.

The number of collisions with collision diameters between ϱ and $\varrho + d\varrho$ per unit time is given by:

$$F(s)\,ds = 8\sqrt{\pi}\left(\frac{m}{2kT}\right)^{3/2} \exp(-mv^2/2kT)\, v^3\, dv\, \varrho\, d\varrho\, GN \tag{2.161}$$

Let us consider this expression at sufficient length for justification. To begin with the factors preceding dv constitute nothing more nor less than the Maxwell–Boltzmann probability expression for the velocity having value between v and $v + dv$. The term $\varrho\, d\varrho$ is the probability for the occurrence of the optical collision diameter between a ϱ and $\varrho + d\varrho$, a simple enough expression since we assume random probability. Finally the expression G for the probability for the occurrence of a given rotational state will be given by a weighted Maxwell–Boltzmann function.

The materials are now at hand for an evaluation of the shift and shape according to this theory. Mizushima carries out this evaluation for (a) a well type potential and (b) an inverse power potential. For the well type:

$$\Delta p = \gamma \text{ for } r < a; \quad \Delta p = 0 \text{ for } r > a \tag{2.162}$$

so that:

$$\eta = 2(a - p)^{1/2}\, \gamma/v \tag{2.163}$$

Eqs. (2.108) give then:

$$\left.\begin{aligned} \delta &= \frac{1}{2} a^2 \langle v\rangle N - \frac{4}{\sqrt{\pi}} \hbar^2 \left(\frac{m}{4kT}\right)^{3/2} \int_0^\infty \varphi^{-8} \exp\left(-\frac{m}{4kT\varphi^2}\right) \\ &\times \left\{\left(\frac{a\varphi}{2\gamma\hbar}\right) \sin\left(\frac{2\gamma a\varphi}{\hbar}\right) + \frac{\left[\cos\left(\frac{2\gamma a\varphi}{\hbar}\right) - 1\right]}{4\gamma^2}\right\} d\varphi \end{aligned}\right\} \tag{2.164}$$

$$\left.\begin{aligned} \Delta &= \frac{4}{\sqrt{\pi}} \hbar^2 \left(\frac{m}{4kT}\right)^{3/2} \int_0^\infty \varphi^{-8} \exp\left(-\frac{m}{4kT\varphi^2}\right) \\ &\times \left\{\frac{\sin\left(\frac{2\gamma a\varphi}{\hbar}\right)}{4\gamma^2} - \frac{a\varphi}{2\gamma\hbar} \cos\left(\frac{2\gamma a\varphi}{\hbar}\right)\right\} d\varphi \end{aligned}\right\} \tag{2.165}$$

For the case $\gamma = \infty$:

$$\delta = \frac{1}{2} a^2 \langle v\rangle N \text{ and } \Delta = 0. \tag{2.166}$$

For the inverse power potential:

$$\Delta p = \frac{\hbar b}{r^n} \tag{2.167}$$

so that:

$$\left.\begin{aligned} \eta &= \int_{-\infty}^{+\infty} \frac{b}{(\varrho^2 + v^2 t^2)^{n/2}}\, dt \\ &= \frac{b}{v\varrho^{n-1}} \pi \frac{\Gamma(n-1)\, 2^{2-n}}{[\Gamma(n/2)]^2} \end{aligned}\right\} \tag{2.168}$$

Eqs. (2.161) and (2.168) subsequently lead to:

$$\left.\begin{aligned}\delta &= \frac{\pi^{(5-n)/(2n-2)}}{2}\left(\frac{k\,T}{2\,m}\right)^{(n-3)/(2n-2)} \Gamma\left(\frac{2\,n-3}{n-1}\right)\Gamma\left(\frac{n-3}{n-1}\right)\\ &\cdot\left[\frac{\Gamma(n-1)}{\{\Gamma(n/2)\}^2}\right]^{2/n-1} \sin\left\{\pi\,\frac{n-3}{2\,n-2}\right\}\langle|b|^{2/n-1}\rangle\,N\end{aligned}\right\} \qquad (2.169)$$

$$\Delta = \tan\left(\frac{\pi}{n-1}\right)\delta\left\langle\frac{b}{|\Delta\mu|}\right\rangle \qquad (2.170)$$

where the averaging is over G.

Eq. (2.167) leads to the following specializations for two interesting interactions:

$$\delta = 1.06\left(\frac{k\,T}{2\,m}\right)^{1/4}\langle|b|^{1/2}\rangle\,N \quad \text{for} \quad n = 5 \qquad (2.171\text{a})$$

$$\delta = 1.03\left(\frac{k\,T}{2\,m}\right)^{3/10}\langle|b|^{2/5}\rangle\,N \quad \text{for} \quad n = 6 \qquad (2.171\text{b})$$

If the sign of b is common to all collisions we will obtain Foley's Eq. (2.116) for the ratio between shift and width from the present theory. The converse of this can be seen, for example, for the case in which the average value of this quantity is zero and no shift results.

In applying the equations obtained above for the shift and width of the spectral line according to this adiabatic theory, the task, of course, becomes one of evaluating the matrix elements of the intermolecular interaction in order to obtain the values of b which must be used in these evaluations.

2.17. Review and summary

Our Interruption considerations began with the simple concept of the breaking off of the wave train with collision. As the time of collision and the effects of distant collisions were included we strayed further and further from these concepts although one bond of union was maintained in the Fourier integral. Of importance to remark is the fact that the simplest Lenz or Weisskopf theory is applicable within the proper limits, and the very simplicity of their intensity expressions does much to recommend these theories. In general our Interruption Theories are applicable when the pressure is low — probably not greater than an atmosphere — or at positions near line center. The inverse square line shape typical of the common ones is then applicable to the van der Waals forces and other forces occurring between unlike atoms. In the broadening between like atoms we shall see that the Statistical Theory yields the same shape as does the Interruption Theory. At any rate then within these restricted limits we found the Lenz half-width to be given by:

$$\delta = \frac{1}{\pi\,\tau} = \varrho^2\,\langle v\rangle\,N$$

Let us note that the mean free path could include the factor $\sqrt{2}$ depending on what derivation we employed for it. The "effective molecular diameter" or, more commonly, "optical collision diameter", ϱ, poses something of a problem. It is almost never sufficient to take it as gas kinetic, so that what we have is a dubious parameter. Now accepting — and this is rather general — the Weisskopf phase shift of one as collision defining we obtain for the van der Waals case:

$$\varrho = \left(\frac{3\pi^2 C}{4\langle v\rangle}\right)^{1/5} = \left(\frac{3\pi^2 \alpha_1 \alpha_2}{4\langle v\rangle}\right)^{1/5}$$

In this case we need the polarizabilities of the interacting atoms, but this will be sufficient. If such are not available, some experimental determination of the optical collision diameter is called for. A number of such experiments have been carried out and for a more or less comprehensive report on these see Ch'en and Takeo.

In the simple Lindholm treatment we saw the effects of distant collisions included. For van der Waals forces ($n = 6$) this yielded:

$$\delta = 2.71\, C_6^{2/5}\, v^{3/5}\, N$$

$$\Delta = 0.98\, C_6^{2/5}\, v^{3/5}\, N$$

In the half-width expression we note that the velocity, density, and van der Waals constant C_6 appear to the same powers as they do in the Weisskopf case. Of course, the Lindholm result does yield a line shift.

We recall that for the same case Foley obtained the following width-shift ratio:

$$\frac{\delta}{\Delta} = \frac{1}{2} \cot\left(\frac{n-3}{2(n-1)}\pi\right)$$

Finally, we obtained the Mizushima result for the following two cases:

$$\delta = 1.06\left(\frac{kT}{2m}\right)^{1/4} \langle |b|^{1/2}\rangle N \quad \text{for} \quad n = 5$$

$$\delta = 1.03\left(\frac{kT}{2m}\right)^{3/10} \langle |b^{2/5}|\rangle N \quad \text{for} \quad n = 6$$

Here again the van der Waals case has the same dependence on the factors density and van der Waals constant.

In toto then, we can hardly expect to improve much on the line shape for low densities and frequencies near line center. For the intermediate cases numerical evaluations of the Correlation Function, a procedure which is surely much simplified by the booming business in large computers, are required for the particular case of interest. For high pressures and wing frequencies we await the results of the next chapter. As to the line width, although the tools are at hand for the calculations, one must be able to compute, generally, the polarizabilities for the molecule in question. Since *a priori* polarizability calculations are questionable, some experimental evidence on interaction constants is generally required.

Two subjects more or less connected with our considerations of this chapter are detailed balancing and a Maxwell–Boltzmann distribution of dipole moments after collision. The former has been included as Appendix I, and the latter has been incorporated in Chapter 6.

CHAPTER 3

STATISTICAL BROADENING

3.1. The Franck–Condon Principle

IN THE previous chapter we considered broadening events which occurred with such rapidity that they could be considered as instantaneous phenomena. The point of view of this chapter will be the direct antithesis of this. We now slow down the broadening action to a virtual standstill. We then investigate the physical consequences of placing an emitting atom in the midst of an assemblage of broadeners under these static assumptions. The result will be the perhaps more physically obvious distortion of the energy levels of the emitter so that the frequency of emission will be perturbed with the resulting spectral line broadening. Finally, of course, we shall see the position of limiting validity which this theory occupies for the case of high pressures of perturbing gas and/or frequencies in the line wing. In order to begin our tale, however, let us recall the paper that Franck presented to the Faraday Society in 1926[48] in which he sketched for the first time the tenets of what is now the well-known Franck–Condon principle.

In order to consider his conception, let us imagine a molecule which for convenience we may take as diatomic. Let us further neglect the rotational motion and consider only the electronic and vibrational energies. We suppose this molecule to be initially in a vibrational ground state. An electronic transition now takes place to, say, some higher electronic state. Now let us consider Fig. 3.1.

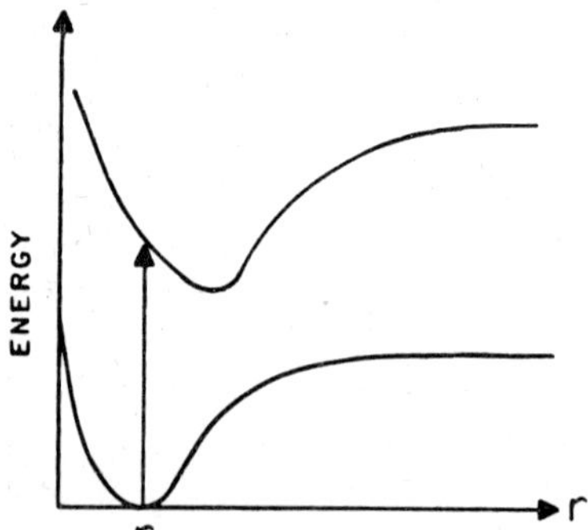

FIG. 3.1. The transition between two electronic states according to Franck

In Fig. 3.1 the curves for two electronic states — between which our transition is assumed to take place — are given in which the molecular energy as a function of nuclear separation has been plotted. In essence, Franck hypothesized that the transition takes place vertically between these two curves. This means that the nuclei have the same separation immediately after the electronic transition as they did before even though the potential curve which governs the nuclear behavior has been changed by the transition. Since in general, as is illustrated by Fig. 3.1, the new separation of the nuclei will not correspond to the equilibrium position of this electronic state, the molecule will no longer be in a vibrational ground state, and it will begin to vibrate, or it may be dissociated.

Shortly after the publication of these views, Condon[25] suggested the manner in which this principle could be used to predict intensities, one might say qualitatively. Let us roughly consider his graphical method by considering Fig. 3.2. We are utilizing the quantum mechanical energy levels of a vibrator, but we shall consider the intensity question classically.

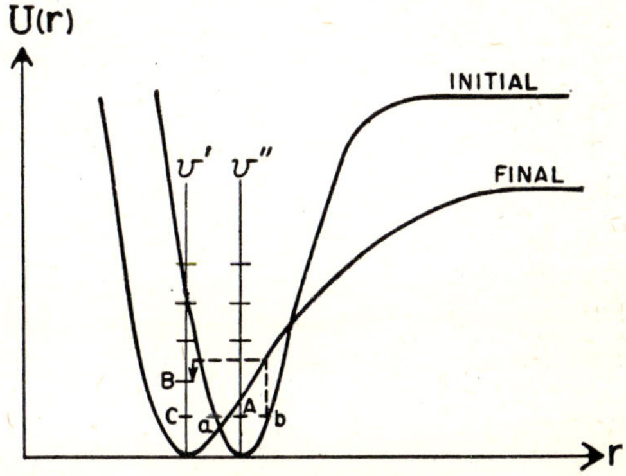

FIG. 3.2. The electronic transition with the indication of the most probable (intense) accompanying vibrational transitions. (After Condon[25])

Consider vibrational level "A" of the initial state. Classically speaking, the internuclear separation during the vibration (corresponding to an energy of "A") could neither be less than "a" nor greater than "b". Further, the molecule will most often be found with an internuclear separation in the neighborhood of "a" or "b", since we may recall that "b" is the separation at which stretching has ceased and contraction is to begin. Thus we may say that the most probable, and hence, the most intense transitions correspond to "a" and "b". If at the time of the electronic transition the internuclear distance corresponds to "b", the transition will proceed vertically to the curve for the upper electronic state, and then, since the vibrational energies are, of course, quantized, would proceed to the closest vibrational level "B". Thus, the most intense transitions from level "A" will be to levels "B" and "C".

Condon[26] later modified this principle to strictly include the wave mechanics and essentially this modification has the effect of adding an indeterminacy to, the principle so that if, for example, $A \rightarrow B$ is favored in Fig. 3.2, we will now have a band around the frequency corresponding to this frequency rather than the frequency itself.

The foundation for a theory of spectral line broadening had thus been laid by Franck and Condon and the theory itself was first qualitatively developed by Jablosnki.[90]

3.2. First application of Franck–Condon to line broadening

For the interaction energy of two atoms in a molecule Jablonski stated the Franck–Condon principle as: "At the moment of the electron jump (1) the reciprocal separation of the atomic nuclei and (2) the velocity of the nuclei are not noticeably changed."[90] Then he considered the emitter and broadener as occupying an analagous position to which the Franck–Condon principle could also be applied. Although only the electronic state of the emitter is changed by the radiating transition — not a molecular orbital — one can consider with due legitimacy the emitter state change to correspond to a state change on the part of the emitter-broadener molecule.

Let us consider, with Jablonski, the potential curves for two free molecules as given in Fig. 3.3. These curves are repulsion curves, but for the principles involved the curve type is of no import. The potential energy of the two particle system is given as a function of the molecular separation. If the broadening molecule is an infinite distance from the absorber, the unperturbed spectral line of frequency corresponding to $E'_\infty - E_\infty$ would be absorbed. In the event that the molecular separation were r_a at the time of absorption, the frequency of the line would have some different value corresponding to $E'_{r_a} - E_{r_a}$. With the aid of Fig. 3.3 let us consider this absorbing transition in slightly greater detail.

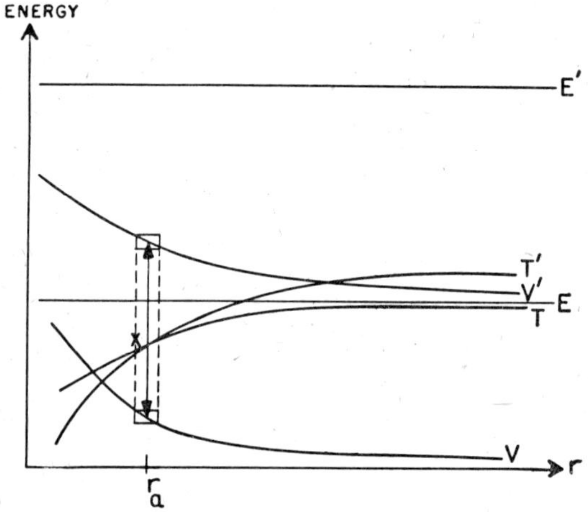

Fig. 3.3. An electronic transition in which translational velocity is unchanged during transition

Firstly, we have assumed that energy is conserved so that we have the T, V, and E curves as shown for the two molecules in the initial state. Now the transition takes place at the separation r_a. In accordance with the Franck–Condon principle, the relative velocity of the molecules is not changed by this transition so that immediately after the transition the point "x" on the T-curve still gives the relative velocity. Now, however, the total energy is E' and this too is to be conserved. Hence, as shown by the T'-curve the relative kinetic energy of the particles after separation will be increased, and this increase is at the expense of the incident radiation. The reverse effect could also occur, that is, the "effective" incident radiation could be increased at the expense of the final kinetic energy. These interactions between collisions and radiation had been considered by Oldenberg[161, 162] and later by Minkowski[157].

Thus, we see radiation differing in frequency from that characteristic for the molecular transition may be absorbed. Qualitatively at least, it follows directly that "... the intensity distribution (and the line width) is dependent on the potential curves and the probability of different r values".[90] Thus, the most probable r values would correspond to the most intense frequencies of the spectral line. Since the probability of a given r would depend on the relative velocity of the "collision partners", the variable relative velocity which may be inferred from the T-curve in Fig. 3.3 must needs be considered.

It is apparent that almost any form of asymmetry of the spectral line could be obtained, depending on the relative profiles of the potential curves. Before considering the matter of line shift, it might be well to emphasize that the line frequency (corresponding to a specific r_a) rapidly approaches the unperturbed line frequency with increasing r regardless of the type of interaction curves under consideration. Thus, it follows that, with increasing probability of small r, there will be an increasing displacement or shift of the line intensity maximum*. Small r would become more probable with increasing pressure.

In what has preceded we have considered the perturbing effect of only one molecule. In "reality" many molecules would contribute to the disturbance, but Jablonski's ruminations do give a qualitative idea as to how this type of spectral line broadening and shift are brought about. He envisioned it as applicable to any type of interaction curves.

3.3. Calculation of the van der Waals forces involved in broadening

Although Jablonski laid the basis for the Statistical Theory, it was Margenau[134] who first developed an applicable mathematical form for it. In so doing he incidentally developed the quantum theory of van der Waals forces†, the forces to which he correctly assigned the task of foreign gas broadening by neutral molecules. As we are aware, these forces are of a great deal of importance to the broadening, sufficient indeed so that their treatment is deserving of something more than appendix relegation.

Margenau first applied himself to the case of foreign gas broadening in which the absorbing atom is in its lowest p-state and the broadening atom is in its ground state.

Let us consider Fig. 3.4.

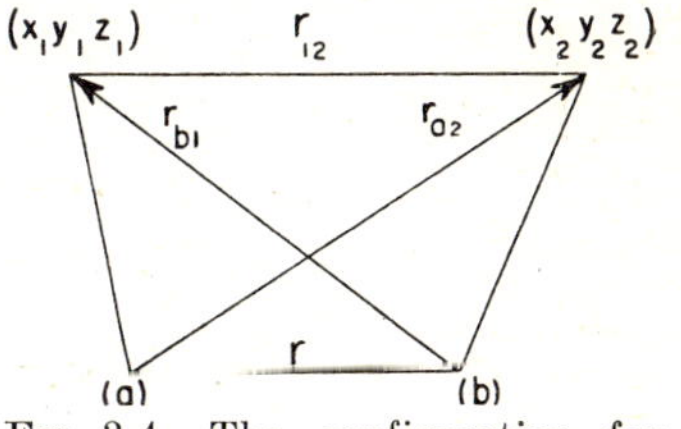

FIG. 3.4. The configuration for the Margenau–van der Waals force calculation. The nuclei are at (a) and (b) and the photoelectrons at $(x_1y_1z_1)$ and $(x_2y_2z_2)$

The atomic nuclei are at (a) and (b), and the centers of negative charge are at $(x_1y_1z_1)$ and $(x_2y_2z_2)$ where the x-axis of the coordinate system to which the system is referred coincides with r, the separation of the two nuclei. We here assume that r is large so that no electron exchange effect need be considered. For example, one could take as eigenfunctions for, say, two H atoms $\psi = \psi_a(1)\,\psi_b(2)$ instead of $\psi = \psi_a(1)\,\psi_b(2) - \psi_a(2)\,\psi_b(1)$, etc.

* This is not true for resonance forces.

† Margenau[216] has cryptically but delightfully defined the van der Waals force as "... that force which gives rise to the constant a in van der Waals equation ..."

The interaction potential energy of the system will be given by:

$$V = e^2\left(\frac{1}{r} + \frac{1}{r_{12}} - \frac{1}{r_{a2}} - \frac{1}{r_{b1}}\right)$$

$$= \frac{e^2}{r} + \frac{e^2}{r}$$

$$\left[\frac{1}{\sqrt{\dfrac{(y_1 - y_2)^2}{r^2} + \dfrac{(z_1 - z_2)^2}{r^2} + \dfrac{(x_1{}^2 + x_2{}^2 - 2x_1 x_2\, 2\, x_1 x_2 - 2x_1 r + 2x_2 r)}{r^2} + 1}}\right]$$

$$- \frac{e^2}{r}\left[\frac{1}{\sqrt{\dfrac{y_2{}^2}{r^2} + \dfrac{z_2{}^2}{r^2} + \dfrac{(x_2{}^2 + 2x_2 r)}{r^2}}}\right] - \frac{e^2}{r}\left[\frac{1}{\sqrt{\dfrac{y_1{}^2}{r^2} + \dfrac{z_1{}^2}{r^2} + \dfrac{(x_1{}^2 - 2x_1 r)}{r^2} + 1}}\right]$$

$$= \frac{e^2}{r^3}(y_1 y_2 + z_1 z_2 - 2\,x_1 x_2) + \frac{3}{2}\frac{e^2}{r^4}[r_1{}^2 x_2 - r_2{}^2 x_1$$

$$+ (2\,y_1 y_2 + 2z_1 z_2 - 3\,x_1 x_2)(x_1 - x_2)]$$

$$+ \frac{3}{4}\frac{e^2}{r^5}[r_1{}^2 r_2{}^2 - 5\,r_2{}^2 x_1{}^2 - 5\,r_1{}^2 x_2{}^2 + 2(y_1 y_2 + z_1 z_2 + 4x_1 x_2)^2] + \cdots$$

since r has been assumed large. The first term is the dipole–dipole interaction term, the second the dipole-quadrupole, and so on. Margenau utilized only the first term, so that the perturbing potential for the system is:

$$V = \frac{e^2}{r^3}(y_1 y_2 + z_1 z_2 - 2x_1 x_2) \tag{3.1}$$

so that:

$$H = E_0 + V_0 + V = H_0 + H'$$

The first and second order equations from ordinary perturbation theory would be:

$$H^0 \psi_n{}' + H' \psi_n{}^0 = E_n{}^0 \psi_n{}' + E_n{}' \psi_n{}^0$$

$$H^0 \psi_n{}'' + H' \psi_n{}' = E_n{}^0 \psi_n{}'' + E_n{}' \psi_n{}' + E_n{}^0 \psi_n{}''$$

We now replace $\psi_n{}'$ by $\sum_k a_{nk} \psi_k{}^0$ and $\psi_n{}''$ by $\sum_k b_{nk} \psi_k{}^0$, replace $H^0\psi_k{}^0$ by $E_k{}^0\psi_k{}^0$, multiply both equations on the left by $\psi_m{}^0$, and integrate over all space to obtain:

$$a_{nm} E_k{}^0 \delta_{mk} + \int \bar{\psi}_m{}^0 H' \psi_n{}^0 \,\mathrm{d}\tau = E_n{}^0 a_{nm} \delta_{mk} + E_n{}' \delta_{mn} \tag{3.2a}$$

$$b_{nm} E_m{}^0 \delta_{nk} + \sum_k a_{nk} \int \bar{\psi}_m{}^0 H' \psi_k{}^0 \,\mathrm{d}\tau = E_n{}^0 b_{nm} \delta_{mk} + E_n{}' a_{nm} \delta_{mk} + E_n{}'' \delta_{nm} \tag{3.2b}$$

These equations yield the first and second order energy corrections:

$$E_{10}' = \int \bar{\psi}_m{}^0 \, H' \, \psi_m{}^0 \, d\tau = 0$$

$$E_{10}'' = \sum_k a_{nk} \int \bar{\psi}_n{}^0 \, H' \, \psi_k^0 \, d\tau$$

$$= \sum_k{}' \frac{\int \psi_1{}^0(a)\, \bar{\psi}_0{}^0 \, V \, \psi_A{}^0(a)\, \psi_B{}^0(b)\, d\tau \int \bar{\psi}_A{}^0(a)\, \bar{\psi}_B{}^0(b) \, V \, \psi_1{}^0(a)\, \psi_0{}^0(b)\, d\tau}{(E_n{}^0 - E_k{}^0)} \qquad (3.3)^*$$

$$= -\sum{}' \frac{|V_{10,AB}|^2}{E_A - E_1 + F_B - F_0} \qquad (3.4)$$

In Eq. (5.4) the E_A are the energy levels of the absorbing atom, E_1 being the energy of the initial p-state, and the F_B are the energy levels of the broadening atom, F_0 being the energy of the initial s-state. $V_{10,AB}$ may now be written out as:

$$V_{10,AB} = \int \bar{\psi}_{AB} \, V \, \psi_{10} \, d\tau = \int \bar{\psi}_A(1)\, \bar{\psi}_B(2)\, V \, \psi_1(1)\, \psi_0(2)\, d\tau_1 \, d\tau_2 \qquad (3.5)$$

since, as mentioned earlier, no electron exchange is considered. If, in the symbolism of Margenau, we let, for example, $x_{1A}(1) = \int \bar{\psi}_A(1)\, x\psi_1(1)\, d\tau_1$, $|V_{10,AB}|$ may be written as:

$$\left.\begin{aligned} |V_{10,AB}| = \frac{e^4}{r^6} [&y_{1A}{}^2(1)\, y_{0B}{}^2(2) + z_{1A}{}^2(1)\, x_{0B}{}^2(2) + 4x_{1A}{}^2(1)\, x_{0B}{}^2(2) \\ &+ 2\, y_{1A}(1)\, z_{1A}(1)\, y_{0B}(2)\, z_{0B}(2) \\ &- 4\, x_{1A}(1)\, y_{1A}(1)\, x_{0B}(2)\, y_{0B}(2) - 4\, x_{1A}(1)\, z_{1A}(1)\, x_{0B}(2)\, z_{0B}(2)] \end{aligned}\right\} \qquad (3.6)$$

Margenau utilized hydrogen-like wave functions for both atoms which, in general, may be represented as:

$$\psi_{nlm} = R_{nl}(r)\, P_l^m(\vartheta)\, e^{im\varphi} \qquad (3.7)$$

in which l is the orbital angular momentum quantum number; m is the magnetic quantum number, and n is the orbital quantum number.

The following relations exist among the matrix elements associated with the broadening atom:

$$\sum_m x_{0B}\, y_{0B} = \sum_m x_{0B}\, z_{0B} = \sum_m y_{0B}\, z_{0B} = 0 \qquad (3.8a)$$

$$\sum_m x_{0B}{}^2(2) = \sum_m y_{0B}{}^2(2) = \sum_m z_{0B}{}^2(2) = \frac{1}{3} r_{0B}(2)\, \delta_{l1} \qquad (3.8b)$$

We have summed over the quantum number m in Eqs. (3.8) since it has no effect on the level energies, but remains a spatial degeneracy parameter.

* $E_{10}' = \int \bar{\psi}_1{}^0(a)\, x\, \psi_1{}^0(a)\, d\tau_1 \int \bar{\psi}_0{}^0(b)\, x\, \psi_0{}^0(b)\, d\tau_2 + \cdots = \int \bar{\psi}_1{}^0(a)\, x\, \psi_1{}^0(a)\, d\tau_1 \times \iiint [R_{n0}]^2 r^3 \sin^2\vartheta \cos\varphi \, dr\, d\vartheta\, d\varphi + \cdots = \int \{\}\, d\tau_1 \iint [R_{n0}]^2 r^3 \sin^2\vartheta \, dr\, d\vartheta \cdot 0 + \cdots = 0$, since $\int_0^{2\pi} \cos\varphi \, d\varphi = 0$, and so on.

$r_{0B}(2)$ is to be defined below. Let us prove a portion of Eq. (3.8a) as an example.

$$\left.\begin{aligned}
\sum_m x_{0B}\, y_{0B} &= \sum_m \int \bar{\psi}_B(2)\, x\, \psi_0(2)\, d\tau \int \bar{\psi}_B(2)\, y\, \psi_0(2)\, d\tau \\
&= \sum_{m=-l}^{+l} \iiint R_{10}\, R_{nl}\, e^{im\varphi}\, P_l^m\, r^3 \sin^2\vartheta \cos\varphi\, dr\, d\vartheta\, d\varphi \\
&\quad \times \iiint R_{10}\, R_{nl}\, P_l^m\, e^{im\varphi}\, r^3 \sin^2\vartheta \sin\varphi\, dr\, d\vartheta\, d\varphi \\
&= \sum_{m=-l}^{+l} \frac{1}{4i} \iiint R_{10}\, R_{nl}\, P_l^m \sin^2\vartheta\, (e^{i(m+1)\varphi} \\
&\quad - e^{i(m-1)\varphi})\, r^3\, dr\, d\vartheta\, d\varphi \iiint \ldots
\end{aligned}\right\} \quad (3.9)$$

The integral over φ in Eq. (3.9) is zero unless $m = \pm 1$, while, when $m = \pm 1$, the sum of all such integrals over m goes to zero. The remaining portion of Eqs. (3.8) may be evaluated in a similar manner. An inspection of Eq. (3.9) also tells us that, in Eq. (3.8b):

$$r_{0B}^2(2) = \left[\int_0^\infty R_{n0}(r)\, R_{nB}(r)\, r^3\, dr\right]^2 \quad (3.8c)$$

Certain relations may also be written down among the matrix elements associated with the absorbing atom. Here the state is the p-state and hence $m_1 = 0, \pm 1$. For the emitter then:

For $m_1 = 0$:

$$\sum_m x_{1A}^2(1) = r_{1A}^2(1)\left[\frac{1}{3}\delta_{10} + \frac{4}{15}\delta_{12}\right] \quad (3.10a)$$

$$\sum_m y_{1A}^2(1) = \sum_m z_{1A}^2(1) = r_{1A}^2(1)\frac{1}{5}\delta_{l2} \quad (3.10b)$$

For $m_1 = \pm 1$

$$\sum_m x_{1A}^2(1) = r_{1A}(1)\frac{1}{5}\delta_{l2} \quad (3.10c)$$

$$\sum_m y_{1A}^2(1) = \sum_m z_{1A}^2(1) = r_{1A}^2(1)\left[\frac{1}{6}\delta_{l0} + \frac{7}{30}\delta_{l2}\right] \quad (3.10d)$$

Eqs. (3.8) and (3.10) may now be utilized to obtain $|V_{10,AB}|^2$ summed over all values of the magnetic quantum numbers for the states A and B. The result may be written as:

For $m_1 = 0$:

$$\sum_{m_A m_B} |V_{10,AB}|^2 = \frac{e^4}{r^6} r_{1A}^2(1)\, r_{0B}^2(2)\left[\frac{22}{45}\delta_{lA2} + \frac{4}{9}\delta_{2l}\right]\delta_{lB1} \quad (3.11a)$$

For $m_1 = \pm 1$:

$$\sum_{m_A m_B} |V_{10,AB}|^2 = \frac{e^4}{r^6} r_{1A}^2(1)\, r_{0B}^2(2)\left[\frac{19}{45}\delta_{lA2} + \frac{1}{9}\delta_{lB0}\right]\delta_{lB1} \quad (3.11b)$$

Margenau expressed the interaction energy in terms of "dispersion-f-values"* which are proportional to the absorbing transition probability. If we denote

* Another name for oscillator strengths.

the f-values for the emitting atom by f_{1A} and for perturbing atom by g_{0B}, we obtain:

$$\left.\begin{aligned} f_{1A} &= \frac{8\pi^2 m}{3h^2}(E_A - E_1)\sum_{m_1}[x_{1A}^2(1) + y_{1A}^2(1) + z_{1A}^2(1)] \\ &= \frac{8\pi^2 m}{3h^2}(E_A - E_1)\, r_{1A}^2(1)\left[\frac{2}{3}\delta_{lA2} + \frac{1}{3}\delta_{lA0}\right] \end{aligned}\right\} \qquad (3.12)$$

$$g_{0B} = \frac{8\pi^2 m}{3h^2}(F_B - F_0)\, r_{0B}^2(2)\,\delta_{lA0} \qquad (3.13)$$

If we now let f_{1A}' be the f-value for the d-states where $l_A = 2$ and f_{1A}'' be the f-values for those states where $l_A = 0$, the substitution of Eqs. (3.12) and (3.13) into Eqs. (3.11) and the subsequent substitution of Eqs. (3.11) into Eq. (3.4) yields:

For $m_1 = 0$:

$$E_{10} = -\frac{1}{r^6}\frac{3}{4m^2}\left(\frac{he}{2\pi}\right)^4\left\{\frac{11}{5}\sum_{A'B}\frac{f_{1A'}g_{0B}}{D_{A'B}} + 4\sum_{A''B}\frac{f_{1A''}g_{0B}}{D_{A''B}}\right\} \qquad (3.14a)$$

For $m_1 = \pm 1$:

$$E_{10} = -\frac{1}{r^6}\frac{3}{4m^2}\left(\frac{he}{2\pi}\right)^4\left\{\frac{19}{10}\sum_{A'B}\frac{f_{1A'}g_{0B}}{D_{A'B}} + \sum_{A''B}\frac{f_{1A''}g_{0B}}{D_{A''B}}\right\}. \qquad (3.14b$$

It is thus apparent that the interaction energy between one atom in a p-state and another in an s-state depends on the orientation of the former through the value of the magnetic quantum number for the p-state.

Eqs. (3.14) may be rewritten as:

$$E_{10} = -\frac{1}{r^6}\frac{3}{4m^2}\left(\frac{he}{2\pi}\right)^4\left\{\frac{22 - 3m_1^2}{10}\sum_{A'B}\frac{f_{1A'}g_{0B}}{D_{A'B}} + (4 - 3m_1^2)\sum_{A''B}\frac{f_{1A''}g_{0B}}{D_{A''B}}\right\}$$

and if, ignoring the dependence of the energy on the orientation, we sum over m_1 and divide by three, we obtain, as average E_{10}:

$$\langle E_{10}\rangle = -\frac{1}{r^6}\frac{3}{2m^2}\left(\frac{he}{2\pi}\right)^4\left\{\sum_{A'B}\frac{f_{1A'}g_{0B}}{D_{A'B}} + \sum_{A''B}\frac{f_{1A''}g_{0B}}{D_{A''D}}\right\}.$$

This equation obviously has the same coefficients for the $f_{1A'}$ and $f_{1A''}$ so that we may write our expression for the interaction energy as follows:

$$\langle E_{10}\rangle = -\frac{1}{r^6}\frac{3}{2m^2}\left(\frac{he}{2\pi}\right)^4\sum_{AB}\frac{f_{1A}g_{0B}}{D_{AB}}. \qquad (3.15)$$

This corresponds to the more general form for the interaction energy which had been obtained by London[129] as:†

$$\langle E_{A'B'}\rangle = -\frac{1}{r^6}\frac{3}{2m^2}\left(\frac{he}{2\pi}\right)^4\sum_{A'B}\frac{f_{A'A}g_{B'B}}{D_{ABA'B'}} + S. \qquad (3.15')$$

† We shall discuss London's interaction considerations in some detail in Chap. 6.

In this equation S refers to the terms which would enter if interactions of higher order than the dipole were included in the perturbing potential. Margenau[133] had investigated the effect of these higher order terms on the van der Waals forces.

Eq. (3.15) may be simplified by applying the London approximation[129] that the summation of g_{0B} over B may be replaced by the multiplicative factor $\alpha\Delta F$. Here α is the polarizability of the broadening atom, and ΔF is the mean energy difference for a transition with the resultant absorption of electric dipole radiation by the broadener. Eq. (3.15) thus becomes:

$$\langle E_{k0}\rangle = -\frac{1}{r^6}\frac{3}{2m}\left(\frac{he}{2\pi}\right)^2 \alpha\,\Delta F \sum_A \frac{f_{kA}}{(E_A - E_k)(\Delta F + E_A - E_k)} \tag{3.16}$$

If we consider the sign of $\langle E_{k0}\rangle$, we find that it is dependent on the sign of $(\Delta F + E_A - E_k)$, that is, on whether the transition is an absorbing or emitting one, since $f_{kA}/(E_A - E_k)$ is always positive and the sign of the remaining factors is obviously always positive. This fact leads to (a) a negative value for E and hence an attractive interaction force for $k = 0$, the absorbing atom initially in the ground state, or (b) the appearance of negative terms in the summation for E_{k0} when $k \neq 0$. Let us assume, however, that $\langle E_{00}\rangle$ and $\langle E_{10}\rangle$ are negative so that the ground and first-excited states of the absorber — we consider here only this transition — are distorted into energy curves which are functions of the atomic separation r and similar to those shown in Fig. 3.5.

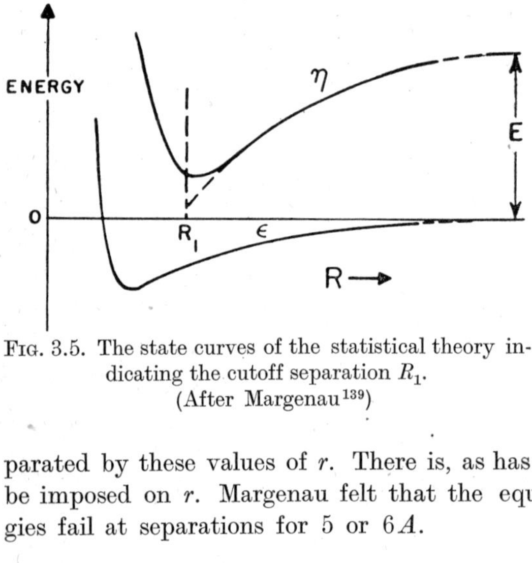

FIG. 3.5. The state curves of the statistical theory indicating the cutoff separation R_1. (After Margenau[139])

In Fig. 3.5, E would correspond to the frequency of the absorbed radiation when the absorber is not perturbed by the broadener. The separation of the two curves at lesser values of r would correspond to the frequency of radiation which is absorbed when the absorbing and broadening atoms are separated by these values of r. There is, as has been mentioned, a lower limit to be imposed on r. Margenau felt that the equations for the interaction energies fail at separations for 5 or 6 Å.

3.4. Early developments of the statistical theory

Let us state what amounts to the fundamental hypothesis of the Statistical Theory: the intensity of a particular frequency in a spectral line will be proportional to the probability for the existence of that spatial configuration of broadeners whose distorting effect leads to the emission of that particular frequency[135]. If, for a moment, we consider the lower level in a radiating transition to be uninfluenced by the broadeners, then the action of the broadeners in lowering or raising the upper level will result in the emission of frequencies to the red or the blue respectively. In a general way then the problem is one of finding a mathematical expression describing the probability for a particular spatial distribution of perturbers. Although the Maxwell–Boltzmann distribution will govern the perturber energies in fine, the static situations *à propos* of description by the Statistical Theory lead to very small values for these energies. Therefore the exponential involved in this Maxwellian description will be unity except for a very small sphere in the neighborhood of the broadener. As a consequence, we may quite legitimately neglect the Maxwell–Boltzmann factor in our probability expression.

Now let the radial coordinates — referred to the origin of coordinates which has been established at the absorber — be given by $r_1, r_2, \ldots, r_n$ for the n perturbing atoms. Thus:

$$I(V)\,dV = c \int_{r_1\sim V} \cdots \int_{r_n \sim V} \int_{\vartheta_1} \cdots \int_{\vartheta_n} \int_{\varphi_1} \cdots \int_{\varphi_n} r_1^2 \ldots r_n^2 \sin\vartheta_1 \ldots \sin\vartheta_n \, dr_1 \ldots dr_n \, d\vartheta_1 \ldots d\vartheta_n \, d\varphi_1 \ldots d\varphi_n \tag{3.17a}$$

$$= c(4\pi)^n \int \cdots \int_V r_1^2 \ldots r_n^2 \, dr_1 \ldots dr_n$$

where $r_1 \ldots r_n$ are integrated over that range which gives the desired dV. dV now refers to the difference between the perturbed energy and the unperturbed energy V_∞. We may take C as the normalizing factor v^{-n} so that we obtain:

$$I(V)\,dV = \left(\frac{4\pi}{v}\right)^n \int \cdots \int_V r_1^2 \ldots r_n^2 \, dr_1 \ldots dr_n \tag{3.17b}$$

Let us refer back to Eq. (3.16). For the lower state we let $V = -a/r_i^6$, and for the upper state we let $V = -b/r_i^6$. Then for the interaction energy change due to the i-th perturber we obtain $(b-a)/r_i^6$ so that the total line perturbation becomes:

$$V = (b-a) \sum_{i=1}^{n} \frac{1}{r_i^6} = \beta \sum_{i=1}^{n} \frac{1}{r_i^6} \tag{3.18}$$

It would be convenient to rewrite Eq. (3.17b) as an integral over all configuration space. We may accomplish this by multiplying it by a Dirichlet factor which has the value one for $-(1/2)\,dV < V < (1/2)\,dV$ and zero for all other values of V. After the suitable Dirichlet factor has been chosen, Eq. (3.17b) becomes:

$$\left.\begin{aligned} I(V)\,dV &= \frac{1}{\pi}\left(\frac{4\pi}{v}\right)^n \int \cdots \int_{\infty} \int_{-\infty}^{+\infty} d\Upsilon \frac{\sin\left(\frac{1}{2}\Upsilon\, dV\right)}{\Upsilon} \\ &\quad \times \exp\left[-iV\Upsilon + i\beta\Upsilon \sum \frac{1}{r_j^6}\right] r_1^2 \ldots r_n^2 \, dr_1 \ldots dr_n \\ &= \frac{1}{\pi}\left(\frac{4\pi}{v}\right)^n \int_{-\infty}^{+\infty} \frac{\sin\left(\frac{1}{2}\Upsilon\, dV\right)}{\Upsilon} e^{-iV\Upsilon} \left[\int_{R_1}^{R} e^{i\beta\Upsilon/r^6} r^2 \, dr\right]^n d\Upsilon \end{aligned}\right\} \tag{3.19}$$

The limits which have been placed on r in Eq. (3.19) call for clarification. We have mentioned the necessity for putting a lower limit on r, and this is, of course, R_1, since we cannot expect our potential curves to be valid at very small r. R is defined by $v = (4/3)\,\pi R^3$, the volume of gas under consideration. In Eq. (3.19) we have also rewritten the product of the n integrals as the product of n identical integrals.

By virtue of the "smallness" with which the definition of a differential vests it, we may replace $\sin(1/2\,\Upsilon\,dV)$ by $(1/2\,\Upsilon\,dV)$ in Eq. (3.19). If we let:

$$E = \frac{\beta}{r^6}; \qquad u(E) = -\frac{2}{3}\frac{\pi}{v}\frac{\beta^{1/2}}{E^{3/2}}$$

Eq. (3.19) becomes:

$$I(V)\,dV = \frac{dV}{2\pi}\int_{-\infty}^{+\infty} e^{-iV\Upsilon}\left[\int_{\beta/R^6}^{\beta/R^6} e^{i\Upsilon E}\,u(E)\,dE\right]^n d\Upsilon \tag{3.20}$$

If we now let:

$$f(\Upsilon) = -\frac{2\pi}{3v}\beta^{1/2}\int_{\gamma}^{\beta/R^6}\frac{e^{i\Upsilon E}}{E^{3/2}}\,dE \tag{5.21}$$

where $\gamma = \beta/R_1^{\,6}$, we obtain:

$$I(V) = \frac{1}{2\pi}\int_{-\infty}^{+\infty} e^{-iV\Upsilon}\,[f(\Upsilon)]^n\,d\Upsilon \tag{3.22}$$

Now let $\Upsilon\gamma = x$, $v_1 = (4/3)\,\pi R_1^{\,3}$, and $u = \Upsilon E$. We now integrate by parts to obtain:

$$\left.\begin{aligned} f(x) &= \frac{2\pi}{3v}\left(\frac{\beta x}{\gamma}\right)^{1/2}\left\{\frac{e^{iu}}{u^{1/2}}\Bigg|_{x}^{xR_1^{\,6}/R^6} - 2i\int_{x}^{xR_1^{\,6}/R^6}\frac{e^{iu}}{u^{1/2}}\,du\right\} \\ &\doteq 1 - \frac{v_1}{v}e^{ix} - i\frac{v_1}{v}x^{1/2}\int_{x}^{0}\frac{e^{iu}}{n^{1/2}}\,du \end{aligned}\right\} \tag{3.23}$$

The remaining integral in Eq. (3.23) is a combination of Fresnel integrals which Margenau[135] obtained from Jahnke and Emde[231], so that:

$$f(x) = 1 - \frac{v_1}{v}\left[\sqrt{2\pi x}\,S(x) + \cos x\right] + i\frac{v_1}{v}\left[\sqrt{2\pi x}\,C(x) - \sin x\right] \tag{3.24}$$

From Eq. (3.23), it is apparent (we recall that v_1 is a small sphere of radius R_1) that $|f(x)| < 1$. This fact justifies the replacement of Eq. (3.24) by $\exp\left[-\frac{n v_1}{v}\left\{\sqrt{2\pi x}\,S(x) + \cos x - i\sqrt{2\pi x}\,C(x) + i\sin x\right\}\right]$ since the right side of Eq. (3.24) may be considered as the first two terms of the McLaurin expansion of the exponential before raising it to the n-th power. Eq. (3.22) thus becomes:

$$I(V) = \frac{1}{2\pi\gamma}\int_{-\infty}^{+\infty}\exp\left[-\frac{n v_1}{v}\left\{\sqrt{2\pi x}\,S(x) + \cos x - i\sqrt{2\pi x}\,C(x) + i\sin x\right\} - i\frac{v}{\gamma}x\right]dx$$

Since $C(-x) = iC(x)$ and $S(-x) = -iS(x)$, this equation becomes:

$$\left.\begin{aligned} I(V) &= \frac{1}{\pi\gamma}\int_0^\infty \cos\left[\frac{n v_1}{v}\left[\sqrt{2\pi x}\, C(x) - \sin x\right] - \frac{v}{\gamma}x\right] \\ &\times \exp\left[-\frac{n v_1}{v}\left\{\sqrt{2\pi x}\, S(x) + \cos x\right\}\right] \mathrm{d}x \end{aligned}\right\} \quad (3.25)$$

Eq. (3.25) gives the intensity distribution in our asymmetrically broadened, shifted spectral line, but it has the undesirable trait of responding only to graphical integration.

However, the intraction curves and resultant line shapes which Margenau was able to obtain are of some interest.

First, we consider Eq. (3.19) in a one-dimensional form for simplicity. Then we assume the lower curve in Fig. 3.5 to be flat, so that we may replace $\beta \sum 1/r_i^6$ by $\Sigma f(x_i)$ where $f(x_i)$ is the upper curve for the i-th particle. Eq. (3.19) then becomes:

$$I(V)\,\mathrm{d}V = \frac{L^{-n}}{\pi}\int_{-\infty}^{+\infty} \mathrm{d}\Upsilon \left[\frac{\sin\left(\frac{1}{2}\Upsilon\,\mathrm{d}V\right)}{\Upsilon}\right] \mathrm{e}^{-iv\Upsilon} \int_0^L \ldots \int_0^L \mathrm{e}^{i\Upsilon\Sigma f(x_i)}\,\mathrm{d}x_1 \ldots \mathrm{d}x_n \quad (3.26)$$

It may be noted that in Eq. (3.26) the lower limits of the r-integration are zero. This may be interpreted as the distance of closest approach of the broadener rather than separation zero.

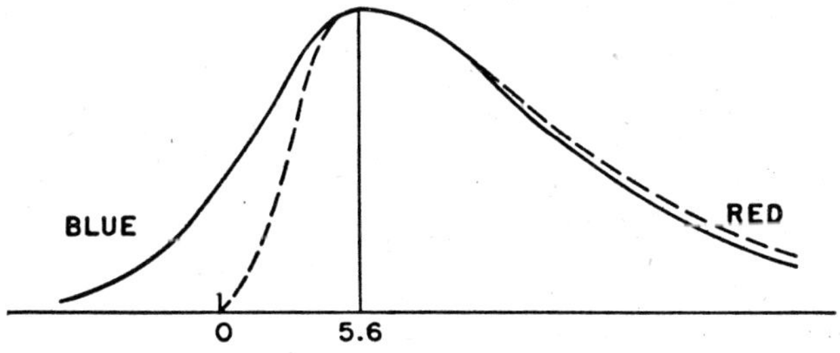

FIG. 3.6. The Hg 2537 line. The solid curve was obtained by Fuchtbauer, Joos, and Dinckelacker[45] for 50 atmospheres N_2. The dotted line is the Margenau result. (After Margenau[135])

Figs. 3.7a, 3.7c, and 3.7e give the $f(x)$ curves which allow the evaluation of Eq. (3.34) in a closed form to yield the line shapes given by Figs. 3.7b, 3.7d, and 3.7f.[136]

3.5. The statistical line shape

In Eq. (3.19) let us, with Margenau[131], replace V by $\Delta\nu$, where $\Delta\nu$ is the frequency separation from the line center, to obtain:

$$I'(\Delta\nu)\,\mathrm{d}(\Delta\nu) = \frac{1}{\pi}\left(\frac{4\pi}{V}\right)^n \int \ldots \int r_1^2\,\mathrm{d}r_1 \ldots \mathrm{d}r_n \int_{-\infty}^{+\infty} \mathrm{d}\Upsilon \frac{\sin\left(\frac{1}{2}\Upsilon\,\mathrm{d}(\Delta\nu)\right)}{\Upsilon} \times \mathrm{e}^{-i(\Delta\nu)\Upsilon + i\beta\Upsilon\Sigma r_i^{-6}} \quad (3.27)$$

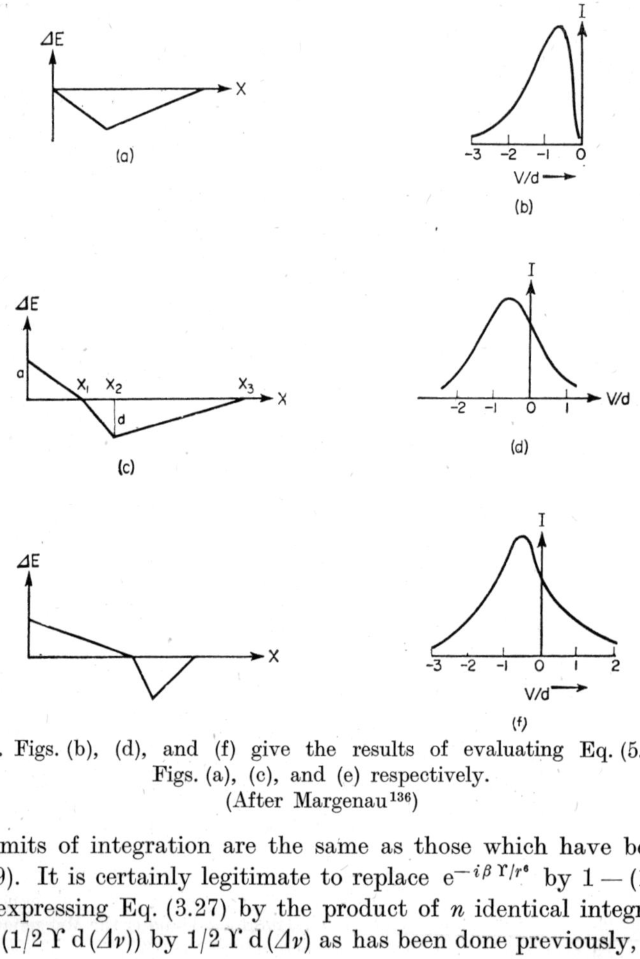

FIG. 3.7. Figs. (b), (d), and (f) give the results of evaluating Eq. (5.34) for Figs. (a), (c), and (e) respectively. (After Margenau[136])

Here the limits of integration are the same as those which have been applied to Eq. (3.19). It is certainly legitimate to replace $\mathrm{e}^{-i\beta\Upsilon/r^6}$ by $1-(1-\mathrm{e}^{i\beta\Upsilon/r^6})$, and, after expressing Eq. (3.27) by the product of n identical integrals and replacing $\sin(1/2\,\Upsilon\,\mathrm{d}(\Delta\nu))$ by $1/2\,\Upsilon\,\mathrm{d}(\Delta\nu)$ as has been done previously, this yields:

$$I'(\Delta\nu) = \frac{1}{2\pi}\left(\frac{4\pi}{V}\right)^n \int_{-\infty}^{+\infty} \mathrm{d}\Upsilon\,\mathrm{e}^{-i(\Delta\nu)\Upsilon}\left\{\int [1-(1-\mathrm{e}^{i\beta\Upsilon/r^6})]\,r^2\,\mathrm{d}r\right\}^n \quad (3.28)$$

Now let us consider the term in braces in Eq. (3.28). The integration should be carried from the closest separation R_1 to a maximum separation d which is, of course, related to the volume of gas under consideration by $V = (4/3)\pi d^3$.

Since R_1 is small, it is apparent that only a very small error will be introduced by integrating $r^2\,dr$ from the lower limit zero. On the other hand, the error which we shall introduce by taking lower limit zero for the remainder of the integrand in the braces will require further discussion. We may write $(1 - e^{i\beta\Upsilon/r^6})$ as $-\frac{i\beta\Upsilon}{r^6} + \frac{\beta^2\Upsilon^2}{2\,r^{12}} - \cdots$, from which it is apparent that, since d is large, we may take infinity as the upper limit of $\int (1 - e^{i\beta\Upsilon/r^6})\,r^2\,dr$ with no appreciable error. Thus we obtain:

$$\int [1 - (1 - e^{-i\beta\Upsilon/r^6})]\,r^2\,dr = \frac{V}{4\pi}\left(1 - \frac{4\pi V'}{V}\right) \tag{3.29}$$

where $V' = \int_0^\infty (1 - e^{i\beta\Upsilon/r^6})\,r^2\,dr$. In raising the braced expression in Eq. (3.28) to the n-th power Margenau allows the volume of the gas to increase indefinitely while maintaining $N = n/V$ constant. This amounts to the same procedure as allowing d to approach infinity, while keeping the density constant. Since $\lim (1 - 4\pi\,NV'/n)^n = e^{-4\pi NV'}$, Eq. (3.28) then becomes:

$$I'(\Delta\nu) = \frac{1}{2\pi}\int_{-\infty}^{+\infty} d\Upsilon\, e^{-i(\Delta\nu)\Upsilon}\, e^{-4\pi NV'(\Upsilon)} \tag{3.30}$$

when the factor $\left(\frac{4\pi}{V}\right)^n$ in Eq. (3.27) has been cancelled by the reciprocal of this factor in Eq. (3.29).

Now let $x = \beta\Upsilon/r^6$ in V' and rewrite V' slightly as:

$$V'(\Upsilon) = -\frac{(\beta\Upsilon)^{1/2}}{6}\int_0^\infty \left(\frac{\cos x}{x^{3/2}} + i\frac{\sin x}{x^{3/2}}\right)dx = \frac{(\beta\Upsilon)^{1/2}}{3}\int_0^\infty \left(\frac{\sin x}{x^{1/2}} - i\frac{\cos x}{x^{1/2}}\right)dx$$
$$= \frac{\sqrt{2\pi\beta\Upsilon}}{6}(1 - i) \tag{3.31}$$

where an integration by parts followed by the reasonably intelligent use of an integral table yields the result. Substitute Eq. (3.31) into (3.30) to obtain:

$$\left.\begin{aligned}
I'(\Delta\nu) &= \frac{1}{2\pi}\int_{-\infty}^{+\infty} d\Upsilon\,\{e^{-i(\Delta\nu)\Upsilon}\, e^{-2/3\,N\pi^{3/2}\beta^{1/2}\Upsilon^{1/2}}\, e^{i\,2/3\,N\pi^{3/2}\beta^{1/2}r^{1/2}}\}\\
&= \frac{1}{2\pi}\int_{-\infty}^{0} d\Upsilon\,\{\} + \frac{1}{2\pi}\int_0^\infty d\Upsilon\,\{\}\\
&\quad = \frac{1}{2\pi}\int_0^\infty d\Upsilon\, e^{i(\Delta\nu)\Upsilon}\, e^{-i\,2/3\,N\pi^{3/2}\beta^{1/2}\Upsilon}\, e^{-2/3\,N\pi^{3/2}\beta^{1/2}\Upsilon^{1/2}}\\
&\quad + \frac{1}{2\pi}\int_0^\infty d\Upsilon\, e^{i(\Delta\nu)\Upsilon}\, e^{-2/3\,N\pi^{3/2}\beta^{1/2}\Upsilon^{1/2}}\, e^{i\,2/3\,\pi^{3/2}\beta^{1/2}\Upsilon^{1/2}}\\
&= \frac{1}{\pi}\int_0^\infty \exp\left[-\frac{2\pi}{3}N\sqrt{2\pi\beta\Upsilon}\right]\cos\left[(\Delta\nu)\Upsilon - \frac{2\pi}{3}N\sqrt{2\pi\beta\Upsilon}\right]d\Upsilon\\
&= \frac{2}{3}\pi\,\beta^{1/2}N(\Delta\nu)^{-3/2}\exp\left[-\frac{4}{9}\frac{\pi^3\beta N^2}{\Delta\nu}\right]
\end{aligned}\right\} \tag{3.32}$$

from the tables.

If we let $\gamma = 2/3\,\pi\beta^{1/2}\,N$ for convenience of notation, this equation becomes:

$$I'(\Delta\nu) = \gamma(\Delta\nu)^{-3/2}\,e^{-\pi\gamma^2/\Delta\nu} \tag{3.33}$$

It should be made clear here that positive $\Delta\nu$ refers to a displacement from the unperturbed line position toward the red, that is, negative frequency displacements. As Kuhn[112] had predicted, the intensity in the wing of the line where the exponential has become small varies as $(\Delta\nu)^{-3/2}$. We may further note from Eq. (3.33) that the poor correspondence between the predicted and the experimental curves on the blue side of the shifted line remains, and that the intensity again goes to zero at the position of the unperturbed line as in Eq. (3.25).

3.6. The statistical shift and half-width

The shift of the intensity maximum of the line for van der Waals forces may be found by equating the first derivative of Eq. (3.42) to zero as:

$$\Delta = \frac{2}{3}\pi\gamma^2 = \left(\frac{2}{3}\pi\right)^3 \beta\,N^2 \tag{3.34}$$

The noteworthy feature in Eq. (3.34) is the dependence of the shift on the square of the density in agreement with Kuhn[112].

Also from Eq. (3.33) or its plot, we may obtain the half-width of the shifted line as:

$$\delta = 0.82\,\pi\gamma^2 \tag{3.35}$$

Before leaving Eq. (3.33) let us briefly consider the effect of taking lower limit zero in Eq. (3.29). This effect is indirectly connected with the simple assumption of a potential curve of the form r^{-6} which is increasingly inaccurate as the optical impact separation (distance of closest approach) falls farther below $10\,A$. The effect of allowing these atomic separations of such small magnitude is to overly lengthen the tail of the line by permitting unwarrantedly large values of $\Delta\nu$. It is apparent that the intensity should be practically zero for $\Delta\nu > \Delta\nu_1 = \beta/R_1^6$ for the true statistical distribution. Thus the ratio $\Delta\nu_1/\Delta\nu_{max}$ should give us a rough measure of the validity of Eq. (3.33). If we assume $R_1 \sim 5A$,

$$\frac{\Delta\nu_1}{\Delta\nu_{max}} = \frac{1}{(5A)^6\left(\frac{2}{3}\pi\right)^3 N^2} \doteq \frac{10^4}{N'^2} \tag{3.36}$$

where N' is measured in units of relative density. Thus, the ratio varies from 10^4 at relative density unity to 10 at relative density 30, so that the validity of the theoretical line wing decreases in this manner. At relative density 30, the intensity corresponding to $\Delta\nu_1$ is 10 per cent of the maximum so that the error due to our lower limit of integration will be less than 10 per cent for this pressure.

3.7 The Jablonski theory

Prior to the Kivel, Bloom and Margenau[104] treatment of broadening by free electrons in 1955, there had only been one treatment of line broadening which did not appeal to the classical path. This, of course, was the theory developed by Jablonski in a series of papers. The development of this theory we shall follow in a moment; the reason for our choosing this chapter for presenting Jablonski's approach is the relation to the Franck–Condon principle which appears throughout. Before doing so, however, a brief word on the classical path.

If we suppose there to be a force acting between two molecules, and if we suppose it to be dependent on their separation — as, indeed, we usually do suppose — then the question arises as to whether we can describe this separation classically. One may consider the problem as follows: Basically, both particles are described by a wave packet. Now if both of these wave packets are of spatial extent small compared to the particle separation, then we may consider them as concentrated at two points, and the separation of the two points or particles is meaningful. In actuality the classical path may be assumed except where the broadeners are electrons; such an assumption was not made in what is to follow. We shall therefore defer more detailed discussion of the criteria for such a path until Chapter 4.

We begin, with Jablonski[91], by assuming that our absorber (emitter) and the N_1 perturbers, whose perturbing influence shall be responsible for the broadening of the spectral line, go to make up a very large, $(N_1 + 1)$-atomic molecule. We are interested in the two eigenfunctions of this molecule which are associated with the two states between which the radiating transition is to take place. Let these eigenfunctions be designated $\psi(E, q, r)$ and $\psi'(E', q, r)$. E and E' are the energies of the two states; q represents the totality of all electron coordinates, and r is the aggregate of the nuclear coordinates. In order to avoid the use of eigendifferentials, the atomic system is restricted to a finite volume V so that $\psi(E, q, r) = \psi'(E', q, r) = 0$ for $r \geq R$, where we have referred all r_i to a coordinate system at the emitter and have placed the emitter at the center of a sphere. This restriction to a finite volume, of course, quantizes the atomic translational (molecular vibrational) motions although our level density will be great enough so that we shall closely approach the continuous curves of our earlier considerations.

The first approximation, that of the separability of electronic and nuclear motions, is now introduced.

$$\psi(E, q, r) = \psi_e(E_e, q)\, \psi_k(E_k, r); \quad E = E_e + E_k \tag{3.37a}$$

$$\psi'(E', q, r) = \psi_e'(E_e', q)\, \psi_k'(E_k', r); \quad E' = E_e' + E_k' \tag{3.37b}$$

If we consider an electronic transition, Eqs. (3.37) yield for the matrix element of $E \leftrightarrow E'$:

$$\left.\begin{aligned} D^{EE'} &= \int \bar{\psi}'(E_e', q)\, D\psi_e(E_e, q)\, \mathrm{d}\tau_q \int \bar{\psi}_K'(E_k', r)\, \psi_k(E_k, r)\, \mathrm{d}\tau_r \\ &= M \int \bar{\psi}_k'(E_k', r)\, \psi_k(E_k, r)\, \mathrm{d}\tau_r \end{aligned}\right\} \tag{3.38}$$

In the first approximation M is independent of r, and we may note that this is merely a restatement of the Franck–Condon principle in that this infers that the nuclear* motion is momentarily unaffected by the electronic transition. Were we to proceed to the second approximation, we would there find M a linear function of r.

We must now find these nuclear eigenfunctions and, in order to do this, we first recall an assumption which has already been utilized. It is assumed that the perturbing forces are additive, which means that the nuclear eigenfunction for the system may be represented as a product of nuclear eigenfunctions as follows:

$$\psi = \prod_i^{N_1} \psi_i \tag{3.39}$$

The reason that this product results may be found in the separability of the Schrödinger equation which results from a potential energy of the form $U(r) = \sum_i U_i(r_i)$, where r_i is the radius vector linking the emitter and the i-th disturber. Thus, Eq. (3.39) would yield N_1 equations of the form:

$$\left[\frac{\hbar^2}{2\mu} \nabla_i^2 + E_i - U_i(r_i)\right] \psi_i = 0 \tag{3.40}$$

where i refers to the i-th perturber. Let us obtain Eq. (3.49) after the manner of Bethe[9].

We consider a system consisting of the emitter and one perturber. Let the mass of the emitter be M and that of the perturber m. Let the coordinates of the emitter be $(x_1^{(1)}, x_2^{(1)}, x_3^{(1)})$ and of the disturber $(x_1^{(2)}, x_2^{(2)}, x_3^{(2)})$. Then the emitter-broadener separation is $r^2 = \sqrt{x_1^2 + x_2^2 + x_3^2}$ where:

$$x_i = x_i^{(2)} - x_i^{(1)} \tag{3.41a}$$

and the coordinates of the center of mass of the system are:

$$X_i = \frac{\mu}{m} x_i^{(1)} + \frac{\mu}{M} x_i^{(2)} \tag{3.41b}$$

The Hamiltonian of the system is:

$$H = \frac{p_1^2}{2M} + \frac{p_2^2}{2m} + U(r)$$

which yields the Schrödinger equation:

$$\frac{\hbar^2}{2M} \nabla_1^2 \psi + \frac{\hbar^2}{2m} \nabla_2^2 \psi + (E - U(r))\psi = 0 \tag{3.42}$$

where $\nabla_i^2 = \sum_{j=1}^{3} \frac{\partial^2}{\partial x_j^{(i)^2}}$. The eigenfunction ψ is a function of the six $x_i^{(j)}$. Utilizing Eqs. (3.41), we may easily show that, since $\psi = \psi(X_1(x_1^{(1)}, x_1^{(2)}, x_1^{(3)}), x_1(x_1^{(1)}, x_1^{(2)}) \ldots)$:

$$\frac{\partial^2}{\partial x_j^{(1)^2}} = \left(\frac{\mu}{m_j}\right)^2 \frac{\partial^2}{\partial X_j^2} + (-1)^j\, 2\,\frac{\mu}{m_j} \frac{\partial^2}{\partial x_j\, \partial X_j} + \frac{\partial^2}{\partial x_j^2}; \quad m_j = \begin{matrix} m \text{ for } j = 1 \\ M \text{ for } j = 2 \end{matrix} \tag{3.43}$$

* We shall use the word "nuclear" extensively, but the reader should keep in mind that it is the group of broadening and emitting molecules which is acting as this group of nuclei.

where $\mu = \dfrac{mM}{m+M}$. Utilizing Eq. (3.43) we may now transform Eq. (3.42) to the following:

$$\frac{\hbar^2}{2(M+m)} \sum_{j=1}^{3} \frac{\partial^2 \psi}{\partial X_j^2} + \frac{\hbar^2}{2\mu} \sum_{j=1}^{3} \frac{\partial^2 \psi}{\partial x_j^2} + [E - U(r)]\,\psi = 0 \tag{3.44}$$

If we neglect the center of mass motion as given by the first term in Eq. (3.44) — separability allows us to do so, and, in addition, this has no effect on our problem — we obtain Eq. (3.40), and this equation describes the behavior of one of our broadeners with respect to the emitter.

Now let $k^2 = \dfrac{2\mu E}{\hbar^2}$ and $U(r) = \dfrac{2\mu}{\hbar^2} V(r)$ in Eq. (3.40) to obtain:

$$\nabla^2 \psi + [k^2 - U(r)]\,\psi = 0 \tag{3.45}$$

The solution* to this equation, after setting it up in spherical polar coordinates, is:

$$\psi_k = \sum_{l=0}^{\infty} A_l\, P_l(\cos\Theta)\, L_{kl}(r) \tag{3.46}$$

where $rL_{kl}(r) = G_{kl}(r)$ satisfies the equation:

$$\frac{d^2 G_{kl}}{dr^2} + \left[k^2 - U(r) - \frac{l(l+1)}{r^2}\right] G_{kl} = 0 \tag{3.47}$$

Each radial eigenfunction L_{kl} refers to a broadening atom of energy $\dfrac{k^2\hbar^2}{2\mu}$ and relative angular momentum $\sqrt{l(l+1)}\,\hbar$. In solving a Schrödinger equation for a given $U(r)$, one obtains a set of orthogonal eigenfunctions, but in the case of this system we have $U(r)$ and $U'(r)$ for the two states between which our transition is to take place, and the L_{kl} and $L'_{k'l'}$ resulting from these different nuclear potential functions for the two states are not generally orthogonal. We may then write:

$$\int \psi_k \psi_k'\, d\tau = 2\pi \int_0^{\pi} \int_0^{\infty} \sum_l \sum_{l'} A_l A_l'\, P_l(\cos\Theta)\, P_{l'}(\cos\Theta)\, L_{kl}(r)\, L'_{k'l'}(r) \cdot \sin\Theta\, d\Theta\, dr \tag{3.48}$$

In Eq. (3.48) the integral is zero unless $l = l'$ due to the orthogonality of the Legendre polynomials which amounts to annother restatement of the Franck–Condon principle in that the relative angular momentum remains constant during the radiating transition.

The Wentzel–Kramers–Brillouin (WKB) approximation yields the following solutions for Eq. (3.47):

$$rL_{kl} = \begin{cases} \dfrac{c}{\sqrt{p_{kl}(r)}}\, e^{-\frac{1}{\hbar}\int_r^{r_0} |p_{kl}(r)|\, dr} & \text{for } r < r_0 \\[2ex] \dfrac{c}{\sqrt{p_{kl}(r)}}\, 2\cos\left[\dfrac{1}{\hbar}\displaystyle\int_{r_0}^{r} p_{kl}(r) dr - \dfrac{\pi}{4}\right] & \text{for } r > r_0 \end{cases} \tag{3.49}$$

* See *infra*, Eq. (7.34) and subsequent.

where $p_{kl}(r) = \sqrt{k^2 - U(r) - \frac{l(l+1)}{r^2}}$, the radial component of the relative momentum. r_0 is the collision partner separation for which $p_{kl}(r) = 0$.

Now let us consider Eq. (3.47a) for the case of large r where $k^2 \ll U(r) + \frac{l(l+1)}{r^2}$. In this asymptotic case we should expect a solution of the form $G_{kl} = A(kr + \Upsilon)$, the solution to:

$$\frac{d^2 G_{kl}}{dr^2} + k^2 G_{kl} = 0 \qquad (3.47b)$$

Upon the evaluation of A and Υ[210] one obtains, for $r \gg r_0$:†

$$L_{kl} = \frac{c}{r} \sin\left(kr - \frac{l\pi}{2} + \eta_l\right) \qquad (3.50)$$

where η_l is the phase shift. We may now apply our boundary condition $\psi_k = 0$ for $r = R$ to obtain:

$$\frac{C}{R} \sin\left(kR - \frac{l\pi}{2} + \eta_l\right) = 0$$

or

$$kR - \frac{l\pi}{2} + \eta = n\pi \qquad (3.51)$$

where n is an integer. From Eq. (3.51) we may evaluate k and subsequently the energy:

$$E_l = \frac{k^2 \hbar^2}{2\mu} = \frac{\left(n\pi + \frac{l}{2}\pi - \eta_l\right)^2}{2\mu R^2} \hbar^2 \qquad (3.52)$$

Let us determine from Eq. (3.52) the energy level density. Differentiation of Eq. (3.52) yields:

$$\frac{dn}{dE_l} = \frac{\sqrt{2\mu} R}{2\hbar\pi\sqrt{E_l}} + \frac{1}{\pi}\frac{d\eta_l}{dE_l} \qquad (3.53)$$

as the number of levels per unit of energy appropriate to the angular momentum $\sqrt{l(l+1)}\,\hbar$. Eq. (3.53) is valid for the level density for R large compared to the range in which $U(r) + \frac{l(l+1)}{r^2}$ is not small compared to k^2.

We now form the overlap integral:

$$\int_0^R \cdots \int_0^R \prod_i^{N_1} L_{inl} L'_{in'l} r_1^2 \, dr_1 \ldots dr_n = \prod_i^{N_1} \int_0^R G_{inl} G'_{in'l} \, dr_i \qquad (3.54)$$

If we now let

$$\prod_i^{N_1} P_{inn'l} = \prod_i^{N_1} \left| \int_0^R G_{inl}(r) G'_{in'l}(r) \, dr \right|^2 \qquad (3.55)$$

† Jablonski chose the lower limit of r as $\sim 10^{-7}$ cm.

$P_{inn'l}$, the square of the matrix element for the i-th broadener, is the probability that the nuclear state will proceed from n, l to n', l as a result of the electronic radiating transition of the emitter. The product given by Eq. (3.55) is the probability that all perturbers will undergo this change in nuclear state due to the radiating transition. We normalize the probability of finding a perturber somewhere in the region $0 \leq r \leq R$ in the usual manner:

$$\int_0^R |G_{inl}(r)|^2 \, dr = 1 \tag{3.56a}$$

Since except for a small region $(r < 10^{-7}$ cm) $G_{inl}(r) = rL_{inl} = c \sin(kr - l\pi/2 + \eta_l)$, we satisfy Eq. (3.56a) as follows:

$$c^2 \int_0^R \sin^2\left(kr - \frac{l\pi}{2} + \eta_l\right) dr = c^2 \frac{R}{2} = 1 \tag{3.56b}$$

thus evaluating c. We now let:

$$P_{inn'l} = \frac{A_{inn'l}}{R^2} \tag{3.57}$$

If we multiply Eq. (3.57), the relative transition probability, by Eq. (3.53), the level density, in which we neglect η_l as small, we obtain:

$$P_{iEE'l} = \frac{A_{iEE'l}}{R} \frac{\sqrt{\mu}}{\hbar\pi\sqrt{2E_l}} \tag{3.58}$$

Eq. (3.58) gives us the probability distribution for the various energy changes. It is now of importance to determine the statistical weight of an initial state E, l.

Let ϱ be the closest distance of approach of a broadening atom. Classically then our angular momentum at closest approach, where $|\mathbf{r} \times \mathbf{v}| = |\mathbf{r}| |\mathbf{v}|$, will be given by $\sqrt{2\mu E_l}\,\varrho$, and the quantum equivalent $\sqrt{l(l+1)}\,\hbar$*. If we equate, square, and differentiate these two expressions for the angular momentum, we obtain:

$$\hbar(2l + 1)\, dl = 2\mu E_l \cdot 2\varrho \, d\varrho \tag{3.59}$$

The Maxwell–Boltzmann distribution yields as the number of disturbers whose energy lies between E and $E + dE$:

$$N_E = N \frac{\sqrt{E}}{(\pi k T)^{3/2}} e^{-E/kT} \, dE \tag{3.60}$$

where $N = N_1/V$. We now desire the number of perturbing atoms of the above mentioned energy whose angular momentum is between $\sqrt{2\mu E}\,\varrho$ and $\sqrt{2\mu E}\,(\varrho + d\varrho)$. Consider Fig. 3.8.

* This equating of the quantum and classical angular momentum is predicated on the preponderant importance of large angular momenta (high l) where the Bohr Correspondence Principle applies.

In Fig. 3.8 the emitter is at 0. dΩ is a solid angle and d$V = r^2$drdΩ is a volume element, a distance r from the emitter. Then the number of molecules of energy between E and $E +$ dE in this volume element is $N_E r^2$dr dΩ. Now the number of molecules in this volume element of optical collision diameter between ϱ and $\varrho +$ dϱ — that is, the number which will pass the emitter with separation between ϱ and $\varrho +$ dϱ — is the fraction $\frac{2\pi\varrho\,\mathrm{d}\varrho}{r^2\,\mathrm{d}\Omega}$. We integrate the latter denominator for agreement with the Jablonski result to obtain the number of molecules in this volume element which are moving inward with impact parameters between ϱ and $\varrho +$ dϱ as $N_E r^2$drdΩ $2\pi\varrho$ d$\varrho/4\pi r^2$. Jablonski used this value for N_E in his first paper[91] while he felt later[94] that he should have utilized twice this value since there will be an equal number moving outward. In following the development of the first paper we shall make this correction.

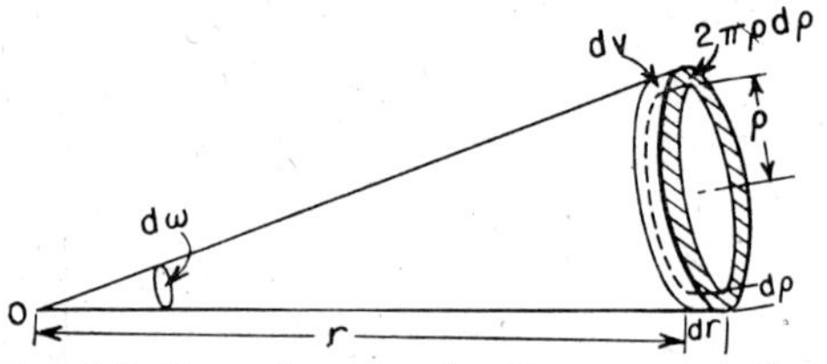

FIG. 3.8. The volume involved in a computation of the random probability (dV/V) of molecules having energies E to $E +$ dE and optical collision diameters ϱ to $\varrho +$ dϱ

We then obtain from Eq. (3.60):

$$N_{E\varrho} = N\frac{\sqrt{E}}{(\pi k T)^{3/2}}\mathrm{e}^{-E/kT}\frac{4\pi\varrho\,\mathrm{d}\varrho}{4\pi r^2}\mathrm{d}E\,\mathrm{d}V \tag{3.61}$$

Eq. (3.61) is now integrated over the entire volume of the container with the exception of as small excluding sphere of radius r' ($\sim 10^{-7}$ cm). The result of this integration is $(R - r')$, and we simply drop r' as small. Substitution for ϱdϱ from Eq. (3.59) into this result yields:

$$m_{(E, E+\Delta E), l} = \pi N \int\limits_E^{E+\Delta} \frac{\mathrm{e}^{E/kT}}{(\pi kT)^{3/2}}\frac{(2l+1)}{\mu\sqrt{E}}\hbar R\,\mathrm{d}E \tag{3.62}$$

Eq. (3.62) gives the number of broadeners whose energy lies between E and $E + \Delta E$ and whose angular momentum quantum number is l or the statistical weight of E_l in Eq. (3.58). Eq. (3.62) then gives us the number of factors in:

$$\prod_i^{N_1} P_{inn'l} = \prod_i^{N_1} P_{iEE'l}(E - E') \tag{3.55}$$

Let us neglect consideration of the restrictions which we must impose on Eq. (3.62) temporarily. We divide the terms in Eq. (3.55) into groups, each group having the same l and energies restricted to a small range ΔE. Now the distribution of the total energy change in each group may be computed, and as a result the distribution corresponding to all E, l groups may be obtained.

3.8. A particular probability distribution

The following problem is posed: The probability distributions $P_1(x_1), P_2(x_2), \ldots, P_{N_1}(x_{N_1})$ of certain quantities $x_1, x_2, \ldots, x_{N_1}$ being given, calculate the probability distribution of:

$$X = \sum_{i=1}^{N_1} x_i. \tag{3.63}$$

The "quantities" here are $x_i = E_i' - E_i$, and the method of solution of this problem will be taken from a later paper of Jablonski.[96]

We assume that the $P_i(x_i)$ are independent of each other in accordance with our earlier assumption of independence of the broadening molecules. The form of the probability distribution $P_i(x_i)$ is:

$$\int P_i(x_i)\,dx_i = \int_{\Delta x_i} [(1-\varepsilon)\,\delta(x_i) + W(x_i)]\,dx_i. \tag{3.64}$$

Eq. (3.64) is based on the assumption of a probability $1-\varepsilon$ for $x_i = 0$*, and for $x_i \neq 0$, the probability of x_i being within Δx_i is:

$$\int_{\Delta x_i} W(x_i)\,dx_i \doteq W(x_i)\,\Delta x_i$$

Also from Eq. (3.64), recalling the delta function behavior:

$$\int_{-\infty}^{+\infty} P_i(x_i)\,dx_i = \int_{-\infty}^{+\infty} [(1-\varepsilon)\,\delta(x_i) + W(x_i)]\,dx_i = 1$$

so that:

$$\varepsilon = \int_{-\infty}^{+\infty} W(x_i)\,dx_i.$$

We consider $P_1(x_1) = P_2(x_2) = \cdots = P_{N_1}(x_{N_1})$, and we now desire:

$$\int_{\Delta X} P_{E,l}(X)\,dX \tag{3.65}$$

where X is given by Eq. (3.63). In order to find Eq. (3.65) Jablonski essentially utilizes mathematical induction.

We assume the probability distribution of a sum of $N_1 - 1$ quantities to be known:

$$P_{1,2,\ldots(N_1-1)}\left(\sum_{i=1}^{N_1-1} x_i\right)$$

Then the probability distribution $P(X)$ of the N_1 quantities:

$$X = \sum_{i=1}^{N_1} x_i = \sum_{i=1}^{N_1-1} x_i + x_{N_1}$$

is surely:

$$P_{E,l}(X) = \int_{-\infty}^{+\infty} P_{1,2,\ldots,(N_1-1)}(X - x_{N_1})\,P_{N_1}(x_{N_1})\,dx_{N_1} \tag{3.66†}$$

* $\delta(x_i)$ is the delta function where by definition $\int_{-\infty}^{+\infty} \delta(x_i)\,dx_i = 1$. This familiar use of δ should not be confused with the use of this symbol for half-width.

† The x_i have been assumed continuous with a range $-\infty$ to $+\infty$.

Now let us return to Eq. (3.65). We let $W^{(1)}(X) \equiv W(x_1)$ and $\int_{\Delta X} W^{(m)}(X)\,\mathrm{d}X$ is the probability that the sum of m quantities, none of which vanish, lies between X and $X + \mathrm{d}X$. Then in analogy to Eq. (3.66):

$$W^{(m)}(X) = \int_{-\infty}^{+\infty} W^{(M-1)}(X - x_i)\, W^{(1)}(x_i)\,\mathrm{d}x_i \tag{3.67}$$

or for the case $N_1 = 2$:

$$W^{(2)}(X) = \int_{-\infty}^{+\infty} W^{(1)}(X - x_i)\, W^{(1)}(x_i)\,\mathrm{d}x_i. \tag{3.68}$$

From Eqs. (3.66) and (3.68) and in analogy with Eq. (3.69) we obtain:

$$\left.\begin{aligned}
\int_{\Delta X} P_{E,l}(X)\,\mathrm{d}X &= \int_{\Delta X}\int_{-\infty}^{+\infty} P_1(X - x_i)\, P(x_i)\,\mathrm{d}x_i\,\mathrm{d}X \\
&= \int_{\Delta X}\int_{-\infty}^{+\infty} [(1-\varepsilon)\,\delta(X - x_i) + W^{(1)}(X - x_i)] \\
&\qquad \times [(1-\varepsilon)\,\delta(x_i) + W^{(1)}(x_i)]\,\mathrm{d}X\,\mathrm{d}x_i \\
\int_{\Delta X} P_{E,l}(X)\,\mathrm{d}X &= \int_{\Delta X}\int_{-\infty}^{+\infty} [\delta(X - x_i)\,\delta(x_i)\,(1-\varepsilon)^2 \\
&\qquad + W^{(1)}(X - x_i)\,\delta(x_i)\,(1-\varepsilon) \\
&+ \delta(X - x_i)\,(1-\varepsilon)\,W^{(1)}(x_i) + W^{(1)}(X - x_i)\,W^{(1)}(x_i)]\,\mathrm{d}X\,\mathrm{d}x_i \\
&= \int_{\Delta X}\left[\delta(X)\,(1-\varepsilon)^2 + \binom{2}{1} W^{(1)}(X)\,(1-\varepsilon) + W^{(2)}(X)\right]\mathrm{d}X
\end{aligned}\right\} \tag{3.69}^*$$

since $\int_{-\infty}^{+\infty} [\delta(X - x_i)\,\delta(x_i)\,(1-\varepsilon)^2]\,\mathrm{d}x_i = \delta(X)\,(1-\varepsilon)^2$, etc., again by virtue of the delta function definition.

We may repeat the procedure for successively higher values of N_2 to obtain for N_1 quantities:

$$\int_{\Delta X} P_{E,l}(X)\,\mathrm{d}X = \int_{\Delta X}\left[\delta(X)\,(1-\varepsilon)^{N_1} + \sum_{m=1}^{N_1}\binom{N_1}{m} W^{(m)}(X)\,(1-\varepsilon)^{N_1-m}\right]\mathrm{d}X \tag{3.70a}$$

Jablonski's inductive proof is completed by showing that Eq. (3.70a) holds for $N_1 + 1$ quantities. No useful purpose would be served here by writing out this final phase of the proof.

For $\varepsilon = 1$, all the terms save the last in Eq. (3.70a) vanish, and for $\varepsilon \ll 1$ we obtain:

$$\int_{\Delta X} P(X)\,\mathrm{d}X = \int_{\Delta X}\left[\delta(X)\,\mathrm{e}^{-\varepsilon N_1} + \sum_{m=1}^{N_1}\binom{N_1}{m} W^{(m)}(X)\,\mathrm{e}^{-\varepsilon(N_1-m)}\right]\mathrm{d}X \tag{3.70b}$$

* The symbol $\binom{n}{m}$ is generally defined as $\frac{n!}{m!\,(n-m)!}$ as in $\binom{2}{1}$.

Now let us consider more carefully what Eq. (3.70a) tells us about our atomic system.

To begin with $\int_{\Delta X} P_{E,l}(X)\, dX$ is the probability distribution for the translational energy of the $N_1 + 1$ perturbers in the E, l group plus the radiator, and $\int_{\Delta x_i} W(x_i)\, dx_i$ is the probability of a change, $x_i = E_i' - E_i$ in the energy of the i-th perturber. We next consider the various terms in the bracketed integrand. The first term is the probability for no translational energy change in the system; the first term in the sum gives the contribution to the probability distribution of a collision of one perturber with the emitter (two body collision), the second term the contribution by a three body collision, etc.†

It now is necessary to construct $P(X)$ for all l and all energy increments ΔE — we have obtained $P_{l,E}(X)$ for the various $l, \Delta E$ groups. The intensity distribution in an absorption line then becomes:

$$P(X) = I(\hbar\,\omega_0 + X) \tag{3.71a}$$

and in an emission line:

$$P'(X) = I'(\hbar\,\omega_0 - X) \tag{3.71b}$$

From the foregoing considerations we might have seen led to expect the same intensity distribution in an absorption line as in an emission line, but this is not the case. $A_{EE'}(X)$ is equal to $A_{E'E}(-X)$ but due to the different distributions of the energy of nuclear motion before and after collisions $P(X)$ is only approximately equal to $P(-X)$.

In Eqs. (3.71a) ω_0 denotes the frequency of the unperturbed spectral line.

This would appear to be an apropos point in the development for a resume of just what we have obtained.

3.9. Review of the Jablonski theory to this point

After taking for our system a large "molecule" made up of N_1 perturbers and an emitter (absorber) restricted to a volume V, we applied Condon's first approximation to separate the electronic motion whose state changes result in our radiation (absorption) from the nuclear motion (Approximation I). We further considered the perturber motions to be independent of one another (Approximation II). Next Eq. (3.46) was obtained for the eigenfunctions of this motion, and the overlap integral Eq. (3.48) resulted. This overlap integral, as a result of the Condon approximation, yielded the selection rule $l = l'$ for the angular momentum, quantum number (Approximation III). From the asymptotic solution (Approximation IV) to Eq. (3.47) we obtained the density of our nuclear energy states Eq. (3.53). As a result of Approximation III the matrix element of a nuclear transition is given in terms of the radial eigenfunctions. The square of the matrix element of a nuclear transition — and hence

† In his first paper Jablonski[91] obtained the equation corresponding to Eq. (3.70a) for the case $P_1(x_1) \neq P_2(x_2) \neq \cdots \neq P_{N1}(x_{N1})$. Since, however, he only utilized the simpler form Eq. (3.70a), we do not reproduce it here.

the probability of the transition — between states n, l and n', l, we found as Eq. (3.57). As a result, the probability of an energy change $E \leftrightarrow E'$ was found as Eq. (3.58), the product of Eq. (3.57) and Eq. (3.53), the nuclear state density. E_l on the right of Eq. (3.58) furnishes the basis for the difference in the absorption and emission lines as given by Eqs. (3.71). It should be kept in mind that Eq. (3.58) is only the intensity or probability due to one perturber. We next found $N_{E\varrho}$ (Eq. 3.61) the number of perturbers with energy between E and $E + \mathrm{d}E$ and angular momentum between $\sqrt{2\mu E\varrho}$ and $\sqrt{2\mu E(\varrho + \mathrm{d}\varrho)}$. In finding this density we assumed the equivalence of classical and quantum angular momentum (Approximation V). We proceeded to find $m_{(E, E+\Delta E), l}$, the number of disturbers with quantum number l and energy between E and $E + \Delta E$, the statistical weight of E_l in Eq. (3.38). This yielded the number of factors in the product Eq. (3.55). This product was divided into E, l groups and $P_{E,l}(X)$ where X as given by Eq. (3.63) was determined from the $P_{iEl}(x_i)$. From these considerations we obtained Eq. (3.70a) for $P_{E,l}(X)$, and subsequently the intensity distribution in the absorption and emission lines. An example might further clarify the meaning of the $W^{(m)}(X)$. As we have mentioned, the first term in the sum in Eq. (3.70a) is the probability distribution resulting from single perturbing atom transits. This term is analogous to Eq. (3.58):

$$W^{(1)}(X) = W^{(1)}(E_{n'} - E_n) = \frac{A^2_{EE'l}}{R^2} \frac{\mathrm{d}n}{\mathrm{d}E_l} m_{(E, E + \Delta E), l} \tag{3.72}$$

The evaluation of $A^2_{E'El}$, the square of the matrix element of the translational motion, now poses the biggest problem. In his later papers[94, 95] on the subject Jablonski carried out a calculation which yielded an approximation for these matrix elements.

3.10. Limiting cases and the matrix element A

Before considering this approximate evaluation of A, we might mention two limiting cases of the theory which Jablonski[92] pointed out. The theory will reduce to the limiting case of a distribution of the type*:

$$I(\omega)_{\omega = \omega_0} = \frac{\Sigma D_i^2}{\hbar(\omega - \omega_0)^2} \tag{3.73}$$

but with no "damping constant" — an additional term in the denominator which prohibits a complete resonance condition — for small differences in the potential curves for the two electronic states involved in the radiating transition.

The other limiting case which essentially corresponds to high densities leads to the statistical distribution smeared out† with a half width:

$$\delta = \frac{1}{2}\sqrt{v_r \frac{\mathrm{d}\nu}{\mathrm{d}r}} \tag{3.74}$$

* See *supra*, Chap. 2.

† By smearing we mean the folding of an Interruption distribution into a Statistical distribution.

This compares favorably with the smearing half-width of Kuhn and London:

$$\delta = \frac{1}{\pi} \sqrt{\pi \frac{d\nu}{dx}} \tag{3.75}$$

Let us now return to the evaluation of A.

For the eigenfunction we take the WKB approximation for $r > r_0$ in the form given by Eq. (3.49) with c as given by Eq. (3.56b). We must insert the additional normalization factor $\sqrt{p_{kl(\infty)}} = \sqrt{2\mu E_{nl}}$. The eigenfunction is then:

$$\left.\begin{aligned} \psi_{nl} &= \left(\frac{2}{R}\right)^{1/2} \left[\frac{2\,m\,E_{nl}}{2m(E_{nl} - V(r)) - \hbar^2 \dfrac{l(l+1)}{r^2}}\right]^{1/4} \\ &\quad \times \cos\left\{\frac{1}{\hbar}\int_{r_0}^{r}\left[2\,m\,(E_{ne} - V\,(r)) - \hbar^2\frac{l(l+1)}{r^2}\right]^{1/2} dr + \Upsilon\right\} \\ &= \left(\frac{2p(\infty)}{Rp(r)}\right)^{1/2} \cos\left[\frac{1}{\hbar}\int_{r_0}^{r} p(r)\,dr + \Upsilon\right] \end{aligned}\right\} \tag{3.76}$$

where, as has been previously noted, $U(r) = (2\mu/\hbar^2)\ V(r)$ and $p(r) = \sqrt{2\,m\,(E_{nl} - V\,(r)) - \dfrac{l(l+1)\,\hbar^2}{r^2}}$. $p(r)$ is, of course, the radial component of the relative linear momentum. The phase Υ has already arisen in the solution of Eq. (3.47b). If we let r_t be the classical turning point or reversal point where $p(r_t) = 0$ then Eq. (3.76) is only valid for $r > r_0 = r_t$.*

By utilizing Eq. (3.76) we may form the matrix element:

$$\left.\begin{aligned} A_{n'n''l} &= \int \bar{\psi}_{n'l}\psi_{n''l}\,d\tau \doteq \frac{2}{R}\int_0^R \left(\frac{p'(\infty)}{p'(r)}\right)^{1/2} \cos\left[\int_r^r \frac{p'(r)}{\hbar}\,dr + \Upsilon'\right] \\ &\qquad \times \left(\frac{p''(\infty)}{p''(r)}\right)^{1/2} \cos\left[\int_{r_t}^r \frac{p''(r)}{\hbar}\,dr + \Upsilon''\right] dr \\ &= \frac{1}{R}\int_{r_t}^R \left(\frac{p'(\infty)\,p''(\infty)}{p'(r)\,p''(r)}\right)^{1/2} \left[\cos\left(\int_{r_t}^r \frac{p'(r) - p''(r)}{\hbar}\,dr + \Upsilon' - \Upsilon''\right)\right. \\ &\qquad \left. + \cos\left(\int_{r_t}^r \frac{p'(r) + p''(r)}{\hbar}\,dr + \Upsilon' + \Upsilon''\right)\right] dr \end{aligned}\right\} \tag{3.77}$$

* "The difficulty arising from the failure of the WKB approximation in the region of the point of closest approach of the colliding atoms (classical turning point) is not surmounted by Weisskopf's treatment, but merely completely camouflaged."[95]

In Eq. (3.77) a single prime denotes a quantity going with one of the electronic states and a double prime a quantity going with the other.

Since the sums involved in the second cosine term in Eq. (3.77) will increase or decrease more rapidly than the differences involved in the first cosine term, the second term will fluctuate much more rapidly than the first with r. Thus, the contribution of the second cosine term to the integral in Eq. (3.77) will be much less in the region of maximum contribution of both terms. On this basis, Jablonski considered only the first. It may also be seen that the region of maximum contribution of this first term is the neighborhood of the so-called "Condon point" where $p'(r_c) = p''(r_c)$.*

It follows immediately that $(A_{n'n''l})^2$ is the probability of change of translational energy by $X = E_{n'} - E_{n''}$:

$$X = V'(r_c) - V''(r_c) = \hbar(\omega_0 - \omega) = \hbar \Delta\omega$$

We now break up the integral within the cosine from one with limits r_t and r into one with limits r_t and r_c and another with limits r_c and r. The second of these two integrals may be expanded in a Taylor series about r_b in a straightforward manner to yield:

$$\left.\begin{aligned} A_{n'n''l} = \frac{1}{R}\int_{r_t}^{R}\left[\frac{p'(\infty)\,p''(\infty)}{p'(r)\,p''(r)}\right]^{1/2} & \\ \times \cos\left\{\frac{1}{\hbar}\int_{r_t}^{r}[p'(r) - p''(r)]\,\mathrm{d}r + \Upsilon' - \Upsilon'' \right. & \\ \left. + \frac{1}{\hbar}\int_{r_c}^{r}\left[p'(r) - p''(r_t) + \left[\frac{\mathrm{d}}{\mathrm{d}r}(p'(r) - p''(r))\right]_{r=r_c}(r - r_c) + \cdots\right]\mathrm{d}r\right\}\mathrm{d}r & \end{aligned}\right\} \quad (3.78)$$

If all terms beyond the second in the series be now neglected; $r - r_c$ be replaced by ξ, and the approximation $p'(r) = p''(r) = p(r_c)$ be made in the denominator of the factor in front of the cosine,† Eq. (3.78) becomes:

$$\left.\begin{aligned} A_{n'n''l} &= \frac{[p'(\infty)\,p''(\infty)]^{\frac{1}{2}}}{Rp(r_c)}\int_{-\infty}^{+\infty}\cos(\eta \pm \beta\xi^2)\,\mathrm{d}\xi \\ &= \frac{[p'(\infty)\,p''(\infty)]^{\frac{1}{2}}}{Rp(r_c)}\left(\frac{\pi}{\beta}\right)^{\frac{1}{2}}\cos\left(\eta \pm \frac{\pi}{4}\right) \end{aligned}\right\} \quad (3.79)$$

where:

$$\eta = \frac{1}{\hbar}\int_{r_0}^{r_c}[p'(r) - p''(r)]\,\mathrm{d}r + \Upsilon' - \Upsilon'' \quad (3.80\,\mathrm{a})$$

* This nomenclature has its origin in the "classical" conception of the *FCP* where it is assumed that the relative nuclear velocities remain constant during an electronic transition.

† Jablonski[95] refers to this approximation as crude. The approximation appears to get "cruder" the closer r_c lies to r_t.

and:

$$\left.\begin{aligned}
\beta &= \frac{1}{2\hbar}\left|\frac{\mathrm{d}}{\mathrm{d}r}\left[p'(r) - p''(r)\right]\right|_{r=r_c} \\
&= \frac{1}{2\hbar}\left|\frac{-m\dfrac{\mathrm{d}V'(r)}{\mathrm{d}r}}{\left[2m\left(E_{n'l} - V'(r)\right) - \hbar^2\dfrac{l(l+1)}{r^2}\right]^{1/2}}\right. \\
&\quad \left. - \left(m\frac{\mathrm{d}V''(r)}{\mathrm{d}r}\right)\Big/\left[2m\left(E_{n'l} - V''(r)\right) - \hbar^2\frac{l(l+1)}{r^2}\right]^{1/2}\right|_{r=r_c} \\
&= m\left|\frac{\mathrm{d}U(r)}{\mathrm{d}r}\right|_{r=r_c}\Big/ p(r)
\end{aligned}\right\} \quad (3.80\,\mathrm{b})$$

where $U(r) = X = V'(r) - V''(r)$.

We may first note that the term $p'(r_c) - p''(r_c)$ in Eq. (3.78) is zero by definition. $\beta\xi$ results from the first integration, since $(r - r_c)^2_{r=r_c}$ is obviously zero, thus simply leaving the upper limit of this integral. In addition, an obvious approximation has been made in Eq. (3.79) by taking the limits $-\infty$ and $+\infty$. Extending the limits in this manner is predicated on the assumption that the integrand only contributes significantly to the integral in the region $r = r_c$.

Substitution of Eq. (3.80b) into Eq. (3.79) yields:

$$\left.\begin{aligned}
A_{n'n''l} = A_X &= \frac{1}{R}\left(\frac{p'(\infty)\,p''(\infty)}{p(r_c)\,m\left|\dfrac{\mathrm{d}U}{\mathrm{d}r}\right|_{r=r_c}}\right)^{1/2}\cos\left(\eta \pm \frac{\pi}{4}\right) \\
&= \frac{1}{R}\left(\frac{p'(\infty)\,p''(\infty)\,2\pi\hbar}{p(r_c)\,m\left|\dfrac{\partial X}{\partial r}\right|_{r=r_c}}\right)^{1/2}\cos\left(\eta \pm \frac{\pi}{4}\right)
\end{aligned}\right\} \quad (3.81)$$

Eq. (3.81) is an approximation to the matrix element which we have been seeking. Before further utilization of Eq. (3.81), let us consider the additional ramifications of the earlier theory which Jablonski[95] introduced in 1945.

3.11. The general system energy change probability

It will perhaps be most straightforward to simply replace Eq. (3.72) by:

$$\left.\begin{aligned}
W^{(1)}(X) = W^{(1)}(E_{n'} - E_{n''}) &= \sum_{l=0}^{l_{\max}} Q(l)\,\frac{D^2_{n'n''l}}{\sum\limits_{n'=1}^{\infty} D^2_{n'n''}}\,\frac{\mathrm{d}n'}{\mathrm{d}E_{n'}} \\
&\doteq \int_0^{l_{\max}} Q(l)\,\frac{D^2_{n'n''l}}{S}\,\frac{\mathrm{d}n'}{\mathrm{d}E_{n'}}\,\mathrm{d}l
\end{aligned}\right\} \quad (3.82)$$

and explain the reason for so doing. $D^2_{n'n''l}$ is defined as before by Eq. (3.38), and $\mathrm{d}n'/\mathrm{d}E_{n'}$ is the energy level density which we have also previously encountered. $Q(l)$ is the probability of occurrence of a certain l which replaced the statistical weight found earlier. The approximation of a high l density has been made so that the summation may be replaced by the integration. If a single prime is taken to mean the upper state, Eq. (3.82) refers to an absorption line. The emission line would replace $\frac{dn'}{dE_{n'}}$ by $\frac{dn''}{dE_{n''}}$.

Now if Eq. (3.82) were substituted into Eq. (3.70a) or (3.70b), the intensity distribution (normalized to unity) in the spectral line — natural and Doppler widths neglected — would be obtained. Needless to remark, this has not been done. The Condon approximation yields:

$$S = M_0^2; \quad D_{n'n''l} = M_0 A_{n'n''l}$$

and Eq. (3.82) may be written:

$$W^{(1)}(X) = \int_0^{l_{\max}} Q(l) A^2_{n'n''l} \frac{dn'}{dE_{n'}} dl \tag{3.83}$$

where $\frac{dn}{dE_n}$ is given by Eq. (3.53), and $A_{n'n''l}$ is given by Eq. (3.81). $Q(l)$, however, remains to be determined.

$Q(l)$ is certainly proportional to the statistical weight due to the spatial degeneracy (m_l) of the state l:

$$Q(l) = g(2l + 1) \tag{3.84}$$

On the basis used for obtaining Eq. (3.59),

$$\varrho = \hbar [l(l + 1)/2mE]^{1/2} \tag{3.85}$$

and

$$Q(l)\, dl \doteq Q'(\varrho)\, d\varrho \tag{3.86}$$

for large l.

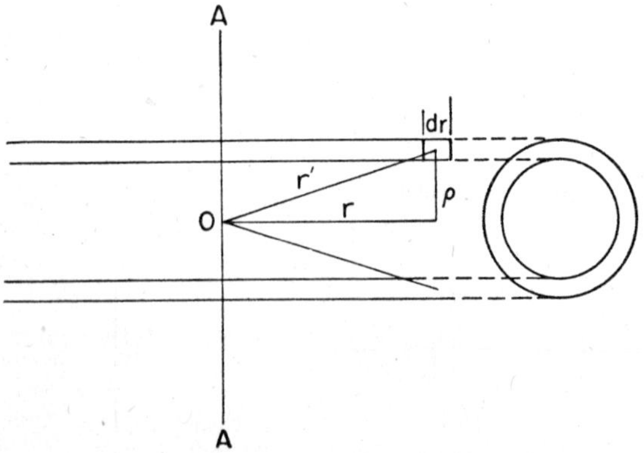

FIG. 3.9. The physical situation involved in the calculation of $Q'(\varrho)$ for Eq. (5.95)

In Fig. 3.9, one-half the molecules may be considered as moving toward (or across) the plane A—A from the left, while the remainder proceed toward the plane from the right. The probability that a collision of optical collision diameter ϱ between ϱ and $\varrho + d\varrho$ occurs, is thus the volume of the tube shell of thickness $d\varrho$ divided by the total volume available to the broadeners. This follows from the assumption of a random distribution in space. We suppose the assemblage to be confined to a sphere of radius R. In consequence:

$$Q'(\varrho)\,\mathrm{d}\varrho = \frac{2\int\limits_0^{r_{\max}} 2\pi\varrho\,\mathrm{d}\varrho\,\mathrm{d}r}{\frac{4}{3}\pi R^3} = \frac{3}{R^3} r_{\max}\varrho\,\mathrm{d}\varrho$$

Now by definition $r = r_{\max}$ when $r' = R$ so that:

$$Q'(\varrho)\,\mathrm{d}\varrho = \frac{3}{R^3}(R^2 - \varrho^2)^{1/2}\varrho\,\mathrm{d}\varrho \tag{3.87a}$$

or for the case under consideration $\varrho \ll R$ so that:

$$Q'(\varrho)\,\mathrm{d}\varrho = \frac{3}{R^2}\varrho\,\mathrm{d}\varrho = \frac{3}{2R^2}\mathrm{d}(\varrho^2) \tag{3.87b}$$

By substituting from Eq. (3.85) for ϱ in Eq. (3.87b) we obtain:

$$Q'(\varrho)\,\mathrm{d}\varrho = \frac{3\hbar^2}{4R^2 mE}(2l+1)\,\mathrm{d}l$$

Now using Eq. (3.86) this yields:

$$Q(l) = \frac{3\hbar^2}{4R^2 mE}(2l+1) \tag{3.88}$$

For convenience, we rewrite Eq. (3.53) slightly, recalling that the equation was obtained for $r = R$, to obtain:

$$\frac{\mathrm{d}n}{\mathrm{d}E_n} = \frac{(m)^{1/2}R}{\pi\hbar(2E_n)^{1/2}} = \frac{mR}{\pi\hbar p_n(\infty)} \tag{3.53}$$

If Eqs. (3.88), (3.53), and (3.81) are substituted into Eq. (3.83) the desired expression for the case of an absorption line is:

$$\left.\begin{aligned} W^{(1)}(X) &= \int\limits_0^{l_1} \frac{3}{4}\frac{\hbar^2}{R^2 m E_{n''}}\frac{(2l+1)\,mR\,p'(\infty)\,p''(\infty)}{\pi\hbar\,p'(\infty)\,R^2\,p'(r)} \\ &\quad\times \frac{2\pi\hbar}{m\left|\frac{\mathrm{d}X}{\mathrm{d}r}\right|_{r=r_c}}\cos^2\left(\eta \pm \frac{\pi}{4}\right)\mathrm{d}l + \zeta(X + l_t) \\ &= \frac{3}{4}\frac{\hbar^2}{mE_{n''}R^3\left|\frac{\mathrm{d}X}{\mathrm{d}r}\right|_{r=r_e}} \\ &\quad\times \int\limits_0^{l_1} \frac{2(l+1)\,2\cos^2(\eta \pm \pi/4)}{\left(1 - \frac{V''(r_c)}{E_{n''l}} - \hbar\frac{2l(l+1)}{2mE_{n''l}r_c^2}\right)^{1/2}}\mathrm{d}l + \zeta(X, l_t) \end{aligned}\right\} \tag{3.89}$$

where

$$\frac{p''(\infty)}{p(r)} = \frac{(2\,m\,E_{n''l})^{1/2}}{\left[2\,m\,(E_{n''l} - V''(r)) - \hbar^2\,\dfrac{l(l+1)}{r^2}\right]^{1/2}}$$

$$= \frac{1}{\left(1 - \dfrac{V''(r_c)}{E_{n''l}} - \hbar^2\,\dfrac{l(l+1)}{2\,m\,E_{n''l}\,r_c^2}\right)^{1/2}}$$

At the classical turning point the following relation holds:

$$2m\,[E_{n''l} - V''(r_c)] - \hbar\,\frac{l_t(l_t+1)}{r^2} = p''^2(r) = 0 \tag{3.90}$$

From Eq. (3.90) we may obtain:

$$1 - \frac{V''(r_c)}{E_{n''l}} = \frac{\hbar^2\,l_t(l_t+1)}{2\,m\,r^2\,E_{n''l}} \tag{3.91}$$

When Eq. (3.91) is substituted into the denominator under the integral sign in Eq. (3.89), the result is:

$$\left.\begin{aligned} W^{(1)}(X) &= \frac{3}{2}\,\frac{r_c^2\left(1 - \dfrac{V''(r_c)}{E_{n''l}}\right)^{1/2}}{l_t(l_t+1)\,R^3\left|\dfrac{dX}{dr}\right|_{r=r_c}} \\ &\times \int_0^{l_1} \frac{(2\,l+1)\,2\cos^2\left(\eta \pm \dfrac{\pi}{4}\right)}{\left[1 - \dfrac{l(l+1)}{l_t(l_t+1)}\right]^{1/2}}\,dl + \zeta(X, l_t) \end{aligned}\right\} \tag{3.92}$$

The $\zeta(X, l_t)$ is a correction term which is added due to the choice of upper limit $l_1 < l_t$. This upper limit must be less than l_t since the *WKB* approximation fails in the region of the classical turning point, and $\zeta(X, l_t)$ is then to correct for this low limit choice. "The more accurate calculations involving eigenfunctions valid in the regions of turning points and outside the classical range of motion would be very tedious unless carried out by the aid of the differential analyzer."[95]

If $|X| = |E_{n'} - E_{n''}|$ is large enough, that is, if we are considering frequencies sufficiently far out in the wing of the line then $\cos(\eta \pm \pi/4)$ is a rapidly oscillating function of l, as we may see from Eq. (3.79) and Eq. (3.80a). In this case we may replace $\cos^2(\eta \pm \pi/4)$ by its average value 1/2. Eq. (3.92) then becomes:

$$
\left.\begin{aligned}
W^{(1)}(X) &= \frac{3}{2}\,\frac{r_c^2\left(1-\dfrac{V''(r_c)}{E_{n''}}\right)^{1/2}}{l_t(l_t+1)\,R^3\left|\dfrac{\mathrm{d}X}{\mathrm{d}r}\right|_{r=r_c}} \\
&\quad\times\int_0^{l_1}\frac{2\,l+1}{\left[1-\dfrac{l(l+1)}{l_t(l_t+1)}\right]^{1/2}}\,\mathrm{d}l+\zeta(X,l_t) \\
&= \frac{3\,r_c^2}{R^3\left|\dfrac{\mathrm{d}X}{\mathrm{d}r}\right|_{r=r_c}}\left(1-\frac{V''(r_c)}{E_{n''}}\right)^{1/2} \\
&\quad\times\left[1-\left(1-\frac{l_1(l_1+1)}{l_t(l_t+1)}\right)^{1/2}\right]+\zeta(X,l_t)
\end{aligned}\right\}\qquad(3.93)
$$

for an absorption line. We simply replace the double prime by a prime in Eq. (3.93) for the emission line.

The asymptotic form ($l_1 = l_t$) for $W^{(1)}(X)$ is:

$$
W^{(1)}(X) = \frac{3\,r_c^2(1-V''(r_c)/E_{n''})^{1/2}}{R^3\left|\dfrac{\mathrm{d}X}{\mathrm{d}r}\right|_{r=r_c}} \qquad (3.94)
$$

and corresponding for an emission line.

Next, let us show that Eq. (3.94) reduces to Kuhn's distribution[112] — Margenau's distribution without the exponential — for certain interactions. Before doing so, however, let us mention a consideration which we have so far neglected.

3.12. Double interaction curves and reduction to the Margenau line shape

It has been tacitly assumed that there is one interaction curve for the upper electronic state and one for the lower. This is not generally the case, for there may well be several interaction curves for each electronic state, and for single encounters we would then have various $W_i^{(1)}(X)$ for the various pairs of curves. Our $W^{(1)}(X)$ would then have the form:

$$
W^{(1)}(X) = a_1\,W_1^{(1)}(X) + a_2\,W_2^{(1)}(X) + \cdots \equiv \Sigma\,a_i\,W_i^{(1)}(X) \qquad (3.95)
$$

where a_i would be related to the probability of transitions between two particular potential curves with which $W_i^{(1)}(X)$ is associated. It is also apparent that, since $A_{n'n''l}$ or $D_{n'n''l}$ would be different for each $W_i^{(1)}(X)$, each $A_{n'n''l}$ would have to comprise a separate calculation for each $W_i^{(1)}(X)$. When one considers that we have here discussed only single encounters, the complexity of the accurate computation for multiple collisions appears rather staggering. Let us return to the limiting case of the statistical distribution.

According to the classical *FCP*:

$$X = V'(r_c) - V''(r_c) = \hbar(\omega - \omega_0) = \hbar \Delta\omega$$

From this equation $\Delta\omega = f(r_c)$ so that $r_c = f(\Delta\omega)$. Let $n = (4/3)\,\pi R^3 N$. Utilizing these relations and Eqs. (3.94) and (3.95) we obtain for an absorption line:

$$I(\omega) = \Sigma a_i\, 4\pi N f^2(\Delta\omega) \frac{\left(1 - \dfrac{V''(r_c)}{E_{n''}}\right)}{\left|\dfrac{d(\Delta\omega)}{dr_c}\right|} \tag{3.96}$$

We now make the approximation $E = 3/2\,kT$, although $I(\omega)$ should be averaged over all E which occur in the gas. Finally, an assumption of one electronic curve for each electronic state is introduced. We let:

$$V'(r_c) = -\frac{c'}{r_c{}^n}; \quad V''(r_c) = -\frac{c''}{r_c{}^n} \tag{3.97}$$

so that:

$$\Delta\omega = \frac{1}{\hbar}\frac{c' - c''}{r_c{}^n} = \frac{K}{r_c{}^n}; \quad r_c = f(\Delta\omega) = \left(\frac{K}{\Delta\omega}\right)^{1/n}; \quad V''(r_c) = \frac{c''}{K}\Delta\omega \tag{3.98}$$

Substitution of Eq. (3.98) into Eq. (3.96) yields:

$$I(\omega) = \frac{4\pi N K^{3/n}}{n(\Delta\omega)^{(n+3)/n}}\left(1 + \frac{2c''\Delta\omega}{3KkT}\right) \tag{3.99}$$

which corresponds to the Kuhn form for $2c''\Delta\omega/(3KkT) \ll 1$. The agreement of this limiting form of the theory with the earlier work of Kuhn[112], and Margenau[137] would tend to imply a corroboration of either this theory or one of the earlier ones depending, of course, on who is pointing out the corroboration.

As Foley[41] has remarked, the divergence of the intensity at line center as given by Eq. (3.99) should hardly come as a surprise when the method of obtaining this equation is considered. The reason for the divergence lies in the fact that the perturber will only spend an infinitesimal portion of its time within the range of the forces, for the case of a large combining volume, when only a single perturbing atom has been initially considered, and then the resulting expression averaged over all possible transitions. This fact results in an infinite probability for the unperturbed line center frequency. The argument is without any particular significance, however, as far as line wing theory validity is concerned.

Insofar as it has been possible to ascertain, this is as far as Jablonski has carried his quantum mechanical broadening theory. Our only concrete results are then given by Eq. (3.99) which had already been given by Margenau's statistical theory (or Kuhn's if you will) with the exception of the small factor in the bracket. This is only a limiting case of the wave mechanical theory and serves no immediately apparent purpose save the wave mechanical verification of the statistical theory. Before a further discussion of this theory, let us review the approximations which we have made in addition to those listed before Eq. (3.72).

3.13. The approximations of the Jablonski theory*

First, let us mention that we have improved Approximation V. Eq. (3.86) certainly holds for very large l. Thus, if we obtain it, as we did, we should have an approximation for $Q(l)$ which is very close to the true state of affairs.

We have replaced our previous Approximation IV by one in which we assume the *WKB* approximate eigenfunction (Approximation IV). The second term in the integral of Eq. (3.77) has been neglected (Approximation VI). We have neglected all terms beyond the second in the series in Eq. (3.78) (Approximation VII). It was then assumed that $p'(r) = p''(r) = p(r_t)$ in the denominator of Eq. (3.78) (Approximation VIII). The limits in Eq. (3.79) have been extended from 0 and R to $-\infty$ and $+\infty$ (Approximation IX). The approximation of Eq. (3.82) — replacing the summation over l by an integration — is probably a trivial one in most cases (Approximation X). Eq. (3.53) is an approximation, but perhaps one of the closest ones (Approximation XI). The limits on Eq. (3.89) are, of course, an approximation, but this one is such a direct result of IV that it may well be included in it. We replaced $\cos^2(\eta \pm \pi/4)$ by its average value (Approximation XII). The obvious approximations resulting in Eqs. (3.94) and (3.99) need not be discussed. We should add, however, that without additional approximations it is apparent that no actual line intensity distribution can be obtained for the case of several potential curves for each state — $\Sigma a_i W_i^{(1)}(X)$ instead of simply $W_i^{(1)}(X)$ — and multiple collisions.

As a result of the approximations which have been introduced, we are left with Eq. (3.94) which is only (a) valid in the wing of the line and (b) valid for heavy broadeners. In addition, no information has been afforded about line shift.

Approximation IV is a serious limitation, but one to which the alternative would be the specific and extensive use of some computing device. As a direct result of this approximation, the asymptotic form Eq. (3.94) of $W^{(1)}(X)$ is larger than the true $W^{(1)}(X)$. The reason for this becomes apparent when we recall that the *WKB* eigenfunctions and, hence, the $A^2_{n'n''l}$ become infinite for $l - l_t$, that is, at the classical turning point. These considerations have the direct effect of making the asymptotic intensity distribution inapplicable to the case of broadening by light particles. Let us investigate the reason for this.

It can be shown† that in order for the *WKB* approximation to be valid,

$$\Delta l = (l_t - l) \gg \frac{[l_t(l_t+1)]^{2/3}}{2\,l_t+1} = \varkappa(l_t) \tag{3.100}$$

In order to make Approximation VI it was necessary for us to assume that there is only a noticeable contribution to Eq. (3.77) in the region of the Condon

* See *supra*, Sec. 3.12.

† An attempt to avoid this phrase has been made, but the calculation which would be involved is not particularly enlightening.

point. This is the same approximation which earlier had restricted the statistical theory to heavy particles (low velocities) by neglecting the intrinsic diffuseness in r_c.

Now let us consider Approximation IX. It is essentially predicated on the assumption of a maximum contribution to the integral for A in the neighborhood of the Condon point. If this is the case we may, with the introduction of small error, extend these limits as has been done. In addition, this means that the phase of the cosine term must increase by at least π as ξ increases from 0 to r. This would automatically require that $\beta r_c^2 > \pi$ which yields:

$$r_c > \left(\frac{2\pi \hbar\, p(r_c)}{m \left. \dfrac{\mathrm{d}X}{\mathrm{d}r} \right|_{r=r_c}} \right)^{1/2} \tag{3.101a}$$

or for $|X| = \hbar K/r^n$

$$r_c < \left(\frac{n\, K\, \mu}{2\pi\, p(r_c)} \right)^{1/(n-1)} \tag{3.101b}$$

Eq. (3.101a) essentially restricts the validity of our solution to the wing where the frequency displacement is relatively large, due to the restrictions imposed by this equation against distant encounters. We might add that Approximation XII also restricts our solution to the line wing.

Jablonski's theory is indeed an interesting, careful, and complete one, and the basic approximation utilizing the Franck–Condon principle should be generally quite good. It appears to have suffered from too much completeness if anything for (1) the needs of those concerned with line broadening and, like its fellows, (2) from the basic paucity of knowledge on potential curves. It is certainly worth our notice, however, and perhaps in the near future it will be worth some gentleman's while to apply a digital computer to a thorough study of its application.

3.14. A mild controversy, Lorentz–Jablonski equivalence

Subsequent developments found Foley[46] investigating the manner in which Jablonski's theory could be shown to lead to a Lorentz type intensity distribution and Jablonski[96] objecting to certain of Foley's demonstrations and concluding that, "One has either to demonstrate rigorously that the Lorentz formula can be obtained from the quantum mechanical theory proposed by the writer or to modify (or to reject) one of these theories."[96] We shall consider this in slightly greater detail.

Initially, Foley remarked that the overlap integral which Jablonski utilized may be written, using WKB functions, as:

$$\frac{1}{R_0} \int_0^{R_0} \mathrm{d}R \cos \left\{ \frac{1}{\hbar} \int_0^{R} [p(R) - p'(R)]\, \mathrm{d}R + \eta - \eta' \right\} \tag{3.102a}$$ *

* Let us note that this is the less exact form which Jablonski utilized in an earlier consideration. The more exact form is, of course, given by Eq. (3.77) from a later work.

Eq. (3.102a) indicates that Foley had already inferred the approximation which negated the second term on the right of Eq. (3.77). Foley next wrote down:

$$\int_{\varrho}^{R_0} \mathrm{d}R \frac{c^2}{p(R)} \cos\left\{\int_{\varrho}^{R} \frac{m}{\hbar} \frac{\Delta E - \Delta U(R)}{p(R)} \mathrm{d}R\right\} \tag{3.102b}$$

In order to obtain this, let us again consider Eq. (3.77). Let us introduce the approximation which we utilized subsequent to Eq. (3.78), namely, $p'(r) \doteq p''(r) \doteq p(r_c)$. This leads us to:

$$\begin{aligned} p'(r) - p''(r) &= \sqrt{2m(E' - U'(r)) - \frac{l(l+1)\hbar^2}{r^2}} \\ &\quad - \sqrt{2m(E'' - U''(r)) - \frac{l(l+1)\hbar^2}{r^2}} \\ &\doteq \frac{p(r_c)}{p(r_c)}\left\{2m(E' - U'(r)) - \frac{l(l+1)\hbar^2}{r^2}\right. \\ &\quad \left. - 2m(E'' - U''(r)) + \frac{l(l+1)}{r^2}\hbar^2\right\} \\ &\doteq 2m\frac{(\Delta E - \Delta U(r))}{p(r)} \end{aligned}$$

This result, with the η's dropped, gives us Eq. (3.102b) wherein the two is neglected. Now at the boundary of the container $p(R_0) = \sqrt{K^2 - U(R_0) - \frac{l(l+1)}{R_0^2}} \to K$ so that Eq. (3.51) becomes:

$$\frac{p(R_0)\cdot R_0}{\hbar} - \frac{l\pi}{2} + \eta = n\pi = \frac{1}{2}(2n' + 1)\pi$$

so that:

$$[p'(R_0) - p''(R_0)]\frac{R_0}{\hbar} + \eta' - \eta'' = q\pi \tag{3.103}$$

Surely, the substitution

$$\mathrm{d}t = \frac{m\,\mathrm{d}R}{p(R)} = \frac{m\,\mathrm{d}R}{\sqrt{2m}\sqrt{E - U(R) - \frac{l(l+1)\hbar^2}{2mR^2}}} \tag{3.104}$$

may be utilized and, in addition, the approximation† $l(l+1)\hbar^2 = m^2v^2\varrho^2$ may be introduced to change the integral within the braces in Eq. (3.102b) to:

$$\frac{2}{\hbar}\int_{\varrho}^{R}(\Delta E - \Delta U)\,\mathrm{d}R \doteq \frac{2}{\hbar}\int_{0}^{T_i}(\Delta E - \Delta U)\,\mathrm{d}R = 2\frac{\Delta e_i}{\hbar}T_i - 2\int_{0}^{T_i}\frac{\Delta U}{\hbar}\,\mathrm{d}t = q\pi$$

† See the discussion of this preceding Eq. (3.59).

at the boundary according to Eq. (3.103) when $Tv = R_0$. Or:

$$2\,\Delta\omega_i\,T_i - 2\int\limits_0^{T_i} \frac{\Delta U}{\hbar}\,\mathrm{d}t = q\pi \tag{3.105a}$$

where now the i refers to the i-th perturber. As a consequence Eq. (3.102b) becomes:

$$\frac{c^2}{m}\int\limits_0^{T_i} \mathrm{d}t_i \cos 2\left[\omega_i\,t_i - \int\limits_0^{t} \frac{\Delta U_i(t)}{\hbar}\,\mathrm{d}t\right] \tag{3.105b}$$

The similarity between Eq. (3.105b) and, say, Eq. (5.5a) would indicate that under the proper manipulations the Lorentz form could probably be obtained from the former equation. Foley proceeds to obtain this form; Jablonski proceeds to criticize the obtention[96]; following which Foley continues to proceed, this time justifying it[47]. Whether or not anything has been proven by this succession of events depends primarily on the justifiability of the approximations involved. Now Foley essentially voiced the opinion that Jablonski's method of evaluating Eq. (3.102b) is only correct in the static or slow motion case. This is as it may be, but Foley had to assume the Franck–Condon principle, a static affair, in order to obtain Eq. (3.102b) in the first place (evidently). Jablonski's theory breaks down (in the sense that generality can no longer be maintained and the development continued) following Eq. (3.77). The flaw in the Jablonski theory seems to lie, not in the fact that it is a special case of the Foley considerations,* but in the obvious difficulties inherent in obtaining results in any but the simplest cases — binary collisions and Franck–Condon.

3.15. A more sophisticated statistical theory[142]

Our basic theory remains, of course, the same, and we again begin by inquiring what the probability is that a configuration of perturbers will exist such that a change in the energy perturbations of $V = \Sigma\, V_i$ between the two levels under consideration will result. This V shift, as we may recall, results in the emission of radiation which is shifted in frequency by this amount. We suppose that the dependence of V on molecular separation will be of the form cr^{-m}, and now we augment this potential with a spin interaction (separable) $u(\xi)$. Thus there results:

$$V_i = c r_i^{-m}\, u(\xi_i) \tag{3.106a}$$

Firstly, the function $u(\xi_i)$ must vanish in the mean. On this basis Margenau chose for $u(\xi_i)$:

$$u(\xi_i) = \begin{matrix} -1 \text{ for } \xi_i \leq 0 \\ +1 \text{ for } \xi_i > 0 \end{matrix}\;;\; -1 \leq \xi \leq 1 \tag{3.106b}$$

which may lead to different numerical factors than would the accurate spin function but nothing else.

* See *supra*, Chap. 2.

On a random basis the occupation probability for r, where $4/3\,\pi\,R^3$ is the volume occupied by the gas, is:

$$\frac{\mathrm{d}V}{V} = \frac{r^2 \sin\vartheta\, \mathrm{d}\vartheta \mathrm{d}\varphi\, \mathrm{d}r}{\frac{4}{3}\pi\, R^3} \leftrightarrow p(r)\, \mathrm{d}r' = \frac{3}{R^3}\, r^2\, \mathrm{d}r \tag{3.107a}$$

and in like manner for the spin:

$$p(\xi)\, \mathrm{d}\xi = \frac{\mathrm{d}l}{l} = \frac{\mathrm{d}\xi}{2} = \frac{1}{2}\, \mathrm{d}\xi \tag{3.107b}$$

Now we have the requisite elements for the determination of $W_n(V)$, the probability that with n perturbers present the shift at the emitter will amount to V. In analogy to Eq. (3.19) and the reasoning which led to it, we may obtain:

$$W_n(V) = \frac{1}{2\pi} \int_{-\infty}^{+\infty} \mathrm{e}^{-i_s V} A_n(s)\, \mathrm{d}s \tag{3.108a}$$

where:

$$A_n(s) = \{\int p(\xi)\, \mathrm{d}\xi \int p(r)\, \mathrm{d}r \exp\,[is\, cr^{-m}\, u(\xi)]\}^n \tag{3.108b}$$

We note that $1/V^n$ fails to appear in Eq. (3.108a) since its equivalent is contained in $p(r)\, p(\xi)$.

Using no subterfuge whatever, Eq. (3.108b) may be rewritten as:

$$A_n(s) = [1 - 3\, B(s)/2\, R^3]^n \tag{3.109a}$$

where:

$$B(s) = \int_{-1}^{+1} \mathrm{d}\xi \int_0^l \{1 - \exp\,[i\, s\, c\, r^{-m}\, u\,(\xi)]\}\, r^2\, \mathrm{d}r \tag{3.109b}$$

Eq. (3.109a) may surely be expanded according to the binomial theorem. In the result we allow R to approach infinity which in turn means n approaches infinity, since $n = 4/3\,\pi\, R^3 N$. This converts our binomial expansion into the infinite MacLaurin series for:

$$A(s) = \exp\,[-2\pi N B(s)] \tag{3.109c}$$

so that our Eq. (3.108a) becomes:

$$W(V) = \frac{1}{2\pi} \int_{-\infty}^{+\infty} \exp\,[-\, isV - 2\pi N B(s)]\, \mathrm{d}s \tag{3.110}$$

Next, Eq. (3.109b) may be integrated over ξ so that:

$$B(s) = 2 \int_0^{+\infty} (1 - \cos s\, v)\, r^2\, \mathrm{d}r \tag{3.111a}$$

where:

$$v = cr^{-m} \tag{3.111b}$$

If we let,

$$sv \equiv t \tag{3.111c}$$

B may be written as:

$$\left.\begin{aligned} B &= \frac{2}{3}(c\,|s|)^{3/m} \int_0^\infty (1-\cos t)\,\mathrm{d}(t^{-3/m}) \\ &= \frac{2}{3}(c\,|s|)^{3/m}\left[t^{-3/m}(1-\cos t)\Big|_{t^{-3/m}=0}^{t^{-3/m}=\infty} - \int_0^\infty t^{-3/m}\sin t\,\mathrm{d}t\right] \\ &= \frac{2}{3}(c\,|s|)^{3/m}\int_0^\infty t^{-3/m}\sin t\,\mathrm{d}t \end{aligned}\right\} \quad (3.112)$$

where integration by parts has been utilized and l'Hospital's rule applied. This application of l'Hospital's rule results in the restriction of m such that $m > 3/2$. This is actually no restriction on the theory, since none of our interaction laws will require a value of less than 3/2.

For convenience, let:

$$\frac{\pi\, 4}{3} c^{3/m} \int_0^\infty t^{-3/m} \sin t\,\mathrm{d}t = g_m \quad (3.113\,\mathrm{a})$$

so that:

$$2\pi B(s) = g_m |s|^{3/m} \quad (3.113\,\mathrm{b})$$

Our probability $W(V)$ is a real affair, and the utilization of this reality and Eq. (3.113b) in Eq. (3.110) yields:

$$\left.\begin{aligned} W(V) &= \frac{1}{2\pi}\int_{-\infty}^{+\infty} \exp[-N g_m s^{3/m}] \cos(s V)\,\mathrm{d}s \\ &= \frac{1}{\pi}\int_0^\infty \exp[-N g_m s^{3/m}]\cos(s V)\,\mathrm{d}s \end{aligned}\right\} \quad (3.114)$$

since the integrand is an even function.

By evaluating Eq. (3.114) for the case $V = 0$ (no shift)

$$W(0) = \frac{m}{3\pi}\,\Gamma\!\left(\frac{m}{3}\right)\frac{1}{(N g_m)^{m/3}} \quad (3.115)$$

Margenau demonstrates the interesting result that the line center intensity decreases as $N^{-m/3}$.

If we define a function I_m as,

$$I_m(X) = \int_0^\infty \exp\left[-\left(\frac{u}{K}\right)^{3/m}\right] \cos u\,\mathrm{d}u \quad (3.116\,\mathrm{a})$$

then Eq. (3.114) may quite readily be written as:

$$W(V) = \frac{1}{\pi\,|V|}\, I_m\!\left(\frac{V}{V_0}\right) \quad (3.116\,\mathrm{b})$$

where:

$$V_0 = (N g_m)^{m/3} \quad (3.116\,\mathrm{c})$$

All of which leads us to the departure point of special cases, and, with Margenau, we shall consider three, namely, broadening by (1) permanent dipoles, (2) quadrupoles, and (3) forces which do not change sign.

In the dipole case we have $m = 3$ so that from Eqs. (3.116c), (3.113), and (3.116a):

$$V_0 = \frac{2}{3}\pi^2 Nc; \quad I_3(X) = \frac{x}{1+x^2}$$

so that Eq. (3.116b) becomes:

$$W(V) = \frac{V_0}{\pi(V^2 + V_0^2)} \tag{3.117a}$$

or, when we recall that $V = \hbar(\omega_0 - \omega)$:

$$W(\omega) = \frac{V_0/\pi\hbar^2}{(\omega_0-\omega)^2 + V_0^2/\hbar^2} \tag{3.117b}$$

This is of some interest since it is the Michelson–Lorentz or dispersion form of the line shape.

A consideration of Eq. (3.113a) and Eq. (3.116a) should suffice to justify the statement that numerical calculations or series expansions are requisite for the evaluation of the integrals in these equations for other values of m. Margenau has carried out this numerical evaluation for the quadrupole case, $m = 5$, and the resulting curve $\frac{I_5(x)}{x}$ is compared to the one for the dipole in Fig. 3.10.

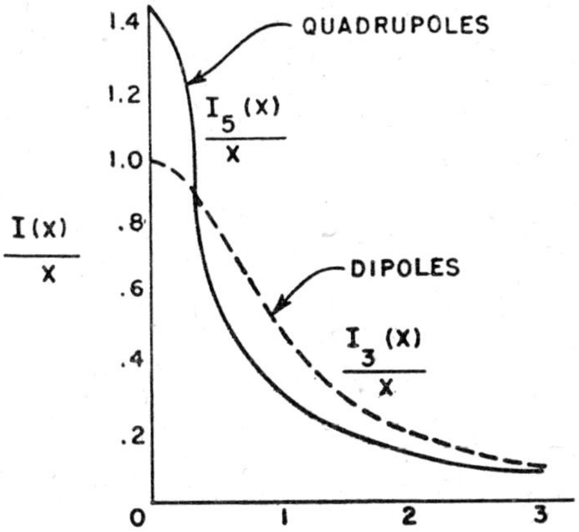

FIG. 3.10. Line shapes according to the Statistical Theory. (After Margenau[119])

Let us note the symmetry displayed by the two line representations of Fig. 3.10 which we have not previously encountered in statistical broadening. The reason for this difference becomes clear when we neglect sign change through spin interaction as we shall now proceed to do. If we simply let $u = 1$, we obtain in place of Eq. (3.111a) the following:

$$B(s) = 2\int_0^\infty (1 - e^{-isv})\, r\, dr \tag{3.118a}$$

from Eq. (3.109b) and

$$B(s) = -i\frac{2}{3}(c\,|s|)^{3/m}\int_0^\infty t^{-3/m} e^{it}\, dt = g \equiv g' + ig'' \tag{3.118b}$$

as in Eq. (3.112).

Then from Eqs. (3.118) and (3.110) we obtain as the form of the intensity (probability) distribution:

$$W(V) = \frac{1}{2\pi}\int_{-\infty}^{+\infty} \exp(-Ng'\, s^{3/m}) \cos(Vs + Ng''\, s^{3/m})\, ds \tag{3.119}$$

Eq. (3.119), of course, will lead us to the type of results we have obtained earlier for van der Waals force, as an example of the same sign forces.

In the case of van der Waals interaction $m = 6$ so that:

$$g' = -g'' = \frac{2}{3}\pi\sqrt{2\pi C}$$

and, *ergo*, Eq. (3.32).

3.16. The reduction of the correlation function

Eq. (2.119) gives the Correlation Function as:

$$C(\tau) = \exp[-N V'(\tau)] \tag{2.119}$$

where:

$$V'(\tau) = 2\pi \int_0^\infty \varrho\,\mathrm{d}\varrho \int_{-\infty}^{+\infty} \mathrm{d}x_0 \left\{1 - \exp\left[i \int_0^\tau \omega\left(\sqrt{(x_0 + \langle v\rangle t)^2 + \varrho^2}\right) \mathrm{d}t\right]\right\} \tag{2.120}$$

If one supposes the interaction between emitter and broadeners to be effectively static so that $\langle v\rangle = 0$, one obtains in place of Eq. (2.120):

$$\left.\begin{aligned} V'(\tau) &= 2\pi \int_0^\infty \varrho\,\mathrm{d}\varrho \int_{-\infty}^{+\infty} \mathrm{d}x_0 \left\{1 - \exp\left[i \int_0^\tau \nu\left(\sqrt{(x_0^2 + \varrho^2)}\right)\mathrm{d}t\right]\right\} \\ &= 4\pi \int_0^\infty R^2\,\mathrm{d}R\,\{1 - \exp i\,\omega(R)\,\tau\} \end{aligned}\right\} \tag{3.120}$$

First we remark the change to a radial coordinate which covers all space and effectively replaces the two former coordinates. Now we compare this with Eq. (3.128a) and we see that the Statistical Theory has indeed been obtained as the static limit of the general Correlation Function treatment.

3.17. Review and summary

The Statistical Theory is based on the Franck–Condon principle which prognosticates vertical transitions between two energy states distorted by the proximity of broadening molecules. Therefore the frequency in a spectral line will depend on the degree of distortion in the two levels at the moment of transition. One describes this mathematically by writing an expression for the probability that the broadeners will be so distributed in space as to distort the natural frequency to a particular value. Then the intensity in the resulting spectral line will be proportional to the probability of distortion to the frequency desired.

The result of application of this theory to van der Waals broadening is a line shape:

$$I(\Delta\nu) = \gamma(\Delta\nu)^{-3/2}\,\mathrm{e}^{-\pi\gamma^2/\Delta\nu}; \quad \gamma = \frac{2}{3}\pi\beta^{1/2} N \tag{3.121a}$$

where in β is the van der Waals constant. The specially defined half-width is given by:

$$\delta = 0.82\,\pi\,\gamma^2 \tag{3.121b}$$

where our "special definition" is necessary for certain asymmetric lines. We define this quantity as the frequency separation of the line intensity maximum and that position on the long wing of the line where the intensity has fallen to half maximum value. The line shift is given by:

$$\Delta = \frac{2}{3}\,\pi\,\gamma^2 \tag{3.121c}$$

We have shown the static nature of the Statistical Theory together with the fact that it is applicable to high densities of perturbers or frequencies well out in the line wing where the effects of close collisions have become important. For these cases the Statistical Theory may be used with complete justification.

In the chapter we have also reviewed the Jablonski Theory of spectral line broadening. Here we suppose the emitter and all the broadeners to make up one large molecule. Then the electronic transition resulting in spectral emission is considered as the molecular electronic behavior while the perturbing effects of the broadeners on the emitter are likened to the vibrational behavior of the molecule. The Franck–Condon principle is applied to this situation to allow separation of the electronic and vibrational effects. The vibrational energies may then be added to the electronic energy to yield any desired frequency in the final spectral line. The problem is then one of evaluation of the vibrational overlap integral which, of course, corresponds to the broadening interaction integral. In a more or less general form the theory is very unwieldy and has not found general application.

CHAPTER 4

STARK BROADENING

SINCE space — not just any space but "outer space" — has become fashionable in the last few years, the laboratory tools associated with related research have become more popular and, of course, more expensive. In these related tools — and here we may include shock tubes and the various species of arcs — plasmas are the rule rather than the exception. Further, in these plasmas the Stark Broadening of spectral lines furnishes a probe for the measurement of various physical quantities of interest to the researcher. Having related the subject matter to the "practical" world of outer space as an interest catcher and in the accepted introductory technique, we may turn our attention to the phenomena involved.

The plasma is, of course, made up of electrons and positive ions, the ratio of their numbers being of no particular interest to us. The results of their presence are quite different for the two species, and, basically, these distinct effects result from the ratio of their masses, a value of two thousand or some multiple thereof. The effects of this mass difference are felt through the relative velocity of the electrons and ions. The ions may be considered as moving slowly and producing nearly static electric fields. In addition their positions may usually be treated classically. On the other hand the electrons move at very high velocity. The electric fields which they produce vary rapidly, and the spatial extent of their associated wave packets often make their quantum decription necessary.

In a general way the Stark Broadening effects of ions and electrons may be considered as four different but not independent phenomena: (1) The slowly varying fields of the ions produce a broadening effect. (2) The rapidly varying fields of the electrons produce a broadening effect. (3) The electrons and ions provide a background which has a shielding effect on the electric field of the ions. (4) The interaction of the ions with each other causes a breakdown of their electric field for high field strengths. In what is to follow we shall first study the individual phenomena and subsequently study their various combinations.

In 1919 and during a few years thereafter Holtsmark developed the Stark Broadening theory which has been applied to the broadening effects of ions ever since. Although its application has been generally so restricted, it was originally meant to cover a broader domain, and, where the broadener was not an ion, the broadening — obviously not Stark in the current sense — was attributed to the fields of the dipole or quadrupole of the perturbing molecule.

4.1. Preliminary considerations

First let us touch briefly on the Stark splitting of a spectral line by an external electric field. It is qualitatively apparent and quantitatively confirmable that, in the absence of a field of force, M the magnetic quantum number has no effect on the level energies but serves only as one of the level degeneracy parameters. There is, in fact, no reason that spatial orientation should affect level energy when no unique direction in space, which may be used to differentiate between various orientations, exists.

Let us now impose an electric field **E** on the gas, thus establishing the requisite unique axis in space. Suppose, for example, the gas to consist of linear dipole molecules. Classically then, the dipoles will be under the influence of a force which tends to align them with the electric field, the magnitude of which force will be dependent on the angle between **E** and the dipole axis. Quantum mechanically, we may consider this angle as a discontinuous function of M. Hence in the case of a linear dipole rotator, the unique J levels will be split into several levels for which the energy displacement from that of the original level will be dependent on the value of M. The spectral lines arising from transitions among levels split in this manner will, of course, also be split. Stark[237] originally suggested applying this principle to the broadening of a spectral line, and later Holtsmark[79,80] carried out the first analytical attack on the problem.*

We begin, with Holtsmark, by considering an emitting atom which is assigned the label 0. This emitter is surrounded by a large number N_1 of disturbing molecules (or ions) with assigned labels 1, . . ., N_1 in general n. Each of these surrounding molecules will contribute to an electric field at the location of 0, and this field will thus be the sum of these N_1 components. Were these perturbing atoms at rest, there would be a constant electric field at 0, and a Stark splitting of the emitted lines would result. The molecular motions change this situation, however. By virtue of these motions of the broadening molecules various molecular system configurations will exist from time to time, and these with varying probabilities. To each of these various configurations will correspond a field strength at the emitter which will occur with the probability of the molecular configuration. Quite obviously then, various Stark splittings may occur with these varying probabilities and, in a manner which will be more apparent after a study of the mathematics of the situation, a line broadening, instead of a simple line splitting, should result.

* In order to alleviate the anguish which this remark may cause some of the more diligent perusers of the literature, a word of clarification would appear to be in order. In several portions of the literature one finds References[79] and/or[80] attributed to Debye. This is simply not correct, but the probable reason for the error may be deduced. Debye wrote a two-page, semiquantitative article[29] on the subject of Stark broadening which immediately preceded Holtsmark's article in the *Physicalische Zeitschrift* for 1919, and it would appear that this juxtaposition has considerably confused the reference situation ever since.

4.2. The probability of an electric field strength at the emitter

Holtsmark[79] began with the assumption that the probability of a particular field strength E is a function of E and set out to determine this probability.

Let the components of the field strength at 0 due to the n-th perturber be X_n, Y_n, Z_n. The components of E will then be given by:

$$X = \sum_1^{N_1} X_n; \quad Y = \sum_1^{N_1} Y_n; \quad Z = \sum_1^{N_1} Z_n \tag{4.1}$$

We desire the probability that

$$\begin{aligned} &X \text{ lies between } X \text{ and } X + \mathrm{d}X \\ &Y \text{ lies between } Y \text{ and } Y + \mathrm{d}Y \\ &Z \text{ lies between } Z \text{ and } Z + \mathrm{d}Z \end{aligned} \tag{4.2}$$

Let the position of the n-th broadener be specified by the m coordinates $x_{1n}, x_{2n}, \ldots, x_{mn}$. Then the X_n, Y_n, Z_n are functions of these m coordinates. It follows that the XYZ of Eq. (4.1) are functions of the $N_1 m$ coordinates belonging to the N_1 surrounding atoms. An $N_1 m$ dimensional space is next set up, and to each point in this space there now corresponds a stipulated field strength. These points will be distributed in space according to some probability law or, what amounts to the same thing, some density function.

The probability that $x_{1n}, x_{2n}, \ldots, x_{mn}$ lie in the range $\mathrm{d}x_{1n}, \mathrm{d}x_{2n} \ldots \mathrm{d}x_{mn}$ may be expressed as:

$$\sigma_n \, \mathrm{d}x_{1n} \, \mathrm{d}x_{2n} \ldots \mathrm{d}x_{mn} \tag{4.3}$$

where σ_n is a function of the $x_{1n}, x_{2n}, \ldots, x_{mn}$. It follows that the probability that all $N_1 m$ coordinates lie in the range $\mathrm{d}x_{11} \, \mathrm{d}x_{12} \ldots \mathrm{d}x_{m N_1}$ is:

$$\sigma_1 \ldots \sigma_{N_1} \mathrm{d}x_{11} \, \mathrm{d}x_{12} \ldots \mathrm{d}x_{1m} \, \mathrm{d}x_{21} \ldots \mathrm{d}x_{N_1 m} \tag{4.4}$$

It should be noted that this coordinate distribution establishes our electric field strength within a certain small range.

In certain portions of our $N_1 m$ dimensional space the range requirements for the field components as given by Eq. (4.2) will be satisfied. On the other hand, it is apparent that these requirements will not be satisfied in certain portions of space. It follows from these considerations that, with Holtsmark, we must now obtain the probability of meeting the requirements of Eq. (4.2).

If Eq. (4.4) is integrated over those portions of space satisfying Eq. (4.2) and this result is divided by the integral of Eq. (4.4) over all space, the field strength probability function is obtained:

$$W(X, Y, Z) = \frac{1}{V_{N_1}} \int_{XYZ} \ldots \int \sigma_1 \sigma_2 \ldots \sigma_{N_1} \mathrm{d}x_{11} \ldots \mathrm{d}x_{m N_1} \tag{4.5a}$$

where:

$$V_{N_1} = \int \ldots \int \sigma_1 \, \sigma_2 \ldots \sigma_{N_1} \mathrm{d}x_{11} \ldots \mathrm{d}x_{m N_1} \tag{4.5b}$$

$W(XYZ)$ may be defined by interpreting $W(XYZ) \, \mathrm{d}X \mathrm{d}Y \mathrm{d}Z$ as the probability that XYZ satisfy Eq. (4.2).

We might consider Eq. (4.5a) more specifically since it is a rather important one in this development. The integrand provides us with the probability that a volume element which satisfies Eq. (4.2) will be occupied by a system point — a point which specifies a certain configuration of the system of perturbers. V_{N_1} serves to normalize this probability, the need for which normalization is apparent. The evaluation of Eq. (4.5a) now becomes of major importance. Holtsmark used what might be considered a modified form of the method developed by Markoff[234] for this evaluation. The first step is the multiplication of Eq. (4.5a) by the proper Dirichlet factors in order to transform this integral into an integral over all space. The Dirichlet factors for this transformation are:

$$H(X) = \frac{1}{\pi} \int_{-\infty}^{+\infty} \frac{\sin \alpha \xi}{\xi} e^{i \Upsilon \xi} d\xi \tag{4.6a}$$

$$H(Y) = \frac{1}{\pi} \int_{-\infty}^{+\infty} \frac{\sin \beta \eta}{\eta} e^{i \varepsilon \eta} d\eta \tag{4.6b}$$

$$H(Z) = \frac{1}{\pi} \int_{-\infty}^{+\infty} \frac{\sin \gamma \zeta}{\zeta} e^{i \chi \zeta} d\zeta \tag{4.6c}$$

where:

$$\alpha = \frac{1}{2} dX; \quad \Upsilon = \sum_{1}^{N_1} X_n(x_{1n} \ldots x_{mn}) - X \tag{4.7a}$$

$$\beta = \frac{1}{2} dY; \quad \varepsilon = \sum_{1}^{N_1} Y_n(x_{1n} \ldots x_{mn}) - Y \tag{4.7b}$$

$$\gamma = \frac{1}{2} dZ; \quad \chi = \sum_{1}^{N_1} Z_n(x_{1n} \ldots x_{mn}) - Z \tag{4.7c}$$

Certain relations among the factors in Eqs. (4.6) may be written as follows:

$$-\alpha < \Upsilon < \alpha \tag{4.8a}$$

$$-\beta < \varepsilon < \beta \tag{4.8b}$$

$$-\gamma < \chi < \gamma \tag{4.8c}$$

If Eq. (4.8a) is valid, Eq. (4.6a) is equal to one. If Eq. (4.8a) is not valid, Eq. (4.6a) is zero. A consideration of Eq. (4.7a) will show that this is simply another way of stating Eq. (4.2). In like manner the validity of Eqs. (4.8b) and (4.8c) control the values of Eqs. (4.6b) and (4.6c) according to Eqs. (4.7b) and (4.7c).

By Eqs. (4.7) α, β, and γ are infinitely small so that:

$$\frac{\sin \alpha \xi}{\xi} = \alpha; \quad \frac{\sin \beta \eta}{\eta} = \beta; \quad \frac{\sin \gamma \zeta}{\zeta} = \gamma \tag{4.9}$$

If Eqs. (4.7) and (4.9) are substituted into Eqs. (4.6) the result is:

$$H(X) = \frac{1}{2\pi} \mathrm{d}X \int_{-\infty}^{+\infty} \mathrm{e}^{i\xi\left(\sum_1^{N_1} X_n - X\right)} \mathrm{d}\xi \tag{4.10a}$$

$$H(Y) = \frac{1}{2\pi} \mathrm{d}Y \int_{-\infty}^{+\infty} \mathrm{e}^{i\eta\left(\sum_1^{N_1} Y_n - Y\right)} \mathrm{d}\eta \tag{4.10b}$$

$$H(Z) = \frac{1}{2\pi} \mathrm{d}Z \int_{-\infty}^{+\infty} \mathrm{e}^{i\zeta\left(\sum_1^{N_1} Z_n - Z\right)} \mathrm{d}\zeta \tag{4.10c}$$

Now multiply Eq. (4.5a) by Eqs. (4.10):

$$W(X, Y, Z)\, \mathrm{d}X \mathrm{d}Y \mathrm{d}Z = \frac{1}{8\pi^3} \mathrm{d}X\, \mathrm{d}Y\, \mathrm{d}Z \iiint_{-\infty}^{+\infty} \mathrm{d}\xi\, \mathrm{d}\eta\, \mathrm{d}\zeta$$

$$\mathrm{e}^{i(\xi X + \eta Y + \zeta Z)} \frac{1}{V_{N_1}} \int_{\text{all space}} \cdots \int \sigma_1 \ldots \sigma_{N_1} \mathrm{e}^{i(\xi \Sigma X_n + \eta \Sigma Y_n + \zeta \Sigma Z_n)} \mathrm{d}x_{11} \ldots \mathrm{d}x_{mN_1} \tag{4.11}$$

The X_n, Y_n, Z_n are functions of the $x_{1n} \ldots x_{mn}$ while the X, Y, Z are constants in all the integrations. It should be stipulated now that all the broadening molecules are considered the same, that is, the Stark broadening is not treated for the case where, for molecules, more than one compound (element) acts as a broadening agent. In this event, identical electric field strength components X_n, Y_n, Z_n will be the same functions of $x_{1n}, x_{2n}, \ldots, x_{mn}$. Let us briefly consider the reason for this. Suppose the broadening molecules may be considered as quadrupoles insofar as the field which they produce is concerned. Then, since the field produced will be a function of the quadrupole moment and the coordinates of the molecule, and since the quadrupole moments for like molecules will be the same, the field strength components will differ only for differing $x_{1n}, x_{2n}, \ldots, x_{mn}$. It also follows from these considerations that for identical X_n, Y_n, Z_n, $\sigma_1, \sigma_2, \ldots, \sigma_{N_1}$ will all be equal.

These equalities among the coordinates and their probability functions lead to a very fortunate simplification of Eq. (4.11). The inner integral can be broken up into the product of N_1 identical integrals, each of which is integrated only over the m coordinates of a single molecule. The result is:

$$J^{N_1} = \frac{1}{V^{N_1}} \left[\int_{\text{all space}} \cdots \int \mathrm{e}^{i(\xi X + \eta Y + \zeta Z)} \sigma\, \mathrm{d}x_1 \mathrm{d}x_2 \ldots \mathrm{d}x_m\right]^{N_1} \tag{4.12}$$

In Eq. (4.12) the principle enumerated in the preceding paragraph has been applied to V_{N_1}. Thus Eq. (4.11) becomes:

$$W(X, Y, Z) = \frac{1}{8\pi^3} \int_{-\infty}^{+\infty} \cdots \int \mathrm{d}\xi\, \mathrm{d}\eta\, \mathrm{d}\zeta\, \mathrm{e}^{-i(\xi X + \eta Y + \zeta Z)} \frac{1}{V^{N_1}}$$

$$\times \left[\int_{\text{all space}} \cdots \int \mathrm{e}^{i(X\xi + Y\eta + Z\zeta)} \sigma\, \mathrm{d}x_1 \ldots \mathrm{d}x_m\right]^{N_1} \tag{4.13}$$

It is at this stage of the development that Holtsmark departed from the Markoff method and proceeded with an integral evaluation scheme which he attributed to Debye.

4.3. The introduction of specific field producers

Let us form the dot product $\mathbf{s} \cdot \mathbf{E}$ where:

$$\mathbf{E} = \mathbf{i}X + \mathbf{j}\,Y + \mathbf{k}Z \tag{4.14a}$$

$$\mathbf{s} = \mathbf{i}\xi + \mathbf{j}\,\eta + \mathbf{k}\,\zeta \tag{4.14b}$$

It might be mentioned that ξ, η, ζ will be constants for the integration over $x_1, \ldots, x_m$.

Eq. (4.12) may now be transformed into spherical polar coordinates and integrated after an expression for $\mathbf{E}$ has been obtained.

A molecule we here consider as something or other which is made up of a number of charges, e_s. Let us establish a moving reference frame at the center of the molecule and determine the potential which would exist at the point x, y, z due to these charges. If $\mathbf{r}$ is the position vector of the point x, y, z; $\mathbf{r}_s$ is the position vector of the s-th charge, and $\mathbf{R}_s$ the sum of these two vectors, then the fact that

$$\begin{aligned} R_s^2 &= (x - x_s)^2 + (y - y_s)^2 + (z - z_s)^2 \\ &= r^2\left[1 + \left(\frac{r^2}{r}\right)^2 - \frac{2\,(r - r_s)}{r^2} + \cdots\right] \end{aligned} \tag{4.15}$$

yields the potential:

$$\begin{aligned} \varphi &= \sum \frac{e_s}{R_s} = \frac{\Sigma e_s}{r} - \frac{(\mathbf{r} \cdot \Sigma e_s \mathbf{r}_s)}{r^3} - \frac{1}{2\,r^3}\left\{\Sigma\, e_s\, r_s^2 - 3\,\Sigma\left[e_s \frac{(\mathbf{r} \cdot \mathbf{r}_s)^2}{r^2}\right]\right\} + \cdots \\ &= \frac{\Sigma e_s}{r} - \frac{1}{r^2}\left[\frac{\mathbf{r}}{r} \cdot \Sigma\, e_s\, r_s\right] - \frac{1}{2\,r^3}\left\{\Sigma\, e_s\, x_s^2\left(1 - \frac{3\,x^2}{r^2}\right) + \Sigma\, e_s\, y_s^2\left(1 - \frac{3\,y^2}{r^2}\right)\right. \\ &\quad + \Sigma\, e_s z_s^2\left(1 - \frac{3\,z^2}{r^2}\right) - \frac{6\Sigma\, e_s x_s y_s x\, y}{r^2} - \frac{6\,\Sigma\, e_s y_s z_s y\, z}{r^2} \\ &\quad \left. - \frac{6\,\Sigma\, c_s z_s x_s z\, x}{r^2}\right\} - \cdots \end{aligned} \tag{4.16}$$

in which the final form of the potential is the most informative. The rapid convergence of the final series in Eq. (4.16) is generally considered as assured by the assumption $r \gg r_s$. When one considers the small portion of space encompassed by the atom, this assumption appears reasonably valid. It is true, however, that the question as to what transpires when the broadener closely approaches the emitter is a rather embarrassing one.

Holtsmark deals with the three cases (a) none of the terms disappears (b) the first term disappears (c) the first and second terms disappear. The rapid convergence of the series tells us that we may neglect all terms but the first in case (a) and all but the second in case (b). Case (c) would appear to be self-explanatory with respect to ignoration. Thus, these three cases yield the following field producers:

1. Ions
2. Dipoles
3. Quadrupoles

An ion may be defined as a charged molecule. A dipole may be defined as a system of charges which gives rise to the second term in Eq. (4.16) but not to the first. A quadrupole is defined by the third term in this equation.

For these three cases we must now specify $\mathbf{s} \cdot \mathbf{E}$ where $\mathbf{E}$ is the electric field strength and $\mathbf{s}$ is the position vector.

For the ion we have:

$$\varphi = \frac{\Sigma e_s}{r} = \frac{\varepsilon}{r} \leftrightarrow \mathbf{E} = \nabla \varphi = -\frac{\varepsilon}{r^3}\mathbf{r}$$

which gives:

$$\mathbf{s} \cdot \mathbf{E} = -\frac{\varepsilon}{r^3}(\mathbf{r} \cdot \mathbf{s}) = -\frac{\varepsilon}{r^2} s \cos \vartheta \tag{4.17}$$

In Eq. (4.17) $s = |\mathbf{s}|$, and ϑ is the angle between $\mathbf{r}$ and $\mathbf{s}$. The dipole potential is

$$\varphi = \frac{\mathbf{r} \cdot \Sigma e_s \mathbf{r}_s}{r^3} = \frac{\mathbf{r} \cdot \boldsymbol{\mu}}{r^3} \tag{4.18}$$

where $\boldsymbol{\mu}$ is the dipole moment, a definition of convenience. Hence:

$$\mathbf{E} = \nabla \varphi = \nabla \frac{\mathbf{r} \cdot \boldsymbol{\mu}}{r^3} = \frac{1}{r^3} \nabla (\mathbf{r} \cdot \boldsymbol{\mu}) \frac{3}{r^5} \mathbf{r} (\mathbf{r} \cdot \boldsymbol{\mu}) = \frac{\boldsymbol{\mu}}{r^3} - \frac{3\mathbf{r}(\mathbf{r} \cdot \boldsymbol{\mu})}{r^5} \tag{4.19}$$

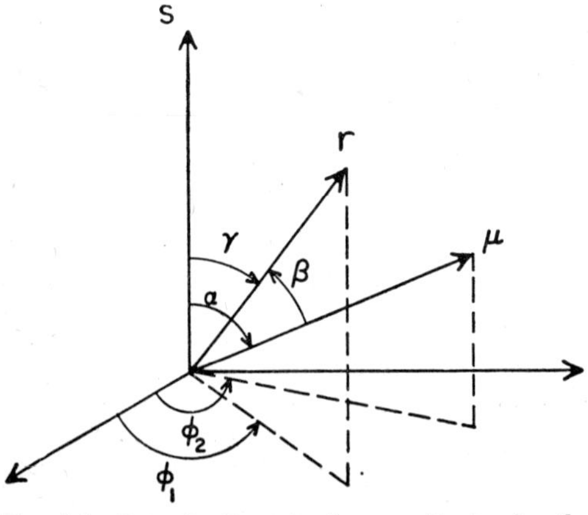

FIG. 4.1. Introduction of polar coordinates for the dipole broadener case. S is the polar axis

So that:

$$\mathbf{s} \cdot \mathbf{E} = \frac{1}{r^2}(\mathbf{s} \cdot \boldsymbol{\mu}) - \frac{3}{r^5}(\mathbf{r} \cdot \boldsymbol{\mu})(\mathbf{s} \cdot \mathbf{r}) \tag{4.20}$$

In Eq. (4.20) we may write:

$$\mathbf{s} \cdot \boldsymbol{\mu} = s\mu \cos \alpha;$$
$$\mathbf{r} \cdot \boldsymbol{\mu} = r\mu \cos \beta;$$
$$\mathbf{s} \cdot \mathbf{r} = sr \cos \gamma$$

Now consider Fig. 4.1.

Spherical coordinates may be introduced as follows. $\mathbf{s}$ is taken as the polar axis. α and γ are then the polar angles ϑ_1 and ϑ_2 respectively, and the azimuth angles φ_1 and φ_2 are as indicated. β, the angle between $\boldsymbol{\mu}$ and $\mathbf{r}$, is given by:

$$\cos \beta = \cos \vartheta_1 \cos \vartheta_2 + \sin \vartheta_1 \sin \vartheta_2 \cos (\varphi_1 - \varphi_2) \tag{4.21}$$

Substitution of Eq. (4.21) into Eq. (4.20) yields:

$$\mathbf{s}\cdot\mathbf{E} = \frac{1}{r^3} s\mu \left[\cos\vartheta_1 - 3\cos\vartheta_2\left(\cos\vartheta_1\cos\vartheta_2 + \sin\vartheta_1\sin\vartheta_2\cos(\varphi_1-\varphi_2)\right)\right] \tag{4.22}$$

Finally, the quadrupole potential must be considered. This potential — the term in braces in Eq. (4.16) — may be simplified by a rotation of the coordinate system which has been attached to the molecule. We may diagonalize the matrix of the quadrupole moment by a rotation of the coordinate axes to which it is referred, thus eliminating terms of the type $\Sigma\, e_s x_s y_s$. We can assume this operation to have been carried out so that only the diagonal constants of the quadrupole remain

$$\Theta_1 = \Sigma\, e_s\, x_s^2;\quad \Theta_2 = \Sigma\, e_s\, y_s^2;\quad \Theta_3 = \Sigma\, e_s\, z_s^2 \tag{4.23}$$

If an atom is symmetric about some axis we obtain $\Theta_1 = \Theta_2$. Hence:

$$\varphi = -\frac{1}{2r^3}\left[\Theta_2\left(1-\frac{3x^2}{r^2}\right) + \Theta_2\left(1-\frac{3y^2}{r^2}\right) + \Theta_3\left(1-\frac{3z^2}{r^2}\right)\right]$$

$$= \frac{1}{2r^5}(\Theta_3-\Theta_2)(-x^2-y^2+2z^2) = \frac{A\beta^2}{2r^5} \tag{4.24}$$

where:

$$A = \Theta_3 - \Theta_2;\quad \beta^2 = -x^2 - y^2 + 2z^2$$

From Eq. (4.24)

$$E_x = \frac{\partial\varphi}{\partial x} = -\frac{A\,x}{2r^5}\left(2-5\frac{\beta^2}{r^2}\right)$$

$$E_y = \frac{\partial\varphi}{\partial y} = -\frac{A\,y}{2r^5}\left(2-5\frac{\beta^2}{r^2}\right);\quad E_z = \frac{\partial\varphi}{\partial z} = \frac{A\,z}{r^5}\left(2-5\frac{\beta^2}{r^2}\right) \tag{4.25}$$

Again spherical polar coordinates are adopted as shown in Fig. 4.2.

c is the molecular figure axis. It follows that:

$$\begin{aligned} s_x &= s\sin\vartheta_1\cos\varphi_1 \\ s_y &= s\sin\vartheta_1\sin\varphi_1 \\ s_z &= s\cos\vartheta_1 \\ x &= r\sin\vartheta_2\cos\varphi_2 \\ y &= r\sin\vartheta_2\sin\varphi_2 \\ z &= r\cos\vartheta_2 \end{aligned} \tag{4.26}$$

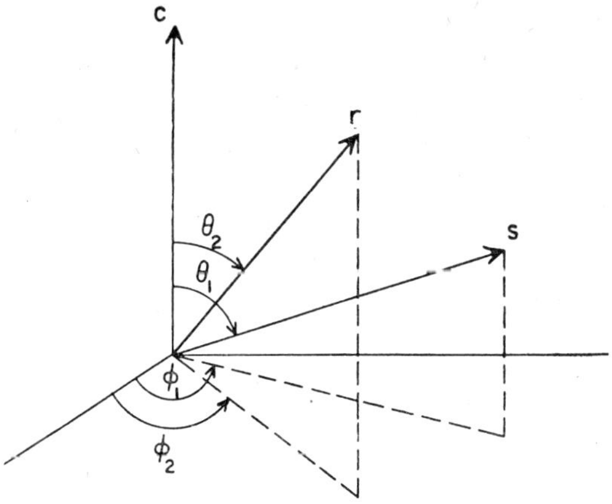

FIG. 4.2. Polar coordinates for the quadrupole broadener

Utilizing Eq. (4.26), we obtain:

$$\mathbf{s}\cdot\mathbf{E} = s_x E_x + s_y E_y + s_z E_z = \frac{A\,s}{2r^4}\{2\cos\vartheta_1\cos\vartheta_2 - \sin\vartheta_1\sin\vartheta_2\cos(\varphi_1-\varphi_2)\}$$
$$\times\,(7 - 15\cos^2\vartheta_2) \tag{4.27}$$

A consideration of Eqs. (4.17), (4.22), and (4.27) indicates that these three cases may all be represented by:

$$\mathbf{s}\cdot\mathbf{E} = \frac{s}{r^p}\,w_p \tag{4.28}$$

where w_p is a function of the angles and p may take on the values 2, 3, or 4.

Eq. (4.28) may now be substituted into Eq. (4.12) to obtain:

$$J = \frac{1}{V}\underset{\text{all space}}{\int\cdots\int} \mathrm{e}^{i\frac{sw_p}{r^p}}\,\sigma\,\mathrm{d}x_1\,\mathrm{d}x_2\ldots\mathrm{d}x_m \tag{4.29}$$

The transformation of the coordinates $x_1, \ldots, x_m$ to the coordinates $r_1, \vartheta_1, \vartheta_2, \varphi_1, \varphi_2$ is in order. After the transformation all positions are referred to the center of the molecule and, further, all directions of a vector from this center are equally probable. Thus, since the molecular distribution is a random one:

$$\sigma\,\mathrm{d}x_1\,\mathrm{d}x_2\ldots\mathrm{d}x_m = r^2\,\mathrm{d}r\sin\vartheta_1\,\mathrm{d}\vartheta_1\sin\vartheta_2\,\mathrm{d}\vartheta_2\,\mathrm{d}\varphi_1\,\mathrm{d}\varphi_2 \tag{4.30}$$

The limits of integration are taken as ϑ from 0 to π, φ from 0 to 2π, and r from 0 to R, thus enclosing the gas in a sphere of radius R.

It follows that:

$$V = \int\cdots\int r^2\,\mathrm{d}r\sin\vartheta_1\,\mathrm{d}\vartheta_1\sin\vartheta_2\,\mathrm{d}\vartheta_2\,\mathrm{d}\varphi_1\,\mathrm{d}\varphi_2 = \frac{16\pi^2}{3}R^3 \tag{4.31}$$

Now the transformation to solid angles is made by utilizing the relations:

$$\mathrm{d}\Omega = \sin\vartheta_1\,\mathrm{d}\vartheta_1\,\mathrm{d}\varphi_1;\quad \mathrm{d}\Omega' = \sin\vartheta_2\,\mathrm{d}\vartheta_2\,\mathrm{d}\varphi_2 \tag{4.32}$$

so that Eq. (4.29) becomes:

$$J = \frac{1}{V}\int_\Omega\int_{\Omega'}\mathrm{d}\Omega\,\mathrm{d}\Omega'\int_0^R r^2\,\mathrm{e}^{i\frac{sw_p}{r^p}}\,\mathrm{d}r \tag{4.33}$$

If we let $u = \dfrac{sw_p}{r^p}$, then $\left(\dfrac{sw_p}{p}\right)^{3/p} du = r^2\mathrm{d}r$ and Eq. (4.33) becomes:

$$J = \frac{3R^3}{16\pi^2}\int_\Omega\int_{\Omega'}\mathrm{d}\Omega\,\mathrm{d}\Omega'\left(\frac{sw_p}{p}\right)^{3/p}\int_\alpha^\infty \mathrm{e}^{iu}\,u^{-\frac{3+p}{p}}\,du \tag{4.34}$$

where $\alpha = \dfrac{sw_p}{R^p}$.

Eq. (4.34) cannot be carried further for the general case, but rather, it is necessary to consider it separately for the ion, the dipole, and the quadrupole. These three cases will now be considered in this order.

4.4. The special case of the ion

In the case of the ion $p = 2$, and the integral in u may be successively integrated by parts to obtain:

$$\int_\alpha^\infty \mathrm{e}^{iu}\,u^{-5/2}\,du = \frac{2}{3}\alpha^{-3/2}\,\mathrm{e}^{iu} + \frac{4}{3}i\,\alpha^{-1/2} + \frac{4}{3}i^2\int_\alpha^\infty \mathrm{e}^{iu}\,u^{-1/2}\,du \tag{4.35}$$

The integral on the right of Eq. (4.35) may be evaluated as:

$$\int_{\alpha}^{\infty} e^{iu}\, u^{-1/2}\, du = \int_{0}^{\infty} e^{iu}\, u^{-1/2}\, du - \int_{0}^{\alpha} e^{iu}\, u^{-1/2}\, du$$

$$= e^{i\frac{\pi}{4}}\, \Gamma\left(\frac{1}{2}\right) - \int_{0}^{\alpha} \left(1 + iu + \frac{1}{2}\, i^2\, u^2 + \cdots\right) u^{-1/2}\, du$$

$$= \sqrt{\pi}\, e^{i\frac{\pi}{4}} - 2\alpha^{1/2} - \frac{2}{3}\alpha^{3/2} - \frac{2}{5}\cdot\frac{1}{2}\, i^2\, \alpha^{5/2} \ldots \quad (4.36)*$$

This series for α appears reasonably well justified in consideration of the small value of α. A straightforward substitution of Eqs. (4.35) and (4.36) into Eq. (4.34) yields:

$$J = \frac{3}{16\pi^3 R^3} \int d\Omega\, d\Omega' \frac{\alpha^{3/2}}{2} R^3 \left[\frac{2}{3}\alpha^{-3/2}\, e^{ia} + \frac{4}{3}\, i\, \alpha^{1/2}\, e^{i\alpha}\right.$$

$$\left. + \frac{4}{3}\, i^2 \left(\sqrt{\pi}\, e^{i\frac{\pi}{4}} - 2\,\alpha^{1/2} - \frac{2}{3}\,\alpha^{3/2} - \cdots\right)\right] \quad (4.37)$$

All terms in Eq. (4.37) with the exception of the first term in the bracket and the first term in the parenthesis are now dropped as small so that, after developing $e^{i\alpha}$ as a power series, one obtains:

$$J = \frac{1}{16\,\pi^2} \int\int d\Omega\, d\Omega' \left(1 + i\,\alpha + 2\, i^2 \sqrt{\pi}\, e^{i\frac{\pi}{4}}\, \alpha^{3/2} + \cdots\right) \quad (4.38)$$

A rather appalling number of terms have been thrown out in the last few operations, but the facts that (a) one is forced to pursue this course in order to solve the problem, and (b) the development has been predicated on large R, do tend to justify the ignoration.

w_2 is a cosine function. Thus, α, which is a function of w_2 integrated over Ω and Ω', is zero. Eq. (4.38) then becomes on partial integration:

$$J = 1 + 2\, i^2 \sqrt{\pi}\, e^{i\frac{\pi}{4}} \frac{1}{16\pi^2} \int\int d\Omega\, d\Omega'\, \alpha^{3/2} \quad (4.39)$$

If we set $\overline{w}_2 = \frac{1}{16\pi^2} \int\int d\Omega\, d\Omega' w_2^{3/2}$, Eq. (4.39) becomes:

$$J = 1 - 2\sqrt{\pi}\, e^{i\frac{\pi}{4}}\, \overline{w}_2\, s^{3/2} \frac{4\,\pi}{3} \frac{N}{N_1} \quad (4.40)$$

N_1 is obviously a large number. Then since $\left(1 - \frac{X}{N_1}\right)^{N_1} = e^{-X}$, we may write:

$$J^{N_1} = \exp\left(-\sqrt{\pi}\, e^{i\frac{\pi}{4}}\, s^{3/2}\, \frac{2}{3}\, 4\,\pi\, N\, \overline{w}_2\right) \quad (4.41)*$$

* By definition $\Gamma(z + 1) = \int_{0}^{\infty} u^z\, e^{-u}\, du = z!$

Also,

$$\overline{w}_2 = \frac{1}{16\pi^2}\iiint \varepsilon^{3/2}\cos^{3/2}\vartheta_1 \sin\vartheta_1\, d\vartheta_1\, d\varphi_1\, d\Omega = \varepsilon^{3/2}\frac{1}{2}\int_0^{\pi}\cos^{3/2}\vartheta_1\sin\vartheta_1\, d\vartheta_1$$

$$+\,\varepsilon^{3/2}\frac{1}{2}\int_{\pi/2}^{\pi}\cos^{3/2}\vartheta_1\sin\vartheta_1\, d\vartheta_1 = \varepsilon^{3/2}\frac{1}{5}(1-i) \tag{4.42}$$

So that, when it is recalled that $e^{i\frac{\pi}{4}} = \frac{i+1}{\sqrt{2}}$, Eq. (4.31) becomes:

$$J^{N_1} = \exp\left(-\frac{8\sqrt{2\pi}}{15}\pi N s^{3/2}\varepsilon^{3/2}\right) = e^{-4.21 N s^{3/2}\varepsilon^{3/2}} \tag{4.43}$$

4.5. The special case of the dipole

We turn now to the calculation of J^{N_1} for the dipole. The integral in u in Eq. (4.34) is:

$$\int_\alpha^\infty e^{iu}u^{-2}\,du = \alpha e^{i\alpha} + i\int_\alpha^\infty e^{iu}u^{-1}\,du$$

$$= \alpha e^{i\alpha} + i\left[\int_0^\infty e^{iu}u^{-1}\,du - \int_0^\alpha e^{iu}u^{-1}\,du\right] \tag{4.44}$$

where here $\alpha = \frac{s\,w_3\mu}{R^3}$

We replace -1 in Eq. (4.44) by $-1+\Lambda$ where Λ is very small and may be set equal to zero in the limit. This gives:

$$f(\alpha,\Lambda) = \int_0^\infty e^{iu}u^{-1+\Lambda}du - \int_0^\alpha e^{iu}u^{-1+\Lambda}du$$

$$= \frac{\Gamma(\Lambda)}{i^\Lambda} - \int_0^\alpha u^{-1+\Lambda}\left[1 + iu + \frac{1}{2}i^2u^2 + \cdots\right]du \tag{4.45}$$

$$= \frac{\Gamma(\Lambda)}{i^\Lambda} - \left[\frac{\alpha^\Lambda}{\Lambda} + i\frac{\alpha^{1+\Lambda}}{1+\Lambda} + \frac{i^2}{2}\frac{\alpha^{2+\Lambda}}{2+\Lambda} + \cdots\right]$$

By MacLaurin expansion:

$$\frac{\alpha^\Lambda}{\Lambda} = \frac{1+\Lambda\log\alpha+\cdots}{\Lambda} = \frac{1}{\Lambda} + \log\alpha + \cdots \tag{4.46}$$

From the relations:

$$\Gamma(\Lambda)\,\Gamma(1-\Lambda) = \frac{\pi}{\sin\Lambda\pi}$$

$$\Gamma(1-\Lambda) = \Gamma(1) - \Lambda\Gamma'(1)$$

$$i^{-\Lambda} = e^{-i\frac{\pi}{2}\Lambda} = \left(1 - i\frac{\pi}{2}\Lambda\right)$$

one obtains:

$$\frac{\Gamma(\Lambda)}{i^{\Lambda}} = \frac{\pi}{\sin \pi \Lambda} \frac{\left(1 - i\frac{\pi}{2}\Lambda\right)}{[\Gamma(1) - \Lambda\Gamma'(1)]} + \frac{1}{\Lambda}\left[\Gamma'(1) - \frac{\pi}{2}i\right] \tag{4.47}$$

Substitute Eqs. (4.46) and (4.47) into Eq. (4.45) and pass to the limit $\Lambda = 0$ to obtain:

$$f(\alpha, \Lambda) = \Gamma'(1) - i\frac{\pi}{2} - \log \alpha - i\,\alpha - \frac{i^2}{2}\frac{\alpha^2}{2} - \cdots \tag{4.48}$$

Eq. (4.48) in conjunction with Eq. (4.45) yields:

$$J = \frac{1}{V}\iint \mathrm{d}\Omega\,\mathrm{d}\Omega'\,\frac{s\,w_3\,\mu}{3}\left[\alpha^{-1}\,e^{i\,\alpha} + i\left(\Gamma'(1) - i\frac{\pi}{2} - \log \alpha - i\,\alpha - \frac{i^2}{2}\frac{\alpha^2}{2}\cdots\right)\right] \tag{4.49}$$

Develop $e^{i\alpha}$ in a power series; insert the value for V and for $\alpha R^3 = \mu s w_3$ and utilize the fact that $\iint \alpha\,\mathrm{d}\Omega\,\mathrm{d}\Omega' = 0$ to rewrite Eq. (4.49) as follows:

$$J = 1 - i\frac{3\,R^3}{16\pi^2\,R^3\,3}\iint \mathrm{d}\Omega\,\mathrm{d}\Omega'\left[\alpha\log\alpha + \left(i + \frac{1}{2}\right)\alpha^2 + \cdots\right] \tag{4.50}$$

The expression for α is substituted into Eq. (4.50) and all terms of higher order than $\alpha \log \alpha$ are dropped as small.

$$J = 1 - \frac{i}{16\pi^2}\iint \frac{\mu\,s\,w_3}{R^3}\,3\left(\log w_3 + \log\frac{\mu\,s}{R^3}\right)\mathrm{d}\Omega\,\mathrm{d}\Omega' \tag{4.51}$$

In Eq. (4.51) $\log\frac{\mu s}{R^3}$ will be a constant for the integration so that this term, due to the periodicity of w_3, will be zero after integration as has occurred in the previous development. Again the number of atoms per unit of volume will be N so that:

$$N_1 = \frac{4}{3}\pi\,R_3\,N \tag{4.52}$$

Also let:

$$\bar{w}_3 = \frac{1}{16\pi^2}\iint w_3 \log w\,\mathrm{d}\Omega\,\mathrm{d}\Omega' \tag{4.53}$$

Using Eqs. (4.52) and (4.53) in Eq. (4.51) and applying the approximation $\left(1 - \frac{x}{N_1}\right)^{N_1} = e^{-x}$ we obtain:

$$J^{N_1} = e^{-\frac{4}{3}\pi\mu s i N \bar{w}_3} \tag{4.54}$$

where $\bar{w}_3$ is defined by Eqs. (4.53) and (4.22). Holtsmark evaluated Eq. (4.53) graphically and obtained $\bar{w}_3 = -0.345\,\pi\,i$. Thus Eq. (4.54) becomes:

$$J^{N_1} = e^{-\frac{4}{3}\pi^2 s\,(0.345)\,\mu N} = e^{-4.54\,s\,\mu\,N} \tag{4.55}$$

4.6. The special case of the quadrupole

Finally, J^{N_1} for the quadrupole where $p = 4$ must be evaluated. For the quadrupole:

$$J = \frac{1}{v} \iint \frac{(A\,s\,w_4)^{3/4}}{2^{3/4} \cdot 4} \int\limits_{\alpha}^{\infty} \mathrm{e}^{i\,u}\, u^{-7/4}\, \mathrm{d}u\, \mathrm{d}\Omega\, \mathrm{d}\Omega' \tag{4.56}$$

The integral in u is evaluated in the same manner as was done for the ion.

$$\int\limits_{\alpha}^{\infty} \mathrm{e}^{i\,u}\, u^{-7/4}\, \mathrm{d}u = \frac{4}{3}\left[\mathrm{e}^{i\,\alpha}\, \alpha^{-3/4} + i\,\mathrm{e}^{i\frac{\pi}{3}}\, \Gamma\left(\frac{1}{4}\right) - 4\,i\,\alpha^{1/4} - \frac{4\,i^2}{5}\,\alpha^{5/4} - \cdots\right] \tag{4.57}$$

where $\alpha = \dfrac{A\,s\,w_4}{2\,R^4}$

Hence:

$$J = \frac{1}{6\pi^2} \iint \mathrm{d}\Omega\, \mathrm{d}\Omega' \left[1 + \mathrm{e}^{i\frac{5}{8}\pi}\, \Gamma\left(\frac{1}{4}\right) \left(\frac{A\,s\,w_4}{2}\right)^{3/4} \frac{1}{R^3}\right] \tag{4.58}$$

In Eq. (4.58) higher order terms have been dropped as in the ion and dipole cases. We then obtain:

$$J^{N_1} = \exp\left[\mathrm{e}^{i\frac{5}{8}\pi}\, \Gamma\left(\frac{1}{4}\right) \frac{4\,\pi}{3}\, N \left(\frac{A\,s}{2}\right)^{3/4} \overline{w_4}^{3/4}\right] \tag{4.59}$$

where, as usual:

$$\overline{w_4}^{3/4} = \frac{1}{16\,\pi^2} \iint \mathrm{d}\Omega\, \mathrm{d}\Omega'\, w_4^{3/4}$$

Utilizing Eq. (4.27), Holtsmark graphically evaluated $\overline{w_4}^{3/4}$ to obtain $\overline{w_4}^{3/4} = 0.70(1 + i^{3/2})$.

Hence:

$$J^{N_1} = \exp\left[\mathrm{e}^{i\frac{5}{8}\pi}\, \Gamma\left(\frac{1}{4}\right) \frac{4\,\pi}{3}\, 0 \cdot 70 \left(1 + \mathrm{e}^{i\frac{3}{4}\pi}\right) N \left(\frac{A\,s}{2}\right)^{3/4}\right]$$

Note that:

$$\mathrm{e}^{i\frac{5}{8}\pi}\left(1 + \mathrm{e}^{i\frac{3}{4}\pi}\right) = \mathrm{e}^{i\frac{5}{8}\pi} + \mathrm{e}^{i\frac{5}{8}\pi} \cdot \mathrm{e}^{i\,2\,\pi} = \mathrm{e}^{i\frac{5}{8}\pi} + \mathrm{e}^{-i\frac{5}{8}\pi} = 2\cos\frac{5\pi}{8}$$

$$= -\,0.7665$$

$$\Gamma\left(\frac{1}{4}\right) = 3.6256$$

Hence:

$$J^{N_1} = \mathrm{e}^{-\,4.87\,N\,(A\,s)\,3/4} \tag{4.60}$$

Eqs. (4.43), (4.55), and (4.60) thus provide J^{N_1} for the three cases under consideration. We may now turn our attention to the evaluation of $W(XYZ)$ as given by Eq. (4.13).

$$W(XYZ) = \frac{1}{8\pi^3} \iiint\limits_{-\infty}^{+\infty} \mathrm{d}\xi\, \mathrm{d}\eta\, \mathrm{d}\zeta\, \mathrm{e}^{-i\,(\xi X + \eta Y + \zeta Z)}\, J^{N_1} \tag{4.61}$$

4.7. The field probability function for the three special cases

For the dipole:

$$W(XYZ) = \frac{1}{8\pi^3} \iiint\limits_{-\infty}^{+\infty} \mathrm{d}\xi\, \mathrm{d}\eta\, \mathrm{d}\zeta\, \mathrm{e}^{-i(\xi X + \eta Y + \zeta Z)}\, \mathrm{e}^{-c_3 s} \tag{4.62}$$

where: $c_3 = 4.54\, \mu N$.

In Eq. (4.62) XYZ can be taken as the components of a vector $\mathbf{E}$ which is independent of s, the vector with components $\xi\,\eta\,\zeta$. Then the exponent in Eq. (4.62) is obviously the dot product $\mathbf{E} \cdot \mathrm{s} = E\, s \cos \vartheta$ where here ϑ is the angle between $\mathbf{E}$ and s. Now transform from the rectangular coordinates $\xi\,\eta\,\zeta$ to the spherical polar coordinates $s\,\vartheta\,\varphi$.

$$\mathrm{d}\xi\, \mathrm{d}\eta\, \mathrm{d}\vartheta = s^2 \sin \vartheta\, \mathrm{d}\vartheta\, \mathrm{d}\varphi\, \mathrm{d}s$$

Eq. (4.62) thus becomes:

$$W(XYZ) = \frac{1}{8\pi^3} \int\limits_0^{\pi} \int\limits_0^{2\pi} \int\limits_0^{\infty} \sin \vartheta\, \mathrm{d}\vartheta\, \mathrm{d}\varphi\, s^2\, \mathrm{d}s\, \mathrm{e}^{-iEs\cos\vartheta}\, \mathrm{e}^{-c_3 s} \tag{4.63}$$

Integrate over ϑ and φ to obtain:

$$W(XYZ) = \frac{1}{4\pi^2} \int\limits_0^{\infty} s^2\, \mathrm{d}s\, \mathrm{e}^{-c_3 s} \frac{2}{E\,s} \sin (E\,s) \tag{4.64}$$

We write $\sin(E\,s)$ as an exponential and take the imaginary part.

$$W(XYZ) = \frac{1}{2\pi^2} \Im \left\{ \int\limits_0^{\infty} \frac{s}{E}\, \mathrm{e}^{(-c + E_i)s}\, \mathrm{d}s \right\} \tag{465}$$

Let $-t = (-c_3 + E\,i)$ so that the integral in Eq. (4.65) becomes a gamma or factorial function.

$$W(XYZ) = \frac{1}{2\pi^2} \Im \left\{ \left[\frac{\Gamma(2)}{(-c_3 + Ei)^2\, E} \right] \right\} = \frac{1}{\pi^2} \frac{c_3}{(c_3{}^2 + E^2)^2} \tag{4.66}$$

Eq. (4.66) gives us for the dipole the field strength probability function which we have been seeking.

A "Normalfeldstarke" may be defined as:

$$E_0 = c_3 = 4.54\, \mu N \tag{4.67}$$

Let β give the relation between this "Normalfeldstarke" and the actual field strength.

$$\beta = \frac{E}{E_0} \tag{4.68}$$

Now we replace the volume element $\mathrm{d}X\, \mathrm{d}Y\, \mathrm{d}Z$ by $4\,\pi\, E^2\, \mathrm{d}E$: From Eq. (4.66):

$$W(XYZ) = \frac{1}{\pi^2} \frac{1}{\left(1 + \frac{E^2}{c_3{}^2}\right)} \frac{1}{c_3{}^3} = \frac{1}{\pi^2} \frac{1}{(1 + \beta^2)^2} \frac{1}{c_3{}^3} \tag{4.69}$$

Hence:

$$\left.\begin{aligned} W(E)\,\mathrm{d}E &= 4\pi\, E^2\, W(XYZ)\,\mathrm{d}E \\ &= \frac{4}{\pi}\,\frac{1}{(1+\beta^2)^2}\,\frac{E^2}{c_3{}^3}\,\mathrm{d}E = \frac{4}{\pi}\,\frac{\beta^2\,\mathrm{d}\beta}{(1+\beta^2)^2} \end{aligned}\right\} \tag{4.70a}$$

Fig. 4.3 furnishes a plot of $W(\beta)$ vs. β as given by Eq. (4.70a).

The case of the ion is taken up next. As in the case of the dipole, let

$$\mathbf{E}\cdot\mathbf{s} = E\, s\cos\vartheta \tag{4.71a}$$

$$\mathrm{d}\xi\,\mathrm{d}\eta\,\mathrm{d}\vartheta = s^2\sin\vartheta\,\mathrm{d}\vartheta\,\mathrm{d}\varphi\,\mathrm{d}s \tag{4.71b}$$

$$J^{N_1} = \mathrm{e}^{-4{\cdot}21\,N\,\varepsilon^{3/2}\,s^{3/2}} = \mathrm{e}^{-c_2 s^{3/2}} \tag{4.71c}$$

After integration over ϑ and φ the equation which corresponds to Eq. (4.64) is obtained as:

$$W(XYZ) = \frac{1}{4\pi^2}\int_0^\infty s^2\,\mathrm{d}s\,\mathrm{e}^{-c_2 s^{3/2}}\,\frac{2}{E\,s}\sin(E\,s) \tag{4.72}*$$

Eq. (4.72) is not directly evaluable. Holtsmark originally derived expressions which subsequent gentlemen corrected and which gave the value of Eq. (4.72) for large β and for small β. We shall not trouble ourselves with these. Instead the reader is referred to Fig. 4.4. This figure is a result of the direct calculation of Eq. (4.72) which we carried out recently on the IBM 704 electronic data processing machine.

Finally, the probability of a field strength E must be calculated for the quadrupole. As in the two preceding cases, integration is first carried out over ϑ and φ to obtain the following:

$$W(XYZ) = \frac{1}{2\pi^2}\int_0^\infty s^2\,\mathrm{d}s\,\mathrm{e}^{-c_4 s^{3/4}}\,\frac{\sin(E\,s)}{E\,s} \tag{4.73}$$

We make the substitutions $v = E\,s$ and $\beta = \dfrac{E}{c_4{}^{4/3}}$ which yield:

$$W(XYZ) = \frac{1}{2\pi^2 E^3}\int_0^\infty \mathrm{e}^{-\left(\frac{v}{\beta}\right)^{3/4}}\, v\sin v\,\mathrm{d}v \tag{4.74}$$

When the exponential is expanded in a series and $\sin v$ is expressed as the imaginary part of an exponential, Eq. (4.74) becomes:

$$\left.\begin{aligned} W(XYZ) = \frac{1}{2\pi^2 E^3}\,\Im\Bigg\{\int_0^\infty v\,\mathrm{e}^{iv}\Bigg[1-\left(\frac{v}{\beta}\right)^{3/4}+\frac{1}{2!}\left(\frac{v}{\beta}\right)^{6/4}-\frac{1}{3!}\left(\frac{v}{\beta}\right)^{9/4} \\ +\frac{1}{4!}\left(\frac{v}{\beta}\right)^{12/4}+\cdots\Bigg]\mathrm{d}v\Bigg\} \end{aligned}\right\} \tag{4.75}$$

* If we now let $v = E_0 s$ we obtain Unsold's Eq. (55)[221] developed by Verweij[200], namely,

$$W(\beta) = \frac{2}{\pi\beta}\int_0^\infty v\sin v\,\mathrm{e}^{-\left(\frac{v}{\beta}\right)^{3/2}}\,\mathrm{d}v.$$

The series in Eq. (4.75) will not be properly convergent for small values of β. Each term in Eq. (4.75) is a gamma function. A straightforward arithmetical

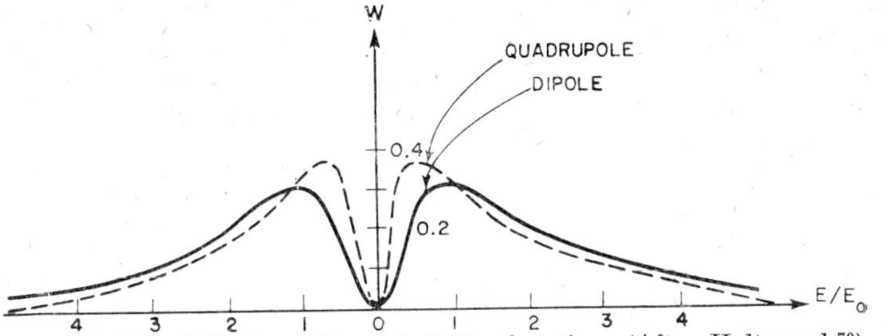

FIG. 4.3. The field strength probability function. (After Holtsmark[79])

calculation after the substitution of β and the gamma functions yields:

$$W(E)\,dE = \frac{4}{\pi}\frac{d\beta}{\beta^{7/4}}\,0.805 \times \left[1 - \frac{0.730}{\beta^{3/4}} - \frac{0.328}{\beta^{3/2}} + \frac{0.621}{\beta^{9/4}} - \frac{0.163}{\beta^{3}} + \cdots\right] \qquad (4.70\,b)$$

The factor $4/\pi$ has been inserted for comparison with the previous Eqs. (4.70). Again $\beta = E/E$ where $E_0 = c_4^{4/3}$.

For $0 < \beta < 1$, Holtsmark utilized graphical integration to obtain:

$$W(E)\,dE = \frac{4}{\pi}\beta^2\,d\beta\,\frac{4}{3}\,(1 - 2.44\,\beta^2 + 11.25\,\beta^4 - 72\,\beta^6 + \cdots) \qquad (4.70\,c)$$

Eqs. (4.70b) and (4.70c) yield the quadrupole curve in Fig. 4.3.

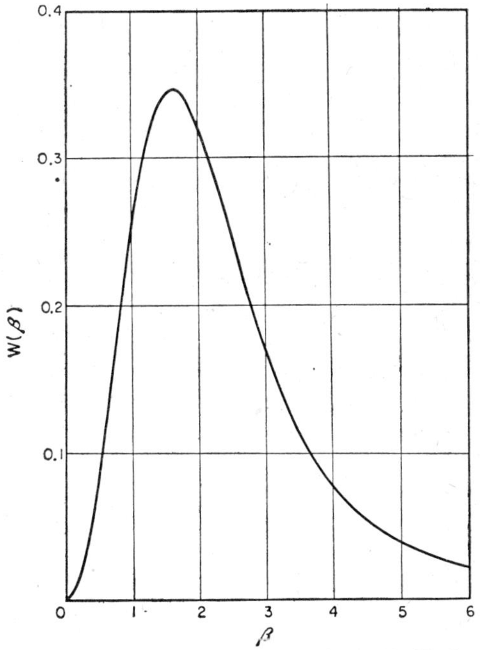

FIG. 4.4. The Holtsmark Field Strength Probability Function

Eqs. (4.70) now give the probability $W(E_0)$ for the existence at a certain time of the field E_0 due to one of three causes. The broadening of the spectral lines due to one of these fields may now be determined.

4.8. General intensity distribution in a Stark broadened line

Let us assume that a spectral line is emitted at the time when the field E exists. E will then split the line according to the Stark effect. Let the intensity in this split line be given by:

$$I\,d\nu = I(E, \nu)\,d\nu \qquad (4.76)$$

I is then the intensity in any relative units for the frequency ν and the field strength E. If we multiply this intensity distribution by the probability for the field strength E and integrate over all E, we obtain the intensity distribution in the broadened line as follows:

$$I\,\mathrm{d}\nu = \mathrm{d}\nu \int_0^\infty I(E,\nu)\,W(E)\,\mathrm{d}E \tag{4.77}$$

Instead of attempting to utilize an accurate but overcomplicated expression for $I(E,\nu)$, Holtsmark makes a rectangular assumption for the broadened spectral line, that is, he assumes $I(E,\nu)$ to be a constant within the limits of the line and zero outside. Thus, for a particular field strength E we have a rectangle of height h and width $2\nu_m$, where $2\nu_m$ is the separation of the outermost Stark components. The area of this rectangle, which is the integrated intensity, is taken as a constant which is independent of E. This is an approximation, but a reasonably good one. We would, of course, obtain rectangles of different width — different $2\nu_m$ — for different values of E. The integrated intensity of the line is:

$$f = 2\nu_m h = 2\nu_m I(E,\nu) \tag{4.78}$$

Thus:

$$I(E,\nu) = \frac{f}{2\nu_m}\ \text{inside}\ 2\nu_m,\ \text{and:}\ I(E,\nu) = 0\ \text{outside}\ 2\nu_m \tag{4.79}$$

Let us take ν_0 as the frequency of the line before splitting. Then it follows that all rectangles for which $\nu_m \geq |\nu - \nu_0|$ will contribute toward the intensity at the frequency ν. In order to obtain the total intensity, integration is carried out for all rectangles which are wider than $2\nu_m$:

$$I\,\mathrm{d}\nu = \mathrm{d}\nu \int_{E'}^\infty \frac{f}{2\nu_m}\,W(E)\,\mathrm{d}E \tag{4.80}$$

E' is that field strength which causes the splitting $2(\nu - \nu_0)$, and ν_m is given by:

$$2\nu_m = c\,E' \tag{4.81}$$

where c is simply a proportionality factor. Eq. (4.80) becomes:

$$I\,\mathrm{d}\nu = \mathrm{d}\nu \int_{E'}^\infty \frac{f}{cE}\,W(E)\,\mathrm{d}E \tag{4.82}$$

The evaluation of Eq. (4.82) for the three cases yields the desired broadening. First, we consider the case of the dipole.

The quadratic Stark effect could be treated in precisely the same manner. Here the separation of the outermost Stark components is given by tE^2. When we replace cE by tE^2 in Eq. (4.82) the result follows.

4.9. Line shape and half widths according to the early Stark theory

When Eq. (4.70a) is substituted into Eq. (4.82), the result is:

$$I\,\mathrm{d}\nu = \mathrm{d}\nu \frac{4}{\pi} \int_{\beta'}^{\infty} \frac{f}{cE_0} \frac{\beta\,\mathrm{d}\beta}{(1+\beta^2)^2} = \mathrm{d}\nu \frac{f}{2cE_0} \frac{4}{\pi} \frac{1}{(1+\beta'^2)} \tag{4.83}$$

in which $\beta' = \dfrac{2(\nu-\nu_0)}{cE_0} = \dfrac{2\Delta\nu}{cE_0}$, so that:

$$I(\nu) = \frac{2f}{\pi} \frac{cE_0}{c^2E_0^2 + 4(\nu-\nu_0)^2} \tag{4.84}$$

The half-width of the line can be seen to be:

$$\delta_{sd} = cE_0 = c\,4.54\,N\mu \tag{4.85}$$

An interesting sidelight seems worthy of introduction at this point. Let us set our integrated intensity f equal to S, a trivial but legitimate move. Then let us substitute Eq. (4.85) into Eq. (4.87) to obtain:

$$I(\nu) = \frac{S}{\pi} \frac{(\delta_{sd}/2)}{(\nu-\nu_0)^2 + (\delta_{sd}/2)^2} \tag{4.86}$$

We turn our attention to the ion and the quadrupole.

Eqs. (4.81) and (4.82) may be rewritten as:

$$2\nu_m = cE_0\beta \tag{4.87}$$

$$I\,\mathrm{d}\nu = \mathrm{d}\nu \frac{f}{cE_0} \int_{\beta'}^{\infty} \frac{W(\beta)\,\mathrm{d}\beta}{\beta} \tag{4.88}$$

Holtsmark evaluated Eq. (4.88) graphically for the cases of the ion and the quadrupole. Figure 4.5, after Holtsmark, represents his results for the intensity

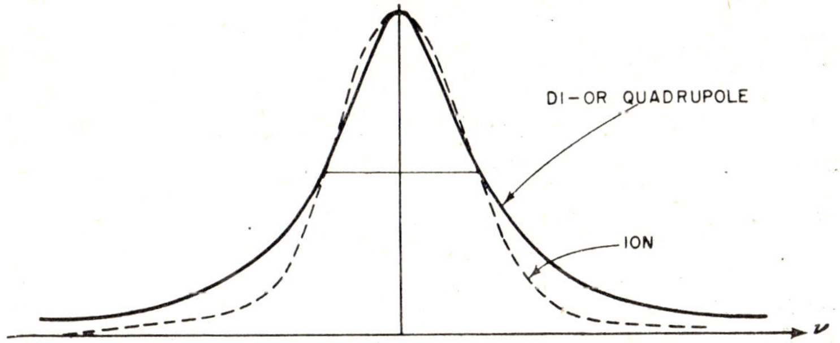

FIG. 4.5. The Stark Theory line shape. (After Holtsmark[79])

distribution for all three cases. In Fig. 4.5 the abscissae have been changed for the three perturbers so that the half-widths coincide, thus giving a curve shape comparison. The half-intensity β's for the ion and the quadrupole may be obtained from the curve as 1.25 and 0.67, respectively. These values of β yield the half-widths:

$$\text{Ion:} \qquad \delta_{si} = 1.25\, c\, E_0 = 3.25\, c\, N^{2/3}\, \varepsilon \tag{4.89a}$$

$$\text{Quadrupole:} \qquad \delta_{sq} = 0.67\, c\, E_0 = 5.53\, c\, N^{4/3}\, A \tag{4.89b}$$

4.10. Review of Holtsmark's early Stark broadening theory

Let us rapidly review this development before beginning its refinement. We began with the assumption that a spectral line is broadened by a Stark effect which arises from a varying field produced at the emitter by (a) ions, (b) dipoles or (c) quadrupoles.* We assumed that the probability $W(E)$, that a field strength E exists at the emitter at a certain time, is a function of E. A rather laborious derivation yielded Eqs. (4.70a) through (4.70c), the equations for these probabilities for the three cases under consideration. A linear Stark effect, $2\nu_m = cE$, was assumed. $I(E, \nu)$ was assumed constant for a given field strength within the outermost Stark components and zero outside. Thus, each field strength yielded a rectangle of different width, but the same area. For a given frequency then, we integrated over all rectangles whose width is such that they contribute to the intensity at this frequency. Eq. (4.82) for the intensity distribution within the Stark-broadened line resulted. For the dipole Eq. (4.82) yielded the line shape:

$$I(\nu) = \frac{S}{\pi} \frac{(\delta_{sd}/2)}{(\nu - \nu_0)^2 + (\delta_{sd}/2)^2} \tag{4.86}$$

which is identical with Eq. (1.78). Eq. (4.82) does not lend itself so readily to evaluation for the ion and quadrupole cases.

$$\delta_{si} = 3{\cdot}25\, c\, N^{2/3}\, \varepsilon \tag{4.89a}$$

$$\delta_{sd} = 4{\cdot}54\, c\, N\mu \tag{4.85}$$

$$\delta_{sq} = 5{\cdot}53\, c\, N^{4/3}\, A \tag{4.89b}$$

Eqs. (4.89a), (4.85), and (4.89b) show the dependence of the half-width on the electrical properties of the perturbing molecule through ε (charge), μ (dipole moment), or A (quadrupole constant) and also on the gas density through N (the number of perturbers per unit of volume).

* Debye[21] obtained approximate expressions for these field strengths and hence, an idea as to the behavior of the half-widths as follows. Electric charge is taken as 5×10^{-10} g$^{1/2}$-cm$^{3/2}$-sec^{-1} and the radius of the molecule as 10^{-8} cm. We assume that (a) the ion has charge 5×10^{-10} g$^{1/2}$-cm$^{3/2}$-sec^{-1} (b) the moment of the dipole is 5×10^{-18} g$^{1/2}$-cm$^{5/2}$-sec^{-1} and (c) the quadrupole constant is 5×10^{-26} g$^{1/2}$-cm$^{7/2}$-sec^{-1}. Since the electric field has the units g$^{1/2}$-cm$^{3/2}$-sec^{-1}, and since we may assume the field to be the product of either (a), (b), or (c) and some power of N (molecules - cm^{-3}), this power of N may be adjusted so that the product has the proper units, thus yielding approximate expressions for the electric field in each of the three cases. Hence, we obtain for (a) $E = \varepsilon N^{2/3}$, for (b) $E = \mu N$, and for (c) $E = \theta N^{4/3}$ or 4500 e. s. u., 135 e. s. u., and 4 e. s. u., respectively. Multiplicative constants would yield the values obtained by Holtsmark.

We shall consider a simplified version of the Holtsmark treatment after our consideration of finite molecular diameters.

4.11. The field strength probability function with finite molecular diameters

It may be recalled that we have essentially considered the molecules involved in the theory as points, in that, when the integration was carried out over the molecular positions in space, no portions of space were excluded on the basis of previous occupancy. In 1920 Debye[30], while in the process of deriving a mean square value for the quadrupole electric field strength due to an atomic collection, commented rather adversely on this punktformig assumption.

As a result Holtsmark attacked the problem again, this time with finite diameters assigned the molecules under consideration. Gans had approached the problem earlier under the assumption of finite diameters for the emitting molecules, but he had retained the point assumption for the broadening molecules. As Holtsmark noted, this would be a good assertion for the case where ions are the field producers, since we would normally expect the ions to make up only a small portion of the total number of molecules present. On the other hand, these assumptions would not appear to be valid for the dipole or quadrupole case. Gans further found a Gaussian distribution for very high field strengths, "... wie zu erwarten war".[81]

In this development, Holtsmark retained the simplifying assumption which classifies the broadening molecules as ions, dipoles, or quadrupoles, that is, he again took only the first term in the series for the potential. Since the assumptions are predicated on large R, it is apparent that when the gas density or pressure is high*, R no longer remains large enough to justify them. For these "high" densities the calculations cannot be carried out, but the field strength distribution in these cases is assumed Gaussian[49]. Holtsmark thus limits himself to those gas densities where the first term in the potential series does give a good approximation of the electric field.

We begin with Eq. (4.11) written in a slightly different form, since the modifications in the development will occur at a later stage.

$$W(XYZ) = \frac{1}{8\pi^3} \iiint_{-\infty}^{+\infty} \mathrm{d}\xi\, \mathrm{d}\eta\, \mathrm{d}\zeta\, \mathrm{e}^{-i(\xi X + \eta Y + \zeta Z)} \int \sigma\, \mathrm{e}^{i(\xi \Sigma X_n + \eta \Sigma Y_n + \zeta \Sigma Z_n)} \mathrm{d}\tau \Big/ \int \sigma\, \mathrm{d}\tau \tag{4.90}$$

Again:

$$\xi X + \eta Y + \zeta Z = s E \cos\vartheta_s; \qquad \mathrm{d}\xi\, \mathrm{d}\eta\, \mathrm{d}\zeta = s^2 \mathrm{d}s \sin\vartheta_s\, \mathrm{d}\vartheta_s\, \mathrm{d}\varphi_s \tag{4.91}$$

where in this case $\mathbf{E}$ has been taken as the polar axis and $\mathbf{s}$ has been referred to it. Substitute Eq. (4.91) into Eq. (4.90) and integrate over ϑ_s and φ_s to obtain:

$$W(E) = \frac{2E}{\pi} \int_0^\infty \mathrm{d}s\, s \sin(Es) \frac{L(s)}{M(s)} \tag{4.92}$$

* In a somewhat similar consideration, Spitzer (see *infra*, this Chap.) takes as a limit a pressure such that $R > 10 r_e$ where r_e is the radius of the Bohr orbit.

where:

$$L(s) = \int d\tau\, \sigma\, e^{i\Sigma(\xi X_n + \eta Y_n + \zeta Z_n)} \tag{4.93}$$

$$M(s) = \int s\, d\tau \tag{4.94}$$

If χ is the angle between s and $\mathbf{E}_n$

$$\Sigma(\xi X_n + \eta Y_n + \zeta Z_n) = s \Sigma E_n \cos\chi_n \tag{4.95}$$

Now in the computation of $L(s)$ and $M(s)$, the finiteness of the molecular diameters is to be taken into account. In the earlier computation, it was possible to transform Eq. (4.16) into Eq. (4.18), a product of identical integrals. This was legitimate, due to the independency of the molecules, in that the motion of one molecular point is not interfered with by the other molecular points. This integral product is no longer admissible after finite diameters have been assigned the molecules, for we may not now allow the center of a molecule to be separated from the center of another molecule by less than this molecular diameter. We let:

$$\Xi_n = e^{isE_n\cos\chi_n} \tag{4.96}$$

so that:

$$\left.\begin{aligned} L(s) &= \int_k d\tau\, \sigma \prod_{k-1}^{N_1} \Xi_n \\ &= \int_{k_1} \Xi_1\, \sigma_1\, d\tau_1 \int_{k_2} \Xi_2\, \sigma_2\, d\tau_2 \ldots \int_{k_{N_1}} \Xi_{N_1}\, \sigma_{N_1}\, d\tau_{N_1} \end{aligned}\right\} \tag{4.97}$$

where the first integral is carried over the space not occupied by the other $(N_1 - 1)$ molecules and the 0-th molecule (emitter). Thus, the first integral is dependent on the coordinates of the other molecules and must be included under the integral sign of the second integral, and so on. This fact obviously does not simplify matters. The difficulty may be eliminated, however, by selecting a suitable initial distribution for molecules. This selection would appear to be justified, since our results should not depend on the arbitrary initial distribution of the molecules. As our initial molecular spatial distribution we place all molecules together in one small region of space, this region being the same during the first $N - 1$ integrations. The position of the first molecule was chosen for this union. After locating the molecules in this manner, we may move molecule N_1 about space in the process of the integration, while keeping the remaining molecules at position one. This process is carried out for $N_1 - 1$ of the molecules. Thus, we see that the first integral contains coordinates which require that it be included only under the integral sign for molecule one — we are working backwards, one might say, from molecule N_1 to molecule one. The happy situation prevails for the first $N_1 - 1$ integrals. Let the integral over all space except that portion occupied by the emitter be:

$$\int_{k_n} \Xi_n\, \sigma_n\, d\tau_n = H \tag{4.98}$$

Further, let the integral over that portion occupied by the N_1 remaining molecules at position one be:

$$\int_{nk} \Xi_n \sigma_n \,\mathrm{d}\tau_n = H_n \tag{4.99}$$

This notation yields for the first integration:

$$(H - H_{N_1-1})$$

After $N_1 - 1$ integrations we have:

$$L(s) = \int \sigma_1 \,\mathrm{d}\tau_1 \,\Xi_1 \prod_{n=1}^{N_1-1} (H - H_n) \tag{4.100}$$

where Eq. (4.100) is to be integrated over the entire coordinate range excepting that portion occupied by the emitter.

Substitute Eq. (4.30) into Eq. (4.99) to obtain:

$$\left.\begin{aligned} H_n &= \int_{nk} \Xi_n r^2 \sin\vartheta_1 \,\mathrm{d}\vartheta_1 \sin\vartheta' \,\mathrm{d}\vartheta' \,\mathrm{d}\varphi_1 \,\mathrm{d}\varphi' \,\mathrm{d}r \\ &\doteq \Xi_n n \int_0^d \int_0^\pi \int_0^\pi \int_0^{2\pi} \int_0^{2\pi} r^2 \,\mathrm{d}r \sin\vartheta_1 \,\mathrm{d}\vartheta_1 \sin\vartheta' \,\mathrm{d}\vartheta' \,\mathrm{d}\varphi_1 \,\mathrm{d}\varphi' = n\,\Xi_1\, 4\pi k \end{aligned}\right\} \tag{4.101}$$

where k is eight times the volume of a molecule; n is the number of molecules at position one; Ξ_1 is the value of Ξ at the position of the first molecule, and d is the diameter (closest separation of two molecular centers) of a molecule. This approximation is only a good one if $n\,k \ll V$, the total volume; and also only at some distance from the emitter, since Ξ could not be taken as the constant Ξ_1 for the integration for position one close to the emitter.

If we let V differ from its previous value by a factor of 4π we may write an analogy to Eq. (4.29) as:

$$H = 4\pi\, V\, J \tag{4.102}$$

We substitute Eqs. (4.101) and (4.102) into Eq. (4.100) to obtain:

$$L(s) = (4\pi\, V\, J)^{N_1-1} \int_{kn} \Xi_1 \sigma_1 \,\mathrm{d}\tau_1 \prod_{n=1}^{N_1-1} \left(1 - n\frac{k\,\Xi_1}{VJ}\right) \tag{4.103}$$

$M(s)$ may now be calculated in a similar manner. Integration over the N_1-th atom yields $4\pi(V - (N_1 - 1)\,k)$, and so on up to the last integration. We obtain:

$$M(s) = (4\pi\, V)^{N_1} \prod_{n=1}^{N_1-1} \left(1 - \frac{n\,k}{V}\right) \tag{4.104}$$

Now let:

$$K(s)\, J^{N_1-1} = \frac{L(s)}{M(s)} = \frac{1}{4\pi V}\, J^{N_1-1} \frac{\int_{kn} \Xi_1 \sigma_1 \,\mathrm{d}\tau_1 \prod_{n=1}^{N_1-1} \left(1 - n\frac{k\,\Xi_1}{VJ}\right)}{\prod_{n=1}^{N_1-1} \left(1 - \frac{n\,k}{V}\right)} \tag{4.105}$$

We have assumed $n\,k \ll V$ so that $N_1 \frac{k\,\Xi_1}{VJ} = N_1\,\varepsilon \ll 1$. This fact allows the approximation:

$$\log \prod_{n=1}^{N_1-1} (1 - n\,\varepsilon) \doteq \sum_{1}^{N_1-1} n\,\varepsilon \doteq -\frac{N_1^2}{2}\,\varepsilon \leftrightarrow \prod_{n=1}^{N_1-1} (1 - n\,\varepsilon) = \mathrm{e}^{-\frac{N_1}{2\varepsilon}} \tag{4.106}$$

Hence:

$$\left.\begin{aligned} K(s) &= \frac{1}{k_n\,k_n} \int \Xi_1\,\mathrm{d}\tau_1\,\mathrm{e}^{-\left(\frac{N_1^2 k}{2}\frac{\Xi_1}{V} + \frac{N_1^2 k}{2V}\right)} \\ &= \frac{\mathrm{e}^{\alpha}}{k_n\,k_n} \int \Xi_1\,\sigma_1\,\mathrm{d}\tau_1\,\mathrm{e}^{-\alpha\,\Xi_1} \end{aligned}\right\} \tag{4.107}$$

where: $\frac{2\,\alpha}{N_1} = \frac{N_1\,k}{V}$, eight times the ratio of molecular to total volume.

A transformation of coordinates at this point facilitates continuation of the calculation. Let:

$$\sigma_1\,\mathrm{d}\tau_1 = r_1^2\,\mathrm{d}r_1\,\sin\chi_1\,\mathrm{d}\chi_1\,\sin\gamma_1\,\mathrm{d}\gamma_1 \tag{4.108a}$$

so that:

$$k_n = \frac{4}{3}\,R^3 \tag{4.108b}$$

where r_1 is the radius vector of molecule one; χ_1 is the angle between s and E_1, and γ_1 is the angle between r_1 and the molecular axis. The legitimacy of this transformation is assured, since we may recall that we are now integrating over the entire volume. E_1 is a function of r_1, γ_1, and the electrical constants of the molecule. Holtsmark limited his calculations to dipoles as field producers, since he felt that in the case of finite diameters the complexity of the calculation for ions or quadrupoles was too great, if, indeed, a reasonable result could be obtained.

From Eq. (4.19):

$$E_1 = \frac{\mu}{r^3}\,w = \frac{\mu}{r^3}\sqrt{1 + 3\cos^2\gamma} \tag{4.109}$$

Let:

$$z = s\,E_1 = s\,\frac{\mu}{r^3}\,w \tag{4.110a}$$

Hence:

$$K(s) = \frac{3\,\mathrm{e}^{\alpha}}{4\,R^3} \int_0^R \int_0^{\pi} \int_0^{\pi} \mathrm{e}^{iz\cos x}\,\mathrm{e}^{-\alpha\,\mathrm{e}^{iz\cos x}}\,r_1^2\,\mathrm{d}r_1\,\sin\chi_1\,\mathrm{d}\chi_1\,\sin\gamma_1\,\mathrm{d}\gamma_1 \tag{4.111}$$

where the lower limit $r = 0$ introduces no noticeable errors. Integration over χ yields:

$$\left.\begin{aligned} K(s) &= \mathrm{e}^{\alpha}\,\frac{3}{2\,R^3} \int_0^{\pi} \sin\gamma_1\,\mathrm{d}\gamma_1 \int_0^R r_1^2\,\mathrm{d}r_1\,\frac{1}{\alpha\,z}\,\mathrm{e}^{-\alpha\cos z}\,\sin(\alpha\sin z) \\ &= \frac{\mathrm{e}^{\alpha}}{\alpha}\,\frac{3}{2\,R^3}\,\Im\left\{\int_0^{\pi} \sin\gamma_1\,\mathrm{d}\gamma_1 \int_0^R r_1^2\,\mathrm{d}r_1\,\frac{1}{z}\,\mathrm{e}^{-\alpha(\cos z - i\sin z)}\right\} \end{aligned}\right\} \tag{4.112}$$

where the imaginary part of the exponential has been substituted for sin (α sin z).

From Eq. (4.110a):

$$r_1^2 \,\mathrm{d}r_1 = - s\,\mu\, w\, z^{-2}\,\mathrm{d}z \tag{4.110b}$$

Hence:

$$\left.\begin{aligned} K(s) &= \frac{e^{\alpha}}{2\alpha}\int_0^{\pi} \mathrm{d}\gamma_1 \sin\gamma_1 \cdot z_1 \cdot \Im\left\{\int_{Z_1}^{\infty} \mathrm{d}z\, z^{-3}\, e^{-\alpha u}\right\} \\ &= P_1 \frac{1}{2}\int_0^{\pi} \mathrm{d}\gamma_1 \sin\gamma_1 \end{aligned}\right\} \tag{4.113}$$

where:

$$u = e^{-iz} = \cos z - i \sin z \tag{4.114}$$

Instead of integrating Eq. (4.113) along the axis of reals from z_1 to ∞. Holtsmark integrated around a path in the complex plane ($z = x + i\,y$) — which, it may be noted, includes no singular points — as shown in Fig. 4.6.

We desire only the imaginary portion of the result. The integral along BC is real, so that it contributes nothing to Eq. (4.113). The integrals along CD and DE disappear for $Y = \infty$ and $z_2 = \infty$ respectively. Thus, by Cauchy's theorem,* the integral with limits z_1 and ∞, which we desire, is equal to the integral along AB.

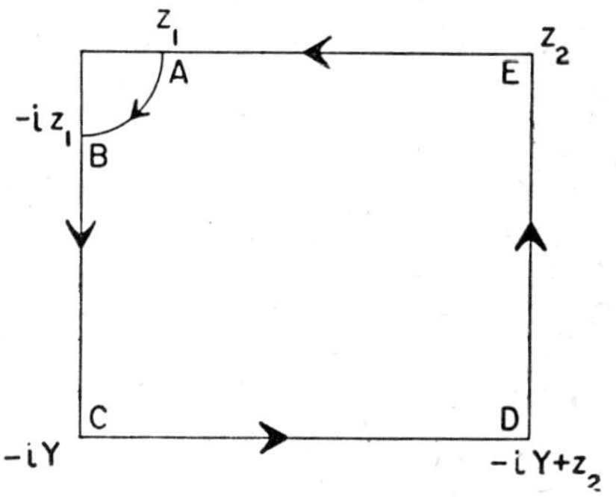

FIG. 4.6. Path in the complex plane used for the evaluation of Eq. (6.113)

Let:

$$z = z_1 e^{-i\varphi} \leftrightarrow \mathrm{d}z = - z_1\, i\, e^{-i\varphi}\,\mathrm{d}\varphi$$

Since z_1, $\alpha \ll 1$,

Let:

$$u \doteq e^{-iz} = 1 - i\,z_1\, e^{-i\varphi} + \cdots \doteq 1 - i\, z_1\, e^{-i\varphi}$$

And:

$$e^{\alpha} e^{-\alpha u} = e^{i\alpha z_1 e^{-i\varphi}} = 1 + i\,\alpha\, z_1 e^{-i\varphi} - \frac{\alpha^2 z_1^2}{2} e^{-2i\varphi} - i\frac{\alpha^3 z_1^3}{6} e^{-3i\varphi} - \cdots \tag{4.115}$$

Eq. (4.115) yields for P_1 after integration:

$$P_1 = 1 + \frac{\pi}{4}\alpha\, z_1 + \frac{\alpha_2 z_1^2}{6} + \cdots \tag{4.116}$$

So that:

$$K(s) = \frac{1}{2}\int_0^{\pi} \mathrm{d}\gamma_1 \sin\gamma_1 \left(1 + \frac{\pi}{4}\alpha\, z_1 + \frac{\alpha^2 z_1^2}{6} + \cdots\right) \tag{4.117}$$

Let:

$$\frac{N_1 k}{2 v} = b; \quad \frac{N_1}{v} = N \leftrightarrow \alpha\, z_1^2 = \frac{4}{3}\pi\, b\, N \mu s\, w \tag{4.118}$$

* Cauchy's theorem states that the integral around a closed path in the complex plane which includes no singular points is zero.

We now substitute Eqs. (4.118) into (4.117) and drop all terms after the second in the series in the resulting equation to obtain:

$$K(s) = 1 + \frac{\pi^2}{6} b N\mu s \int_{-1}^{+1} \sqrt{1+3x^2}\, dx = 1 + 4.54\, b N\mu s \qquad (4.119)$$

where $x = \cos \gamma_1$.

An objection to taking the upper limit of z as infinity ($r = 0$) may well be raised, for this would mean that the molecules are point masses in that the emitter is infinitely close to the broadener for $z = \infty$. Practically, the effect is trivial.

We have assumed large N_1 and, essentially:

$$\frac{L(s)}{M(s)} = K(s) \lim_{N_1 \to \infty} J^{N_1 - 1} \qquad (4.120)$$

The quantity $\lim_{N_1 \to \infty} J^{N_1-1}$ is nearly the same as has been given earlier for the ion, dipole, and quadrupole, respectively, by Eqs. (4.43), (4.55), and (4.60). Instead of integrating from 0 to R to obtain J^{N_1-1}, we must here take into account the finite diameters by integrating from a finite lower limit which is taken as the mean of the gas kinetic diameters of the emitting and broadening atoms. If we let:

$$a = 4\pi N d^2; \qquad z = \frac{K_p s}{d^p} \qquad (4.121)$$

where K_p is ε, μ, or A for p equal to 2, 3, or 4 respectively, we may write Gans[5] result as:

$$J^{N_1-1} = e^{-\alpha G_p(z)} \qquad (4.122)$$

Eq. (4.122) yields Eqs. (4.43), (4.55), and (4.60) as a first approximation with **a** $\ll 1$. In second approximation one obtains for the dipole:

$$J^{N_1-1} = e^{4.54\,\mu N s}\, e^{a/3} \qquad (4.123)^*$$

Thus, a constant factor has been introduced which depends on the mean molecular diameters. This would also be the case for the ion and the quadrupole. Eq. (4.92) now becomes:

$$\left.\begin{aligned} W(E) &= e^{a/3} \frac{2E}{\pi} \int_0^\infty ds\, s \sin(Es)\, e^{-4.54\,\mu N s} (1 + 4.54\, b N\mu s) \\ &= e^{a/3} \frac{2E}{\pi} \Im \left\{ \int_0^\infty ds\, s\, e^{(-4.54\,\mu N s + iEs)} (1 + 4.54\, b N \mu s) \right\} \\ &= \left[\frac{4}{\pi} \frac{E^2 E'}{(E'^2 + E^2)} \right] e^{a/3} \left\{ 1 + b \frac{3E'^2 - E^2}{E'^2 + E^2} \right\} \end{aligned}\right\} \qquad (4.124)$$

where $E' = 4.54\, N\mu$.

* As Debye[30] pointed out, this equation does not hold for $E = \infty$ since $W(E)$ decreases too slowly for very high field strengths, thus yielding an infinite mean value of E. Gans found a Gaussian distribution for these large E.

The factor in brackets in Eq. (4.124) is identical to Eq. (4.70a) the probability for the case of point atoms, while the remaining factors essentially correct it for finite diameters. Gans had already obtained the corrective factor $e^{a/3}$ which, it may be noted, is independent of field strength and, thus, does not take into account any field strength changes, and, as a consequence, this factor will not influence the broadening. We should remember that Eq. (4.124) breaks down for very large field strengths, and the Gaussian probability function must be utilized. Eq. (4.124) may be written as:

$$W(\eta) = \frac{4}{\pi} \frac{\eta^2}{(\eta^2+1)^2} e^{a/3} \left\{1 + b\frac{3-\eta^2}{1+\eta^2}\right\} \tag{4.125}$$

where: $\eta = E/E'$.

A simple calculation shows the maximum of the probability curve to fall at $\eta = 1 - b$, while this maximum is at $\eta = 1$ for the dipole on the uncorrected curve.

4.12. Dipole line shape from refined Stark broadening theory

Let us make the short calculation necessary to obtain an idea of the shape for this corrected case. We shall utilize Eq. (4.88) and assume that for $\eta' \leq \eta \leq \eta''$ Eq. (4.125) yields the correct form of $W(\eta)$ while for $\eta'' \leq \eta \leq \infty$ Erf (η) yields the correct $W(\eta)$. We are assuming η'' very large. Eq. (4.88) becomes:

$$I(\nu)\,d\nu = d\nu\, e^{a/3} \frac{4}{\pi} \frac{f}{c E_0} \int_{\eta'}^{\eta''} \frac{\eta}{(1+\eta^2)^2} \left\{1 + b\frac{3-\eta^2}{(1+\eta^2)}\right\} d\eta + d\nu \frac{4}{\pi}\frac{f}{c E_0} \int_{\eta'}^{\infty} \operatorname{erf}(\eta)\, d\eta$$

$$= d\nu\, e^{a/3} \frac{4}{\pi}\frac{f}{c E_0} \left[\int_{\eta'}^{\infty} \frac{\eta}{(1+\eta^2)^2} \left\{1 + b\frac{3-\eta^2}{(1+\eta^2)}\right\} \right.$$

$$\left. - \int_{\eta''}^{\infty} \frac{\eta}{(1+\eta^2)^2} \left\{1 + b\frac{3-\eta^2}{(1+\eta^2)}\right\} d\eta + \int_{\eta''}^{\infty} \operatorname{Erf}(\eta)\, d\eta \right]$$

$$I(\nu) = \frac{4}{\pi} \frac{f}{cE_0} \frac{e^{a/3}}{4} \frac{(2+5b) + (2+4b)\eta'^2}{(1+\eta'^2)^2} + K(\eta'') \tag{4.126}$$

The spectral line given by Eq. (4.126) neglecting the $K(\eta'')$ term will still be symmetrical about the line center. For small b, and b has been assumed small, the curve shape as given by Eq. (4.126) is very nearly identical with the shape given by Eq. (4.83). A good approximation for values of b up to 0.0020 is:

$$\left.\begin{aligned} I(\nu) &= \frac{4}{\pi} \frac{f}{2cE_0} e^{a/3} \frac{1+2b}{(1+\eta'^2)} = \frac{2f}{\pi} e^{a/3} \frac{(1+2b)\,cE_0}{c^2 E_n{}^2 + 4(\nu-\nu_0)^2} \\ &= \frac{f}{\pi} e^{a/3} \frac{(1+2b)\,(\delta/2)}{(\nu-\nu_0)^2 + (\delta/2)^2} \end{aligned}\right\} \tag{4.127}$$

Holtsmark calculated b for four monatomic and four diatomic gases on the basis of Debye's work[20] and obtained values of $b < 0.0020$ in all but one instance.

4.13. A simplified version of the Holtsmark theory[211]

In obtaining Eq. (4.82) we supposed the field E to split the spectral line into an intensity distribution $I(E, \nu)$. Let us now suppose this field simply to move the spectral line a frequency interval $\Delta\omega$. Then the spectral line is shifted to various frequencies with varying probabilities, and a broadened spectral line results. The probabilities or intensities in the line are directly related to the field probabilities as given by the Holtsmark function $W(\beta)$.

In the case of the linear Stark effect $\Delta\omega/\Delta\omega_0 = E/E_0 = \beta$ where $\Delta\omega_0$ is the angular frequency shift due to E_0, and one obtains:

$$I(\Delta\omega)\,\mathrm{d}(\Delta\omega) = W(\beta)\,\mathrm{d}\beta$$

so that the intensity is:

$$I(\Delta\omega) = W(\Delta\omega/\Delta\omega_0)\,(1/\Delta\omega_0) \tag{4.128}$$

For the quadratic Stark effect $\sqrt{\Delta\omega/\Delta\omega_0} = E/E_0 = \beta$ so that:

$$I(\Delta\omega) = W\left(\sqrt{\Delta\omega/\Delta\omega_0}\right)/2\sqrt{\Delta\omega\cdot\Delta\omega_0} \tag{4.129}$$

4.14. Interaction between broadening electrons and ions

In Section 4.11 we have considered the effect which interaction between broadeners would have on the field strength probability function, but the result is only applicable to the case where the broadening field producers are dipoles. Of more general interest and of concern to us in the present section is the situation where the broadener field producers are electrons and ions. The Holtsmarkian consideration of broadener interaction really treated the broadeners as if they were hard spheres rather than as if they were themselves producers of fields. Here the opposite viewpoint is taken in that it is only the field of the perturbers which is considered.

It is to be supposed that the fields produced by the ions are shielded by the presence of the surrounding electrons and ions, that is to say, the field of a given ion only penetrates so far through the surrounding plasma, and if no emitter happens to be within the penetration range to benefit, its effect is completely shielded out. It is rather generally conceded that the Debye–Hückel potential will yield a good expression for the shielded Coulomb field:

$$\mathbf{E} = \frac{Z e\,\mathbf{r}}{r^2}\left[\frac{1}{r} + \frac{1}{D}\right]\mathrm{e}^{-r/D} \tag{4.130a}$$

$$D = \sqrt{\frac{k\,T}{4\pi N e^2 (1 + Z^2)}} \tag{4.130b}$$

Ecker[38–40] originally chose a simplified version of Eqs. (4.130)

$$\mathbf{E} = \begin{cases} \dfrac{Z e^2\,\mathbf{r}}{r^3} & \text{if} \quad r < D \\ 0 & \text{if} \quad r \geq D \end{cases} \tag{4.131}$$

thus choosing the Debye length as the long range cut-off. That author then applied machine calculations to the simplified Holtsmark Theory of the previous section to obtain the results indicated in Fig. 4.7. In this figure the parameter is $\zeta = \frac{4\pi D^3}{3} N$. It is apparent then that the parameter tells us how many

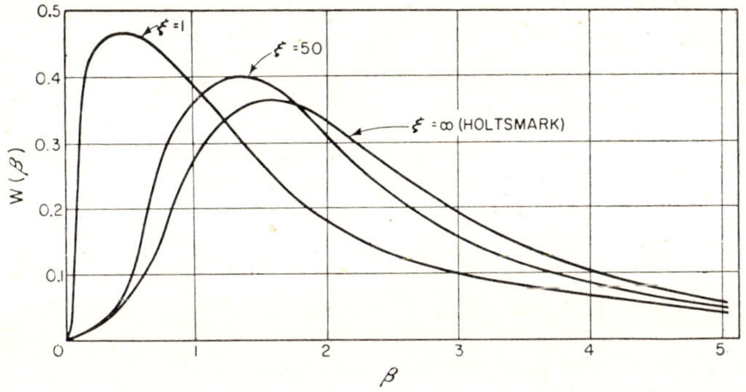

FIG. 4.7. The field strength probability function for various values of the shielding parameter $\xi = (4\pi D^3/3)N$. (After Margenau and Lewis[219])

broadening ions there are in a sphere about the emitter of radius the Debye length. When the Debye length goes to infinity then, of course, there is no shielding and the Holtsmark unshielded result is obtained.

Hoffman and Theimer[78] have carried out the calculation without the introduction of the cut-off, but a later calculation by Ecker using the correct potential leads to just about the same results as given in Fig. 4.7.

Our considerations above have modified the broadening electric field at large separation. Margenau[142] has carried out a treatment which has shown the short range modifying effects of atomic screening. This occurs due to the rare and very close approaches and results in the potential going to a finite value for zero separation. The result is a predicted flattening of the Holtsmark curve at very large separations from line center with a β^{-1} frequency dependence instead of a $\beta^{-5/2}$.

We shall consider this screening effect in somewhat more detail in Section 4.27 where the phenomenon enters into the original theory development.

4.15. The Stark effect in parabolic coordinates

The Holtsmark theory is, of course, a classical study of ionic Stark broadening, and for such broadening it generally suffices. As one would imagine, however, the problem has been treated quantum mechanically, and it is to Spitzer's quantum treatment that we shall turn our attention.

Before doing so, however, let us consider the Stark broadening of the spectral lines of hydrogen from a quantum mechanical point of view. In doing this we shall very rapidly sketch the theory as presented by Epstein[42] and Bethe[9].

To begin with we suppose the electron to be moving in the field of its nucleus of charge e under the influence of the potential $-e^2/r$. A homogeneous field of strength E is imposed along the z-axis resulting in the addition of $-e\,E\,z$ to the potential. Finally, we transform to parabolic coordinates by utilizing the relations:

$$\left.\begin{aligned} &x = \sqrt{\xi\,\eta}\cos\varphi; \quad y = \sqrt{\xi\,\eta}\sin\varphi; \quad z = \frac{\eta-\xi}{2} \\ &0 \le \xi \le \infty; \quad 0 \le \eta \le \infty; \quad 0 \le \varphi \le 2\pi \end{aligned}\right\} \quad (4.132)$$

If we utilize Eq. (4.132) and the potential energy expression which we have obtained, the Schrödinger equation may be written as:

$$\left.\begin{aligned} &\frac{\partial}{\partial\eta}\left(\eta\frac{\partial\psi}{\partial\eta}\right) + \frac{\partial}{\partial\xi}\left(\xi\frac{\partial\psi}{\partial\xi}\right) + \frac{1}{4}\left(\frac{1}{\eta}+\frac{1}{\xi}\right)\frac{\partial^2\psi}{\partial\varphi^2} + \frac{\mu}{2\,\hbar^2} \\ &\times\left[E_n(\eta+\xi) + 2\,e^2 - \frac{e\,E\,(\eta^2-\xi^2)}{2}\right]\psi = 0 \end{aligned}\right\} \quad (4.133)$$

This equation must be solved by the methods of perturbation theory, and, since the carrying out of the solution involves large quantities of mathematical spadework which do not directly contribute to our line broadening considerations, we shall simply give the first and second order energy results. For other information concerning the Stark effect which we shall utilize we may treat the problem in more general terms.

The zeroth-order solution of Eq. (4.133) simply yields the unperturbed energies of the hydrogen atom. The first-order solution yields the energy perturbation of the linear Stark effect:

$$E^{(1)} = \frac{3\,\hbar^2\,E}{2\,\mu\,e}\,n\,(k_2 - k_1) \quad (4.134\text{a})$$

The second-order solution results in the energy perturbation of the quadratic Stark effect:

$$E^{(2)} = -\frac{E^2}{16\,\mu^2}\left(\frac{\hbar}{e}\right)^6 n^4\,[17\,n^2 - 3\,(k_2-k_1) - 9\,m^2 + 19] \quad (4.134\text{b})$$

In Eqs. (4.134) $n = k_1 + k_2 + m$ is the total quantum number, while k_1, k_2, and m are the quantum numbers associated with the parabolic coordinates ξ, η, and φ respectively. We might note that φ corresponds exactly to the φ of spherical polar coordinates. This means that the quantum number m is the magnetic quantum number specifying angular momentum about the atomic figure axis in parabolic as well as in polar coordinates. As a consequence, the selection rules for the m of parabolic coordinates will correspond to those for the m of polar.

The eigenfunctions of hydrogen in parabolic coordinates will be written down at a later point when their utilization will be required.

4.16. General treatment of the Stark effect

Now let us consider the problem in a more general form. To begin with the imposition of the external electric field, which is the Stark effect producer, will present us with a problem in perturbation theory if we consider the case of the isolated hydrogen atom as the unperturbed problem. The perturbation introduced will be $e\,E\,r\cos\vartheta$, if, as in the last section, we take the field along the z-axis and set up the problem in spherical polar coordinates. Let us assume our familiarity with the solution of this unperturbed problem in these coordinates and specify the solution by $\psi_{n\,l\,m}$. The symbols n, l, and m we take to indicate the principal, the orbital angular momentum, and the magnetic quantum numbers, respectively, and, for convenience of notation, we further suppose the numbers n and l to be represented by n and the number m by j. Now a different energy eigenvalue corresponds to each eigenfunction going with a different value of n. For a given value of n, however, there are a set of eigenfunctions corresponding to the various possible values of j to which set there corresponds but one energy eigenvalue, namely, E_n. This means that the energy level of energy E_n is degenerate. The perturbation, which is to be introduced, will tend to remove this degeneracy.

In zeroth-order there are, say, i eigenfunctions which satisfy the Schrödinger equation for energy E_n. It follows then that the complete solution of the differential equation for the state under consideration is a linear combination of these i eigenfunctions:

$$\psi_{nj}^{(0)} = \sum_{j'} c_{nj'}\,\psi_{nj'}^{(0)} \qquad (4.135)$$

with the result:

$$H = H^0 + H' \qquad (4.136\text{a})$$

$$\psi_{nj} = \sum_{j'} c_{nj'}\,\psi_{nj'}^{(0)} + \varepsilon\,\psi_{nj}^{(1)} + \varepsilon^2\,\psi_{nj}^{(2)} + \cdots \qquad (4.136\text{b})$$

$$E_{nj} = E_n^{(0)} + \varepsilon\,E_{nj}^{(1)} + \varepsilon^2\,E_{nj}^{(2)} + \cdots \qquad (4.136\text{c})$$

Eqs. (4.136) are next utilized to obtain from $H\psi_n = E_n\psi_n$:*

$$H^0 \sum_{j'} c_{nj'}\,\psi_{nj'}^{(0)} = E_n^{(0)} \sum_{j'} c_{nj'}\,\psi_{nj'}^{(0)} \qquad (4.137\text{a})$$

$$(H^0 - E_n^{(0)})\,\psi_n^{(1)} = \sum_{j'} c_{nj'}\,(E_n^{(1)} - H)\,\psi_{nj'}^{(0)} \qquad (4.137\text{b})$$

$$H'\,\psi_{nj}^{(1)} + H^0\,\psi_{nj}^{(2)} = E_n^{(0)}\,\psi_{nj}^{(2)} + E_{nj}^{(1)}\,\psi_{nj}^{(1)} + \sum_{j'} c_{nj'}\,E_{nj}^{(2)}\,\psi_{nj'}^{(0)} \qquad (4.137\text{c})$$

We agree that the ψ_{ni} are a complete orthonormal set. This unanimity of viewpoint allows us to introduce the expansions:

$$\psi_{nj}^{(1)} = \sum_{n'j'} a_{njn'j'}\,\psi_{n'j'}^{(0)}; \quad \psi_{nj}^{(2)} = \sum_{n'j'} b_{njn'j'}\,\psi_{n'j'}^{(0)} \qquad (4.138)$$

with the result, after slight rearrangement:

$$\sum_{n'j'} a_{njn'j'}\,(H^0 - E_n^{(0)})\,\psi_{n'j'}^{(0)} = \sum_{j'} c_{nj'}\,(E_{nj}^{(1)} - H')\,\psi_{nj'}^{(0)} \qquad (4.137\text{b}')$$

$$\sum_{n'j'} b_{njn'j'}\,(H^0 - E_n^{(0)})\,\psi_{n'j'}^{(0)} + \sum_{n'j'} a_{njn'j'}\,(H' - E_{nj}^{(1)})\,\psi_{n'j'}^{(0)} = \sum_{j'} c_{nj'}\,E_{nj}^{(2)}\,\psi_{nj'}^{(0)} \qquad (4.137\text{c}')$$

* For the simpler non-degenerate case see *supra*, Chap. 3.

If Eq. (4.137b′) is multiplied through on the left by $\bar{\psi}_{nj}^{(0)}$ and integrated over all space, the result is:

$$\sum_{n'j'} a_{njn'j'} \int \bar{\psi}_{nj}^{(0)} (E_{n'}^{(0)} - E_{n}^{(0)})\, \psi_{n'j'}^{(0)}\, d\tau = \sum_{j'} c_{nj'} \int \bar{\psi}_{nj}^{(0)} (E_{nj}^{(1)} - H')\, \bar{\psi}_{nj'}^{(0)}\, d\tau \qquad (4.138)$$

according to Eq. (4.137a). Orthonormality decrees:*

$$\bar{\psi}_{nj}^{(0)}\, \psi_{n'j'}^{(0)}\, d\tau = \delta_{nn'}\, \delta_{jj'} \qquad (4.139)$$

Thus, if $n \neq n'$, or $j \neq j'$ or if both inequalities hold, the left side of Eq. (4.139) goes to zero. If, on the other hand, $n = n'$, it is apparent that the difference $E_{n'}^{(0)} - E_{n}^{(0)}$ is zero. Under any condition then:

$$\sum_{j'} c_{nj'} \int \bar{\psi}_{nj}^{(0)} (E_{nj}^{(1)} - H')\, \psi_{nj'}^{(0)}\, d\tau = 0 \qquad (4.140)$$

which, according to Eq. (4.139) further reduces to:

$$\sum_{j'} c_{nj'} (\delta_{jj'}\, E_{nj}^{(1)} - H'_{jj'}) = 0 \qquad (4.141)$$

by virtue of the fact that $\int \bar{\psi}_{nj}^{(0)} E_{nj}^{(1)} \psi_{nj'}^{(0)}\, d\tau = E_{nj}^{(1)} \int \bar{\psi}_{nj}^{(0)} \psi_{nj'}^{(0)}\, d\tau$ and where we have introduced the matrix element:

$$H'_{jj'} = \int \bar{\psi}_{nj}^{(0)} H' \psi_{nj'}^{(0)}\, d\tau$$

Eq. (4.141) yields a set of equations from which, in theory at least, the c_{nj} may be obtained.

Let us consider the matrix of $E_{nj}^{(1)} - H'_{jj'}$ for hydrogen:

TABLE 4.1

nlm \ nlm	(100)	(200)	(210)	(211)	(21-1)	...
(100)	$\mathbf{E}^{(1)}$	0	0	0	0	...
(200)	0	$\mathbf{E}^{(1)}$	$-\mathbf{H}_{23}'$	0	0	...
(210)	0	$-\mathbf{H}_{23}'$	$\mathbf{E}^{(1)}$	0	0	...
(211)	0	0	0	$\mathbf{E}^{(1)}$	0	...
(21-1)	0	0	0	0	$\mathbf{E}^{(1)}$	...
.	.	.	.	.	.	
.	.	.	.	.	.	
.	.	.	.	.	.	

First we may consider the reason for the disappearance of the various matrix elements. To begin with we have forbidden† the appearance of the matrix ele-

* δ_{nn} is the Kronecker delta of definition $\delta_{nn'} = \begin{matrix} 0 \text{ for } n \neq n' \\ 1 \text{ for } n = n'. \end{matrix}$

† Failure to forbid this leads to the same result, i. e., in first order levels of different n are still separated by the amount of zeroth-order.

ments between states of different n in obtaining Eq. (4.138). Next the angle φ does not appear in H' so that integrals of the form $\int_0^{2\pi} e^{i(m-m')\varphi}\, d\varphi$ will cause H' to disappear for $m \neq m'$. Finally the presence of $\cos\vartheta$ in H' is responsible through $\int_0^{\pi} \Theta_l(\vartheta)\,\Theta_{l'}(\vartheta)\cos\vartheta \cdot \sin\vartheta\, d\vartheta$ for the disappearance of H' except for $l' = l \pm 1$ by virtue of the orthogonality of the associated Legendre functions.

The matrix tells us several important facts about the Stark effect. To begin with we see that the linear Stark effect has no "effect" on the energy of the ground state. Secondly, the matrix elements for $m \neq m'$ disappear. Finally, a twofold degeneracy corresponding to $m = \pm 1$ still exists in the state with $n = 2$. We might remark that for $n = 3$ the states with $m = \pm 1$ have the same energy, as do the states with $m = \pm 2$, and so on for the higher values of n.

Now let us return to Eq. (4.137b'), multiply through on the left by ψ_{ml} and integrate over all space where we take $n' = m \neq n$ and $l = j'$. The result is:

$$a_{njn'j'} = \frac{\sum_{j''} c_{nj''} H'_{n'j'nj''}}{(E_{n'}^{(0)} - E_n^{(0)})}\,; \quad n' \neq n \tag{4.142}$$

If Eq. (4.137c) is multiplied through on the left by $\bar{\psi}_{nj}^{(0)}$ and integrated over all space, the result is:

$$c_{nj} E_{nj}^{(2)} = \sum_{n'j'} a_{njn'j'} \int \bar{\psi}_{nj}^{(0)} H' \psi_{n'j'}^{(0)}\, d\tau \tag{4.143}$$

From Eqs. (4.142) and (4.143) we obtain:

$$\left.\begin{aligned} E_{nj}^{(2)} &= -\sum_{n'j'} \frac{\sum_{j''} c_{nj''} H'_{n'j'nj''}}{c_{nj}\,(E_{n'}^{(0)} - E_n^{(0)})} H'_{njn'j'} \\ &= -E^2 \sum_{n'j'} \frac{\sum_{n'j'} c_{n'j'} (n'j'\,|er\cos\vartheta|\,nj'')}{c_{nj}(E_{n'}^{(0)} - E_n^{(0)})} (nj\,|er\cos\vartheta|\,n'j') \end{aligned}\right\} \tag{4.144*}$$

Eq. (4.144) tells us the second order energy of our perturbed degenerate system. We may glean a few factors of importance from this equation. First it may be noted that this energy correction is proportional to the square of the electric field, or a quadratic Stark effect has been obtained. Next, matrix elements of H' are still non-vanishing only for $\Delta l = \pm 1$ and $\Delta m = 0$. Finally, a fact which will be of some importance later is that now the matrix elements for changes in the principal quantum number will not disappear.

4.17. Quantum ionic Stark broadening for adiabatic Case I[189]

We suppose the collision potential in this case to change very slowly, sufficiently so that adiabaticity is assured.

* The alternate expression for the matrix element $(a\,|z|\,b) = \int \bar{\psi}_a z \psi_b\, d\tau$ is here introduced.

Let us immediately establish the three basic approximations: (1) The emitter exists in a homogeneous field of strength $Z\,e^2/r$. Since the field is produced by an ion, the homogeneity assumption means we will have to keep it far enough from the emitter so that the assumption is reasonable. Spitzer[189] settled on $R > 8.0 \times 10^{-8}\,\pi^2$ where n is the principal quantum number. (2) The collisions are binary, (this is certainly a familiar proposition). (3) The mass of the colliding particles is very large. This should restrict us to ions.

In the state A the molecule has energy $E_A(t)$ and there are no photons in the radiation field so that the unperturbed eigenfunction for the system molecule plus field is:

$$\Psi_a^{(0)} = \psi_a^{(0)}\,e^{-\frac{i}{\hbar}\int_0^t E_A(t)\,dt} \tag{4.145a}$$

In the state B the molecule has energy $E_B(t)$ and a photon of frequency ν_i, is present in the field. The unperturbed eigenfunction is:

$$\Psi_{b_i}^{(0)} = \psi_{b_i}^{(0)}\,e^{-\frac{i}{\hbar}\int (E_B(t)+h\nu_i)\,dt} \tag{4.145b}$$

If Eqs. (4.145) are substituted into the time dependent form of the Schrödinger equation:

$$H\,\psi = i\,\hbar\,\frac{\partial}{\partial t}\,\psi \tag{4.146}$$

the more familiar form of this equation results:

$$H\,\psi_a^{(0)} = E_A(t)\,\psi_A^{(0)}(t) \tag{4.147a}$$

$$H\,\psi_{b_i}^{(0)} = (E_B(t) + h\,\nu_i)\,\psi_{b_i}^{(0)}(t) \tag{4.147b}$$

In Eqs. (4.147) t is a parameter, and $E_A(t)$ and $E_B(t)$ are the molecular level energies *as perturbed by the linear Stark effect of the ionic field.*

We now wish to introduce the molecule-field interaction and ascertain the solution to Eq. (4.146) under the influence of this perturbation. The result of this perturbation will be to smear out the probability of finding the molecule-field system in state $\Psi_a^{(0)}$, $\Psi_{b_1}^{(0)}$, $\Psi_{b_2}^{(0)}$, etc. This prognostication leads us to assume a solution of the perturbed problem of the form:

$$\psi_n = a(t)\,\Psi_a^{(0)} + \sum_i b_i(t)\,\Psi_{b_i}^{(0)} \tag{4.148a}$$

Eq. (4.147) tells us that, if the system is initially unperturbed and subsequently perturbed for a time t, the probability that the system is in the state $\Psi_a^{(0)}$ is $|a(t)|^2$, in the state $\Psi_{b_i}^{(0)}$ is $|b_i(t)|^2$, and so on. Thus, these coefficients may well be dubbed "state growth coefficients", for this they are. Finally then, the solution of the Schrödinger equation amounts to nothing more nor less than the procuring of these coefficients. Since $\left|a(t)\exp\left(\frac{i}{\hbar}\int_0^t E_A(t)\,dt\right)\right|^2 = |a(t)|^2$ and $\left|b_i(t)\exp\left[\frac{i}{\hbar}\int_0^t (E_B(t)+h\nu_i)\,dt\right]\right|^2 = |b_i(t)|^2$, we prefer to write Eq. (4.146) in the equivalent form:

$$\psi_n = a(t)\, e^{\frac{i}{\hbar}\int E_A(t)\,dt}\, \Psi_a^{(0)} + \sum_i b_i(t)\, e^{\frac{i}{\hbar}\int (E_B(t) + h\nu_i)\,dt}\, \Psi_{b_i}^{(0)} \qquad (4.148\text{b})$$

If the molecule-field interaction Hamiltonian is taken as S, Eq. (4.148b) may be substituted into Eq. (4.146) with the result:

$$a(H^0 + S)\, e^{\frac{i}{\hbar}\int E_A dt}\, \Psi_a^{(0)} + \sum_i b_i (H^0 + S)\, e^{\frac{i}{\hbar}\int (E_B + h\nu_i)\,dt}\, \Psi_{b_i}^{(0)} = i\hbar \dot{a}\, e^{\frac{i}{\hbar}\int E_A dt}\, \Psi_a^{(0)}$$

$$- E_A\, a\, e^{\frac{i}{\hbar}\int E_A dt}\, \Psi_a^{(0)} + i\hbar a\, e^{\frac{i}{\hbar}\int E_A dt}\, \frac{\partial}{\partial t}\Psi_a^{(0)} + i\hbar \sum_i \dot{b}_i\, e^{-\frac{i}{\hbar}\int (E_B + h\nu_i)\,dt}\, \Psi_{b_i}^{(0)}$$

$$- \sum_i (E_B + h\nu_i)\, b_i\, \Psi_{b_i}^{(0)}\, e^{-\frac{i}{\hbar}\int (E_B + h\nu_i)\,dt} + i\hbar \sum_i b_i\, e^{-\frac{i}{\hbar}\int (E_B + h\nu_i)\,dt}\, \Psi_{b_i}^{(0)} \qquad (4.149)$$

Certain terms drop out, since $H^0 \Psi_r = i\hbar \frac{\partial}{\partial t} \Psi_r$. In order to determine $a(t)$, let us then multiply through on the left by $\overline{\Psi}_a^{(0)}$ and integrate over all space to obtain:

$$a\, S_{aa}\, e^{\frac{i}{\hbar}\int E_A dt} + \sum b_i\, S_{a b_i}\, e^{\frac{i}{\hbar}\int E_A dt} = i\hbar \dot{a}\, e^{\frac{i}{\hbar}\int E_A dt} - E_A\, a\, e^{\frac{i}{\hbar}\int E_A dt} \qquad (4.150)$$

The diagonal matrix elements, such as S_{aa}, of the field-molecule interaction disappear.* Eq. (4.150) then becomes:

$$i\hbar \dot{a} = E_A(t)\, a(t) + \sum_i S_{ab_i}\, b_i(t) \qquad (4.151\text{a})$$

The multiplication of Eq. (4.149) by $\overline{\Psi}_{b_i}^{(0)}$ in like manner results in:

$$i\hbar \dot{b}_i = \{E_B(t) + h\nu_i\}\, b_i(t) + S_{b_i a}\, a(t) \qquad (4.151\text{b})$$

Eqs. (4.151) are the so-called "state growth equations", since they tell us the manner in which the state probabilities $|a(t)|^2$ and $|b_i(t)|^2$ change with time.

Let us assume that at time $t = 0$, the atom is in the state A with no radiation present in the field. Thus, $|a(0)|^2$ is unity, and the $|b_i(0)|^2$ are zero. These requirements and Eq. (4.151a) are satisfied by:

$$a(t) = e^{-(1/2)\Gamma t} \exp\left\{-\frac{i}{\hbar}\int_0^t E_A(\tau)\, d\tau\right\} \qquad (4.152)$$

Now let us consider Eq. (4.151b). From the fact that the right side of this equation is a sum of two terms we may infer that $b_i(t)$ is a product of two functions. Hence:

$$b_i(t) = f(t)\, g(t) \qquad (4.153)$$

* See *supra*, Sec. 4.12.

From Eqs. (4.151b) and (4.152) we obtain:

$$g(t)\frac{\mathrm{d}f(t)}{\mathrm{d}t} = \frac{1}{i\hbar}\{E_B(t) + \hbar\nu_i\}\, g(t)\, f(t) \tag{4.154a}$$

$$f(t)\frac{\mathrm{d}g(t)}{\mathrm{d}t} = H_{b_i a}\, a(t)\frac{1}{i\hbar} \tag{4.155a}$$

The solution of Eq. (4.154a) is:

$$f(t) = \exp\left\{-2\pi i\nu_i t - \frac{1}{\hbar}\int_0^t E_B(\tau)\,\mathrm{d}\tau\right\} \tag{4.154b}$$

If we now substitute Eqs. (4.152) and (4.154b) into Eq. (4.155a), we may solve the resulting equation to obtain:

$$g(t) = -\frac{i S_{b_i a}}{\hbar}\int^t \mathrm{e}^{-(1/2)\Gamma T}\mathrm{d}T \exp\left\{2\pi i\nu_i T + \frac{i}{\hbar}\int_0^T \big(E_B(\tau) - E_A(\tau)\big)\,\mathrm{d}\tau\right\} \tag{4.155b}$$

Eqs. (4.154b) and (4.155b) may now be substituted into Eq. (4.153) to yield:

$$\left.\begin{aligned} b_i(t) &= -\frac{i S_{b_i a}}{\hbar}\exp\left\{-2\pi i\nu_i t - \frac{i}{\hbar}\int_0^t E_B(\tau)\,\mathrm{d}\tau\right\} \\ &\quad\times\int_0^t \mathrm{e}^{-(1/2)\Gamma T}\,\mathrm{d}T \exp\left\{2\pi i\nu_i T + \frac{i}{\hbar}\int_0^T \big(E_B(\tau) - E_A(\tau)\big)\,\mathrm{d}\tau\right\}\end{aligned}\right\} \tag{4.156}$$

By time $t = \infty$ the emitter will certainly be in the ground state, and a photon of frequency ν_i will be in the field. The intensity distribution in the spectral line is surely given by $|b_{i(\nu)}(\infty)|^2$, since this will represent the probabilities for the appearance of these various frequencies. If we call $I'(\nu)$ the intensity of the frequency ν there results:

$$I'(\nu) = \frac{(S_{b_i a})^2}{\hbar^2}\left|\mathrm{e}^{-(1/2)\Gamma T}\,\mathrm{d}t \exp\left\{2\pi i\nu_i T + \frac{i}{\hbar}\int_0^T \big(E_B(\tau) - E_A(\tau)\big)\,\mathrm{d}\tau\right\}\right|^2 \tag{4.157}$$

If we let E_{A0} and E_{B0} be the unperturbed energies of the A and B states respectively, we may utilize the equations:

$$\Delta_A(t) = \{E_A(t) - E_{A0}\}\,\hbar \tag{4.158a}$$

$$\Delta_B(t) = \{E_B(t) - E_{B0}\}\,\hbar \tag{4.158b}$$

$$x = 2\pi(\nu - \nu_{AB}) \tag{4.158c}$$

to transform Eq. (4.157) to:

$$I(x) = \frac{\Gamma}{2\pi}\left|\int_0^\infty \mathrm{e}^{i(x+1/2i\Gamma)T}\,\mathrm{d}T \exp\left\{-i\int_0^T \big(\Delta_A(\tau) - \Delta_B(\tau)\big)\,\mathrm{d}\tau\right\}\right|^2 \tag{4.159}$$

where $I(x)\,\mathrm{d}x$ over all x has been normalized to unity.

From Eq. (4.134a):

$$\Delta_A(t) = \frac{3\hbar n_A}{2me}E(k_{2A} - k_{1A}) \tag{4.160}$$

We have supposed the field E to be produced by an ion of, say, charge Ze at distance $r(t)$. We assume straight paths for the perturbing ions, and we now let v be the ionic velocity, R the distance of closest approach and t_0 the time of closest approach. The results:

$$E = \frac{Ze}{r^2(t)} \tag{4.161a}$$

$$r^2(t) = R^2 + v^2(t - t_0)^2 \tag{4.161b}$$

$$\Delta_A(t) - \Delta_B(t) = \frac{g/\hbar}{R^2 + v^2(t - t_0)^2} \tag{4.161c}$$

$$g/\hbar = 3hgZ/2m = 1.73gZ \tag{4.161d}$$

$$g = n_A(k_{2A} - k_{1A}) - n_B(k_{2B} - k_{1B}) \tag{4.161e}$$

Eqs. (4.161c) may be integrated as follows:

$$\int_{t_0}^{T} \{\Delta_A(\tau) - \Delta_B(\tau)\}\, d\tau = \frac{g}{\hbar R v} \tan^{-1} \frac{v(T - t_0)}{R} \tag{4.162}$$

Now surely if we allow T in Eq. (4.162) to have the limits $-\infty$ and $+\infty$, we will obtain the total phase shift of the emitted radiation due to a collision. The result of taking these limits is, of course, a shift of $\pi \Upsilon$ where:

$$\Upsilon = \frac{g}{\hbar R v} \tag{4.163}$$

It may be noted that in Eq. (4.159) we desire the limits 0 and T on the integral in the exponential, whereas, we have taken the limits t_0 and T in Eq. (4.162). A reasonable approximation is to assume that between 0 and t_0 one half of our total phase shift will occur so that if we subtract $1/2\, \pi\Upsilon$ from Eq. (4.162) we shall obtain the desired result for Eq. (4.159). Let us now substitute Eq. (4.162) into Eq. (4.159) and integrate by parts to obtain:

$$I(x) = \frac{\Gamma}{2\pi} \frac{1}{\left(x^2 + \left(\frac{1}{2}\Gamma\right)^2\right)} \left| \left\{ e^{i(x + (1/2) i\Gamma) T} \exp\left[-i\Upsilon \tan^{-1} \frac{v(T - t_0)}{R}\right] \right|_0^\infty \right. \tag{4.164a}$$

$$\left. + e^{-1/2 i\pi\Upsilon} \frac{v\Upsilon}{R} \int_0^\infty \frac{1}{1 + \frac{v^2(T - t_0)^2}{R^2}} e^{i(x + (1/2) i\Gamma) T} \exp\left[-i\Upsilon \tan^{-1} \frac{v(T - t_0)}{R}\right] dT \right|^2$$

Since:

$$e^{i(x + (1/2) i\Gamma) T} \exp\left[-i\Upsilon \tan^{-1} \frac{v(T - t_0)}{R}\right]\Bigg|_0^\infty$$

$$= e^{-1/2\Gamma\infty}\, e^{ix\infty}\, e^{-i\gamma\pi/2} - e^0\, e^{i\gamma\pi} = 0 - 1$$

Eq. (4.164a) becomes:

$$I(x) = \frac{\Gamma}{2\pi} \frac{1}{\left[x^2 + \left(\frac{1}{2}\Gamma\right)^2\right]} \tag{4.164b}$$
$$\times \left| i + \frac{\Upsilon v}{R} e^{-(1/2) i \pi \gamma} \int_0^\infty \frac{e^{i(x+(1/2)i\Gamma)T} \exp\left[-i\Upsilon \tan^{-1} \frac{v(T-t_0)}{R}\right]}{1 + \frac{v^2 (T-t_0)^2}{R^2}} dT \right|^2$$

Now let us take a factor $\exp(x + 1/2\, i\Gamma)\, t_0$ out of the integral and partially complete the absolute square to obtain:

$$I(x) = \frac{\Gamma}{2\pi\left[x^2 + \left(\frac{1}{2}\Gamma\right)^2\right]} \tag{4.164c}$$
$$\times \left\{1 + e^{-\Gamma t_0} \frac{\Upsilon^2 v^2}{R^2} \left| \int_{-t_0}^\infty \frac{e^{i(x+(1/2)i\Gamma)t} \exp\left[-i\Upsilon \tan^{-1} \frac{v\tau}{R}\right] d\tau}{1 + \frac{v^2\tau^2}{R^2}} \right|^2 \right\}$$

where $T = \tau + t_0$. Suppose x is much different from Γ. When such is the case the cross product in Eq. (4.164b) will be removed by the averaging process. Now let us extend the lower limit to $-\infty$, — this will have little effect — assume $x \gg \Gamma$ and neglect the damping factor under the integral to yield:

$$I(x) = \frac{\Gamma}{2\pi x^2} \left\{1 + e^{-\Gamma t_0} \frac{\Upsilon^2 v^2}{R^2} \left| \int_{-\infty}^{+\infty} \frac{e^{ix\tau} \exp\left[-i\Upsilon \tan^{-1} \frac{v\tau}{R}\right]}{1 + \frac{v^2\tau^2}{R^2}} d\tau \right| \right\} \tag{4.164d}$$

We now let:

$$u = \frac{v\tau}{R} \tag{4.165a}$$

$$\xi = \frac{xR}{v} \tag{4.165b}$$

and integrate Eq. (4.164d) over t_0 to obtain, if $\Omega(\Upsilon, \xi) = v/R$ is the number of collisions per second:

$$I(x) = \frac{\Gamma}{2\pi x^2} \left\{1 + \frac{\Omega(\Upsilon, \xi)}{\Gamma} \Upsilon^2 f_\Upsilon^2(\xi)\right\} \tag{4.166}$$

where

$$f_\Upsilon(\xi) = \int_{-\infty}^{+\infty} \frac{e^{i(\xi u - \gamma \tan^{-1} u)}}{1 + u^2} du \tag{4.167}$$

Let us first consider the result of the two limiting cases (1) ξ very large and (2) ξ approaches zero. In order to consider Case (1) let us integrate Eq. (4.167) by parts:

$$\lim_{\xi\to\infty} f_\Upsilon(\xi) = \lim_{\xi\to\infty}\left\{\frac{e^{-i\Upsilon\tan^{-1}u}\, e^{i\xi u}}{i\xi}\Bigg|_{-\infty}^{+\infty} + \frac{\Upsilon}{\xi}\int_{-\infty}^{+\infty}\frac{e^{i\xi u}\, e^{-i\Upsilon\tan^{-1}u}}{(1+u^2)^2}\,du\right\} = 0$$

From this result it is apparent that $f_\Upsilon(\xi)$ goes to zero for sufficiently large ξ, and, as a result, the line profile is independent of the number of collisions for x large enough. For Case (2) straight integration for $\xi = 0$ yields $\Upsilon f_\Upsilon(0) = 2\sin\Upsilon_\pi/2$. Thus $\Upsilon^2 f_\Upsilon{}^2(0) = 4\sin^2\Upsilon_\pi/2$ and the average value of this expression is, of course, two. If $\Omega_{1/\pi}$ is the total number of collisions for which the phase shift is greater than unity or $\Upsilon > 1/\pi$ as required by the Weisskopf theory, we obtain, by neglecting the unity in Eq. (4.166), the interruption line shape for ξ small.

$$I(x) = \frac{\Omega_{1/\pi}}{\pi x^2} \tag{4.168}$$

These are the two limiting results for Eq. (4.166), but Eq. (4.167) may be integrated, if rather laboriously. To begin with:

$$\begin{aligned} e^{-i\tan^{-1}u} &= \cos(\tan^{-1}u) - i\sin(\tan^{-1}u) \\ &= \cos[\tan^{-1}u]\,[1 - iu] = \frac{(1-iu)}{(1+u^2)^{1/2}} \end{aligned} \tag{4.169}$$

Substituting Eq. (4.169) into Eq. (4.167) and expanding $(1 - iu)^\Upsilon$ according to the binomial theorem we obtain:

$$f_\Upsilon(\xi) = \int_{-\infty}^{+\infty}\frac{e^{iu\xi}}{(1+u^2)^{1+(1/2)\Upsilon}}\,du\left[1 - i\Upsilon u + \frac{\Upsilon(\Upsilon-1)(-iu)^2}{2!} + \cdots + (-iu)^\Upsilon\right] \tag{4.170}$$

Now since the first integral in Eq. (4.170) is an even function, we may surely write:

$$\int_{-\infty}^{+\infty}\frac{e^{iu\xi}}{(1+u^2)^{1+(1/2)\Upsilon}}\,du = 2\int_0^\infty\frac{\cos u\xi}{(1+u^2)^{1+1/2\Upsilon}}\,du$$

Basset has shown that:*

$$K_l(xz) = \frac{\Gamma\left(l+\frac{1}{2}\right)(2z)^l}{x^l\,\Gamma\left(\frac{1}{2}\right)}\int_0^\infty\frac{\cos xu\,du}{(u^2+z^2)^{l+1/2}} \tag{4.171a}$$

* See also reference 240.

where $K_l(x\,z)$ is the Bessel function of the second kind. If we let $g\Upsilon(\xi)$ be the first integral in Eq. (4.170), we obtain:

$$g\,\Upsilon(\xi) = \frac{2^{1/2(1-\Upsilon)}\sqrt{\pi}}{\Gamma\left(\frac{1}{2}\Upsilon+1\right)}\,\xi^{1/2(1+\Upsilon)}K_{1/2(1+\Upsilon)}(\xi) \tag{4.171b}$$

Eq. (4.170) then becomes:

$$f_\Upsilon(\xi) = g(\xi) \mp \Upsilon\frac{d}{d\xi}g_\Upsilon(\xi) + \frac{\Upsilon(\Upsilon-1)}{2}\frac{d}{d\xi^2}g_\Upsilon(\xi) \mp \cdots \tag{4.172}$$

If in the expression for the hypergeometric function $F(a, b, c; z)$ we set $b\,z = x$ and allow b to approach infinity, we obtain the confluent hypergeometric function (series):[231]

$${}_1F_1(\alpha, \gamma, x) = 1 + \frac{\alpha}{\gamma}x + \frac{\alpha(\alpha+1)}{2!\,\gamma(\gamma+1)}x^2 + \cdots$$

Now if Υ is an even positive integer, we may use an expression due to Oltramare:†[163]

$$\int_0^\infty \frac{\cos x\,u}{(u^2+z^2)^n}du = \frac{(-1)^{n-1}\pi}{z^{2n-1}(n-1!)}\left[\frac{d^{n-1}\,e^{-xzp}}{dp^{n-1}(1+p)^n}\right]_{\gamma=1}$$

to obtain for Eq. (4.170):

$$f_{2\alpha}(\xi) = 2\pi(-1)^{\alpha-1}\,\xi\,e^{-\xi}\,{}_1F_1(1-\alpha, 2, 2\xi) \tag{4.173}$$

An inspection of Eq. (4.170) for $f_\Upsilon(\xi)$ is sufficient to show that the integral of $f_\Upsilon{}^2(\xi)$ over ξ is independent of Υ. Now surely:

$$\int_{-\infty}^{+\infty} f_\Upsilon{}^2(\xi)\,d\xi = \iiint_{-\infty}^{+\infty}\frac{e^{i\xi(u-u')}}{(1+u^2)(1+u'^2)}du\,du'\,d\xi \tag{4.174}$$

Let us write out the Fourier transform of $f(x) = f_\Upsilon{}^2(\xi)$ as follows:

$$f(x) = \frac{1}{2\pi}\iint_{-\infty}^{+\infty} f(\zeta)\,e^{ik(x-\zeta)}\,dk\,d\zeta \tag{4.175}$$

If in Eq. (4.174) we take $f(u') = \dfrac{1}{(1+u^2)(1+u'^2)}$, then in essence we have, from Eq. (4.177), $2\pi f(u)$ integrated over u. Thus, Eq. (4.174) becomes:

$$\int_{-\infty}^{+\infty} f_\Upsilon{}^2(\xi)\,d\xi = \int_{-\infty}^{+\infty} 2\pi f(u)\,du = 2\int_{-\infty}^{+\infty}\frac{du}{(1+u^2)^2} = \pi^2 \tag{4.176}$$

This then will essentially be a normalizing factor for our spectral line.

† We could probably also use a relation obtained by Malmsten[132]

$$\int_0^\infty \frac{\cos a\,x}{(1+x^2)^{n+1}}dx \sim \frac{\pi\,e^{-n}}{2^{2n+1}n!}\left[(2a)^n + {}_nC_1(n+1)(2a)^{n-1} + {}_nC_2(n+1)(n+2)(2a)^{n-2} + \cdots\right]$$

for the same result.

Figure 4.8 gives the results of Spitzer's calculations for several values of $\Upsilon \cdot x_{max}$ is nothing more than $q/\hbar R^2$. The limiting value for $\Upsilon = \infty$ corresponds, of course, to zero velocity according to Eq. (4.163). Thus, this value of Υ should yield the line shape as given by the Statistical Theory.* Spitzer obtains the limiting value of $\Upsilon f_\Upsilon^2(\xi)$ from the Statistical theory rather than from Eq.(4.167) going back to Eq.(4.161b) to do so. The result is:

$$\lim_{\Upsilon \to \infty} \Upsilon f_\Upsilon^2(\xi) = \frac{2\pi(x-1)^{1/2}}{x_{max}^{1/2}} \quad (4.177)$$

which together with Eq. (4.166) will give the inverse 3/2 dependence on x as predicted by the Statistical Theory.

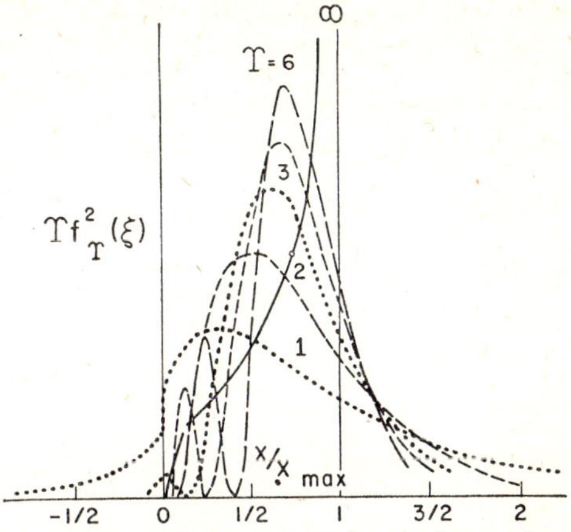

FIG. 4.8. Stark effect line shapes for various collision phase shifts. (After Spitzer[189])

In obtaining Eq. (4.166) we have assumed that the collision may be considered adiabatic, and we shall now proceed to demonstrate that the breakdown of the adiabatic hypothesis results in a change in $f_\Upsilon(\xi)$.

4.18. The effect of a diabatic assumption[189]

We may recall that the interaction between the radiation field and the atom resulted in the appearance of Γ, and, since we have assumed $x > (1/2)\Gamma$ at any rate, we neglect, with Spitzer, this interaction. For this case:†

$$i\hbar \frac{\partial \Psi}{\partial t} = H(t)\Psi \quad (4.178a)$$

where

$$\Psi = \sum_r a_r(t)\,\psi_r(t) \quad (4.178b)$$

We substitute Eq. (4.178b) into Eq. (4.178a), multiply through on the left by $\bar{\psi}_s(t)$, and integrate over all space to obtain:

$$-\dot{a}_s = i\frac{E_s(t)}{\hbar} a_s(t) + \sum_r k_{sr}\,a_r(t) \quad (4.179a)$$

since: $H(t)\psi_r(t) = E_r(t)\psi_r(t)$ instantaneously, where:

$$k_{sr}(t) = \int \bar{\psi}_s(t)\frac{\partial \psi_r(t)}{\partial t}\,d\tau \quad (4.179b)$$

* See *supra*, Chap. 3.

† The derivation carried out here was first performed in an identical fashion for this case by Guttinger[68].

We may obtain k_{sr} in slightly different form by first differentiating Eq. (4.150) to obtain:

$$\left(\frac{\partial H}{\partial t} - \frac{\partial E_r}{\partial t}\right) \psi_r = [E_r(t) - H(t)] \frac{\partial \psi_r}{\partial t} \tag{4.180a}$$

If Eq. (4.180a) is multiplied through on the left by $\bar{\psi}_s(t)$ and integrated over all space, the result is:

$$\left(\frac{\partial H}{\partial t}\right)_{sr} = [E_r(t) - E_s(t)]\, k_{sr}; \quad s \neq r \tag{4.180b}$$

where the $E_s(t)$ may be obtained due to the Hermitian* quality of $H(t)$, that is $\bar{H}_{rs} = H_{sr}$. Since $k_{ss} = 0$, Eq. (4.180b) is sufficient.

Eq. (4.161a) gives the field, and we assume that the ion is moving in the $y_e z_e$ plane where the subscript e refers to the electronic coordinates of the emitter. In addition, the ion is supposed moving parallel to the Y axis. For simplicity, let us consider Fig. 4.9.

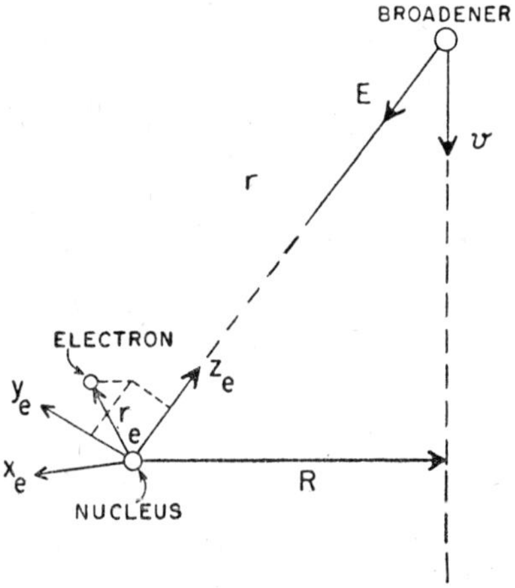

FIG. 4.9. The physical conditions for the non-adiabatic collision. The broadener and the emitter nucleus are in the $y_e z_e$ plane

If μ is the electric dipole moment of the molecule, the interaction is $\boldsymbol{\mu} \cdot \mathbf{E} = \mu_x E_x + \mu_y E_y + \mu_z E_z$. Then from Fig. 4.9, the time dependent portion of this Hamiltonian is $\mu_z E_z = -\dfrac{z_e Z e^2}{r^2(t)}$.

It is further apparent from the figure that:

$$r^2 = R^2 + v^2(t - t_0)^2 \leftrightarrow \frac{dr}{dt} = \frac{v^2(t - t_0)}{r} \tag{4.181a}$$

since R is a constant t. An inertial Y-axis is taken as shown in Fig. 4.9 and an inertial Z-axis is taken as corresponding to R. A study of the figure tells us that.

$$Y_e = z_e \sin\vartheta + y_e \cos\vartheta$$

$$Z_e = z_e \cos\vartheta - y_e \sin\vartheta$$

Differentiation of these relations, multiplication of the first by $\sin\vartheta$ and the second by $\cos\vartheta$, and addition of the resulting expressions yields:

$$\frac{dz_e}{dt} = y_e \frac{d\vartheta}{dt}$$

* We may recall that an Hermitian matrix is self adjoint, that is, $||A||^+ = ||\tilde{\bar{A}}|| = ||A||$.

but:

$$\frac{d\vartheta}{dt} = \frac{v\cos\vartheta}{r} = \frac{v\,R}{r^2}$$

so that:

$$\frac{\partial z_e}{\partial t} = \frac{\gamma_e\,v\,R}{r^2} \tag{4.181b}$$

Utilizing Eqs. (4.181a) and (4.181b) we may find:

$$\left.\begin{aligned}\frac{\partial H}{\partial t} &= -Z\,e^2\,\frac{\partial}{\partial t}\,\frac{z_e}{r^2} = \frac{2\,Z\,z_e\,e^2}{r^3}\,\frac{dr}{dt} - Z\,\frac{e^2}{r^2}\,\frac{dz_e}{dt}\\ &= \frac{Z\,e^2\,v}{r^4(t)}\,\{2\,z_e\,v\,(t-t_0) - y_e\,R\}\end{aligned}\right\} \tag{4.181c}$$

and we need only the $y_e R$ term since, we may recall that, for states of the same total quantum number, the matrix of z_e is diagonal.

From Eq. (4.178a):

$$i\,\hbar\,\frac{\partial\psi_s}{\partial t} = E_s\,\psi_s = H(t)\,\psi_s$$

so that, when we multiply through on the left by $\bar{\psi}_s$, and integrate over all space, we obtain:

$$E_s = \big(H(t)\big)_{ss} = (\boldsymbol{\mu}\cdot\mathbf{E})_{ss} = -\frac{Z\,e^2\,z_{ss}}{r^2(t)} \tag{4.182}$$

Combining Eqs. (4.180b), (4.181), and (4.182) we obtain

$$R_{sr} = \frac{v\,R}{r^2(t)}\,\frac{y_{sr}}{z_{rr}-z_{ss}} \tag{4.183}$$

Now letting,

$$\{v^2(t-t_0)^2 + R^2\}\,dw = vR\,dt \tag{4.184a}$$

$$w = \tan^{-1}\frac{v(t-t_0)}{R} \tag{4.184b}$$

in Eq. (4.179a), and substituting for $r^2(t)$ from Eq. (4.182) we obtain:

$$\frac{da_s}{dw} + i\,\Upsilon_{ss}\,a_s + \sum_{r\neq s}\frac{y_{sr}}{z_{rr}-z_{ss}}\,a_r = 0 \tag{4.185a}$$

where:

$$\Upsilon_{ss} = -\frac{Z\,e^2\,z_{ss}}{\hbar\,v\,R} \tag{4.185b}$$

Eq. (4.185a) for the coefficients a may be solved for any specific case given a solver with sufficient time and patience. Spitzer's time and patience extended through the first excited state of hydrogen, and we shall now consider his example.

In obtaining Table 4.1 we found $\Delta m = 0$ since, in our operator, φ did not occur due to the lone presence of z. In those cases in which our operator contains y or x, however, φ will appear and we find, for example, since $y = r\sin\vartheta\sin\varphi$:

$$\int_0^{2\pi} e^{-im\varphi}\sin\varphi\,e^{im'\varphi}\,d\varphi = \frac{1}{2\,i}\int_0^{2\pi}\left[e^{-i(m^1-m'-1)} - e^{-i(m-m'+1)}\right]d\varphi$$

Thus, the matrix elements of y fail to disappear only for $\Delta m = \pm 1$.

Let us write down the deferred eigenfunctions of hydrogen in parabolic coordinates:

$$\psi_{k_1 k_2 m} = \frac{1}{\sqrt{2\pi}} u_1(\xi)\, u_2(\eta)\, e^{i m \varphi} \quad (4.186\,a)$$

$$u_1(\eta) = \frac{\sqrt{k_1!}}{k_1 + (m!)^{3/2}} e^{-(1/2)\varepsilon\eta}\, \eta^{1/2 m}\, \varepsilon^{(1/2)(m+1)} L^m_{k_1+m} \quad (\varepsilon\eta) \quad (4.186\,b)$$

$$u_2(\xi) = \frac{\sqrt{k_2!}}{k_2 + (m!)^{3/2}} e^{-(1/2)\varepsilon\xi}\, \xi^{(1/2) m}\, \varepsilon^{(1/2)(m+1)} L^m_{k_2+m} \quad (\varepsilon\xi) \quad (4.186\,c)$$

Our first excited state has the possible functions ψ_{100}, ψ_{00-1}, ψ_{010}, and ψ_{001} which we may designate as ψ_1, ψ_3, ψ_4, and ψ_2. The matrix elements of z will be the matrix elements of $\frac{\eta - \xi}{2}$ according to Eq. (4.132). As an example of the type of relation which we desire we may consider the following:

$$\left.\begin{aligned} \left(\frac{\eta-\xi}{2}\right)_{11} &= z_{11} = \frac{A}{2\pi} \int e^{-\varepsilon\eta}\, \varepsilon\eta\, [L_1^0(\varepsilon\eta)]^2\, d\eta \int u_2^2(\xi)\, d\xi \int d\varphi - \frac{A}{2\pi} \\ &\times \int u_1^2(\xi)\, d\xi \int e^{-\varepsilon\xi}\, \varepsilon\xi\, [L_0^0(\varepsilon\xi)]^2\, d\xi \int d\varphi \\ = -\Bigg[\frac{A}{2} &\int e^{-\varepsilon\eta}\, \varepsilon\eta\, [L_0^0(\varepsilon\eta)]^2\, d\eta \int u_2^2(\xi)\, d\xi \int d\varphi \\ - \frac{A}{2\pi} &\int u_1^2(\eta)\, d\eta \int e^{-\varepsilon\xi}\, \varepsilon\xi\, [L_1^0(\varepsilon\xi)]^2\, d\xi\Bigg] = -z_{44} = -\left(\frac{\eta-\xi}{2}\right)_{44} \end{aligned}\right\} \quad (4.187)$$

In this manner we arrive at the relations:

$$z_{11} = -z_{44}; \quad z_{22} = z_{33} = 0$$

$$y_{12} = y_{21} = y_{13} = \cdots = y_{34} = -\frac{1}{2} z_{11}$$

Eqs. (4.163) and (4.161) tell us that for the case where the lower state involved in a transition is unperturbed — which the ground state certainly is — $\Upsilon_{ss} = \Upsilon$. Using this fact together with Eq. (4.187) in Eq. (4.185) we obtain:

$$\frac{da_1}{dw} + i\Upsilon a_1 + \frac{1}{2}\left(a_2 + a_3 + \frac{1}{2} a_4\right) = 0 \quad (4.188\,a)$$

$$\frac{da_2}{dw} - \frac{1}{2}(a_1 - a_4) = 0 \quad (4.188\,b)$$

$$\frac{da_3}{dw} - \frac{1}{2}(a_1 - a_4) = 0 \quad (4.188\,c)$$

$$\frac{da_4}{dw} - i\Upsilon a_4 - \frac{1}{2}\left(\frac{1}{2} a_1 + a_2 + a_3\right) = 0 \quad (4.188\,d)$$

In the usual manner we may assume $a(t) = a(t_0)\exp(i\sigma w)$ to obtain the secular determinant for the a_i:

$$\begin{vmatrix} i(\sigma+\Upsilon) & \frac{1}{2} & \frac{1}{2} & \frac{1}{4} \\ -\frac{1}{2} & i & 0 & \frac{1}{2} \\ -\frac{1}{2} & 0 & i & \frac{1}{2} \\ -\frac{1}{4} & -\frac{1}{2} & -\frac{1}{2} & i(\sigma-\Upsilon) \end{vmatrix} = 0$$

whose solution is $\sigma = 0,\ 0,\ \pm\left(\Upsilon^2 + \frac{17}{16}\right)^{1/2}$. If the coefficients of a_4 in Eq. (4.188a) and a_1 in Eq. (4.188d) were zero the solution would be $\sigma = 0,\ 0,\ \pm\ (\Upsilon^2 + 1)^{1/2}$. The a_i themselves could now be found.

Of these state functions it is quite obvious we shall have three, one corresponding to $\sigma = 0$, the second given by:

$$a(t) = a(t_0)\exp\left[i\sigma w\right] = a(t_0)\exp\left\{i\left(\Upsilon^2 + \frac{17}{16}\right)^{1/2}\tan^{-1}\frac{v(t-t_0)}{R}\right\} \tag{4.189}$$

and the third the complex conjugate of the second. Thus the lowest hydrogen line has been split into three components.

If we take the line profile as the absolute square of the Fourier transform of the state function a comparison of Eq. (4.189) with Eq. (4.164d) tells us that the line profile will still be given by Eq. (4.166) where now $f_\Upsilon(\xi)$ is replaced by $f_\sigma(\xi)$. This then is essentially the effect of the breakdown of the adiabatic hypothesis on the profile of the lowest hydrogen line.

4.19. Quantum ionic Stark broadening for adiabatic Case II

We suppose that an adiabatic collision may occur very rapidly so as to leave the atomic state functions unchanged after the fashion in which a Born approximation leaves the electronic state functions unchanged.

Now let us take as our Hamiltonian:

$$H_a + H_f + V + S = H \tag{4.190}$$

where H_a is the unperturbed atomic Hamiltonian, H_f the unperturbed field Hamiltonian, S the atom-field interaction, and V the ionic field Hamiltonian. We suppose there to be two upper states, A_1 and A_2, of equal unperturbed energy E_{A0} and two lower states, B_1 and B_2, of equal unperturbed energy E_{B0}. We let Γ be the probability coefficient for a transition from A_1 to B_1 per unit time.

We assume that our perturbed eigenfunction may be expanded in terms of the four unperturbed eigenfunctions of the states which we are considering as:

$$\psi = a_1\psi_1 + a_2\psi_2 + b_1\psi_1 + b_2\psi_2 = \sum_i c_i\psi_i \tag{4.191}$$

Although it will prove a somewhat laborious procedure, let us, for clarity's sake, work out the equation for a_1 specifically.

Into the equation:

$$i\hbar \frac{\partial \psi}{\partial t} = H\psi$$

we substitute Eqs. (4.190) and (4.191) to obtain:

$$i\hbar \left(\sum_i \dot{c}_i \psi_i + \sum_i c_i \dot{\psi}_i \right) = \sum_i c_i H \psi_i \tag{4.192a}$$

In the usual manner we multiply through on the left by $\bar{\psi}_1$ and integrate over all space. First the contributions of H_f will disappear either due to eigenfunction orthogonality or due to the fact that in state a_1 we assume no photons in the field so that all n in $\sum_t n_t h \nu_t$ are zero. Secondly the H_a and the E_a diagonal term will obviously disappear. After orthogonality has relieved us of several terms of the form $\dot{c}_i \psi_i$ and $c_i \dot{\psi}_i$ $(i \neq 1)$ we are thus left with:

$$\hbar i \dot{a}_1 = a_1 S_{11} + a_2 S_{12} + b_1 S_{13} + b_2 S_{14} + \sum_i c_i V_{1i} \tag{4.192b}$$

Now $b_1 V_{13}$ and $b_2 V_{14}$ disappear due to the fact that the matrix elements of the ionic field disappear for changes in the total quantum number. Finally, $a_1 S_{11}$ and $a_2 S_{12}$ disappear so that Eq. (4.192b) becomes:

$$i\hbar \dot{a}_1 = a_1 V_{11} + a_2 V_{12} + b_1 S_{13} + b_2 S_{14} \tag{4.192c}$$

Let us notice that Eq. (4.152) infers a value $-(1/2)\, i\hbar \Gamma a(t)$ for the sum on the right of Eq. (4.151a). In essence we again make the same assumption — $b_1 S_{13} + b_2 S_{14}$ corresponds to the sum in question — to obtain:

$$i\hbar \dot{a}_1 = \hbar \Delta_{A1}(t)\, a_1 + \hbar K_A(t)\, a_2 - \frac{1}{2} i\hbar \Gamma a_1 \tag{4.192d}$$

where we have let $\hbar \Delta_{A1}(t)$ be the diagonal and $\hbar K_A(t)$ the off-diagonal matrix elements of V as in Eqs. (4.158). In like manner we let $\hbar \Delta_{A2}(t)$ be the other diagonal matrix element in the upper state, $\hbar \Delta_{B1}(t)$ and $\hbar \Delta_{B2}(t)$ the diagonal matrix elements of the lower state, and $K_B(t)$ the off-diagonal matrix elements of the lower state. Thus, we obtain for a_1 and, analogously, for the time derivatives of the other state growth coefficients, the following:

$$\dot{a}_1 = -i\Delta_{A1}(t)\, a_2 - i K_A(t)\, a_2 - \frac{1}{2} \Gamma a_1 \tag{4.193a}$$

$$\dot{b}_{1_s} = -i(x + \Delta_{B1}(t))\, b_{1_s} - i K_B(t)\, b_{2_s} - i S_{A B_s} a_1 \tag{4.193b}$$

$$\dot{b}_{2_s} = -i(x + \Delta_{B2}(t))\, b_{2_s} - i \overline{K_B(t)}\, b_{1_s} \tag{4.193c}$$

where x is still given by Eq. (4.158c).

From Eq. (4.193a) it is apparent that $a_1(t)$ will be of the form:

$$a_1(t) = \mathrm{e}^{-(1/2)\Gamma(t)}\, h_1(t) \tag{4.194a}$$

and differentiation shows that Eq. (4.193a) is satisfied if we take for $h_1(t)$ the following:

$$\left.\begin{aligned} h_1(t) &= \exp\left(-i\int_0^t \Delta_{A1}(\tau)\,dv\right)\left\{h_1(0) - i\int_0^t K_A(T)\,a_2(T)\right.\\ &\times \exp\left(i\int_0^T \Delta_{A1}(\tau)\,d\tau - \frac{1}{2}\,i\,\Gamma T\right) dT\Big\}\end{aligned}\right\} \quad (4.194\text{b})$$

Eqs. (4.194) surely satisfy the boundary condition $a_1(0) = h_1(0)$.
From Eqs. (4.193b) and (4.193c) there then results:

$$b_{1_s}(t) = -i\,\overline{p(t)}\int_0^t p_1(T)\,dT\,\{H_{AB_s}'\,e^{-(1/2)\Gamma t}\,h_1(T) + K_B(T)\,b_{2s}(T)\} \quad (4.195\text{a})$$

$$b_{2_s}(t) = -i\,\overline{p_2(t)}\int_0^t p_2(T)\,dT\,\overline{K_B(T)}\,b_{1s}(T) \quad (4.195\text{b})$$

where:

$$p_1(t) = \exp\left\{i\,x\,t + i\int_0^t \Delta_{B1}(\tau)\,d\tau\right\} \quad (4.195\text{c})$$

$$p_2(t) = \exp\left\{i\,x\,t + i\int_0^t \Delta_{B2}(\tau)\,d\tau\right\} \quad (4.195\text{d})$$

We shall not go through the succeeding steps in detail, but merely sketch them in. Spitzer next assumed, say, $b_{1s}(t)$ as a series of successive approximations,

$$b_{1s}(t) = b_{1s}^{(0)}(t) + b_{1s}^{(1)}(t) + b_{1s}^{(2)}(t) + \cdots \quad (4.196)$$

$b_{1s}^{(0)}(t)$ is obtained from Eq. (4.195a) by neglecting K_B and integrating by parts. This result is substituted into Eq. (4.195b) to obtain $b_{2s}^{(1)}(t)$, and this latter result finally utilized to find $b_{1s}^{(2)}(t)$. In addition $b_{2s}^{(0)}$, $b_{1s}^{(1)}$, and $b_{2s}^{(2)}$ are shown to vanish. There ultimately results:

$$\left.\begin{aligned} |b_{1_s}(\infty)|^2 + |b_{2_s}(\infty)|^2 &= \frac{|S_{ABs}|^2}{x^2 + \left(\frac{1}{2}\Gamma\right)^2}\left[|h_1(0)|^2\left\{1 + \left|\int_0^\infty k(T)\,dT\right|^2\right.\right.\\ &- 2R\left[\int_0^\infty k(\tau)\,d\tau\int_0^\infty \overline{k(T)dT}\right]\Big\} + \left|\int_0^\infty p_1(T)\,e^{-(1/2)\Gamma T}\,dT\right.\\ &\times\left.\left\{\frac{dh_1}{dT} + i\Delta_{B1}h_1\right\}\right|^2 + \left.\left|\int_0^\infty p_2(T)\,e^{-(1/2)\Gamma T}\,dT\,K_B\,h_1\right|^2\right]\end{aligned}\right\} \quad (4.197\text{a})$$

where:

$$k(t) = K_B(t)\,p_1(t)\,\overline{p_2(t)} \quad (4.197\text{b})$$

It can be shown that the double integral cancels the term immediately preceding it in Eq. (4.197a). In addition, the changes in h_1 and h_2 during a collision are neglected so that these quantities are taken outside the integral signs and the mean square of each equated to unity. Finally $\exp(-(1/2)\Gamma T)$ is taken outside the integral and x is assumed $\gg (1/2)\,\Gamma$. Again we recall that Eq. (4.197a) should give the intensity distribution. Thus:

$$I(x) = \frac{\Gamma}{2\pi x^2}\left[1 + \exp(-\Gamma t_0)\left\{\left|\int_0^\infty e^{ixT}(\Delta_{A1} - \Delta_{B1})\,dT\right|^2 + \left|\int_0^\infty e^{ixT}\,K_A\,dT\right|^2 + \left|\int_0^\infty e^{ixT}\,\bar{K}_B\,dT\right|^2\right\}\right] \quad (4.198)$$

The integrals of Δ_{B1} and Δ_{B2} may be noted to have disappeared, and this is by virtue of their first expected appearance in fourth order. Use has also been made of the fact that the phases of h_1 and h_2 are arbitrary which would lead to the disappearance of their cross product terms in the main (occurrence with equal probability of positive and negative signs).

Now we take the z_e axis as parallel to the direction of rectilinear motion of the perturbing ion and the $y_e z_e$ plane as the plane of perturber motion. If, as usual, ϑ is the angle between the distance of closest broadener approach and the instantaneous emitter-broadener separation, then the z_e component of the field produced by the perturbing ion will be given by $E \sin\vartheta$ and the y_e component by $E\cos\vartheta$. Under these conditions Eqs. (4.161) tell us that the diagonal matrix elements of the ionic field yield:

$$\Delta_{A1}(T) - \Delta_{B1}(T) = q\,\frac{\sin\vartheta}{\hbar\, r^2}$$

On the other hand the matrix of z_e is diagonal. Thus, we should expect the contributions to the off-diagonal elements of the ionic field, K_A and K_B, to arise from the matrix of y_e alone so that, if we let K_A be some constant the off-diagonal element, K_A will be given by $\bar{K}_A \cos\vartheta \hbar\, r^2$. According to the conventions introduced:

$$\sin\vartheta = \frac{v(T - t_0)}{r}\,; \qquad \cos\vartheta = \frac{R}{r} \quad (4.199)$$

so that from Eqs. (4.161):

$$\int_0^\infty e^{ixT}(\Delta_{A1} - \Delta_{B1})\,dT = \Upsilon \int_{-\infty}^{+\infty} e^{i\xi u}\,\frac{u\,du}{(1+u^2)^{3/2}} \quad (4.200\text{a})$$

$$\int_0^\infty e^{ixT}\,K_A\,dT = \frac{K_A}{\hbar R v}\int_{-\infty}^{+\infty} e^{i\xi u}\,\frac{du}{(1+u^2)^{3/2}} \quad (4.200\text{b})$$

Integrals of the form Eqs. (4.200) have already been quite handily disposed of in Eq. (4.170) and subsequent, and when the result of such disposition is substituted into Eq. (4.198), the result is:

$$I(x) = \frac{\Gamma}{2\pi x^2}\left\{1 + 4\exp(-\Gamma t_0)\,\xi^2\left[\Upsilon^2 K_0^2(\xi) + \frac{K_A^2 + K_B^2}{\hbar^2 R^2 v^2}\,K_1^2(\xi)\right]\right\} \quad (4.201)$$

It appears reasonably evident that when the y_e and z_e axes are interchanged, the roles of the K's and the Δ's will be interchanged resulting in the interchange of K_0^2 and K_1^2. We weight $I(x)$ for each component in a hydrogen line by the oscillator strength for the component. When $I(x)$ is then summed over all components, the result must be independent of the choice of axes. This appears physically sensible, and it is to be admitted that, if such is the case, the weighted sum over Υ must equal the weighted sum over K_A^2 and K_B^2 in Eq. (4.207). With the understanding then that the sum in question is to be taken, it is legitimate to replace $(K_A^2 + K_B^2)/\hbar^2 R^2 v^2$ by Υ^2. If we again specify the number of collisions as in Eq. (4.166), we obtain this latter equation for Eq. (4.201) where now $f_\Upsilon(\xi)$ is replaced by:

$$f_{0'}(\xi) = 2\xi\,[K_1^2(\xi) + K_0^2(\xi)]^{1/2} \tag{4.202}$$

From Eq. (4.113):

$$f_1(\xi) = 2\xi\,[K_1(\xi) \pm K_0(\xi)] \tag{4.203}$$

so that we see that $f_0'^2(\xi)$ is nothing more nor less than the average of $f_1^2(\xi)$ over plus and minus values.

The replacement of $f_\Upsilon(\xi)$ by $f_\sigma'(\xi)$ in Eq. (4.166) is somewhat similar to the replacement by $f_\sigma(\xi)$ earlier. Now the change in $f_\sigma(\xi)$ as Υ goes from one to zero is small. On the other hand Υ is of the order of $2K_B/\hbar v R$ which in turn is approximately equal to the square of the integral of $K_B(t)$. This latter quantity tells us approximately the probability of the molecule changing its state from B_1 to B_2, i. e., of invalidating the Born approximation. This line of reasoning leads us to the conclusion that as Υ gets large enough to allow the use of the adiabatic approximation as a good assumption, the Born approximation is no longer valid.

4.20. Inclusion of different types of collisions in the ionic Stark theory[190]

In the last few sections then we have (1) considered the effect on a hydrogen spectral line of an adiabatic collision of specific optical collision diameter and perturber velocity to obtain Eq. (4.166) for the intensity distribution in the broadened line, (2) considered the deviations from this adiabatic hypothesis of (1) for the specific case of the first Lyman lines and found that $f_\Upsilon(\xi)$ must be replaced by $f_\sigma(\xi)$ to account for the diabaticity and (3) utilized what corresponded to the Born approximation to obtain a line shape (as given by Eq. (4.201)) and in this derivation the transitions considered were more general than (2) since they could take place between any two perturbed levels. *In toto* then (2) and (3) have indicated the necessity of replacing Eq. (4.166) by:

$$I(x) = \frac{\Gamma}{2\pi x^2}\left\{1 + \frac{\Omega(\Upsilon, \xi)}{\Gamma}\,\Upsilon^2 f_\Upsilon^2(\xi)\right\} \tag{4.204a}$$

where:

$$\sigma^2 = 1 + \Upsilon^2 \tag{4.204b}$$

and the remainder of the symbols are defined, as before, by Eqs. (4.161), (4.163), and (4.165).

The problem which remains then is one of integrating over all values of Υ, and ξ, that is, all values of v and ϱ, in order to obtain a comprehensive picture of the intensity distribution in the spectral line.

In order to accomplish this integration, we first suppose $\Omega_{m'}(\varrho, v)\,d\varrho\,dv$ to be the number of times per second (essentially a probability) that a hydrogen atom undergoes an optical collision of diameter between ϱ and $\varrho + d\varrho$ with an ion of mass M_m, charge $Z_m e$ and having a relative velocity between v and $v + dv$. Then an integration of Eq. (4.204a) over ϱ and v together with a summation over m, the types of ions, will yield the expression desired.

$$I(x) = \frac{\Gamma}{2\pi x^2}\left\{1 + \frac{1}{\Gamma}\sum_m \int_0^\infty dv \int_0^S \Omega_m'(\varrho, v)\,\Upsilon^2 f_\sigma^2(\xi)\,d\varrho\right\} \qquad (4.205)$$

The upper limit of integration over the collision diameter was taken as:

$$\frac{4}{3}\pi S^3 = \frac{1}{N}$$

where N has its usual meaning. There are two obvious approximations here. One involving the upper and the other the lower limit.

For the distribution of velocities a Maxwell–Boltzmann distribution was assumed, and for the optical collision diameter — distance of closest approach — the simple probability $2\pi\varrho v N_m$ was assumed so that:

$$\Omega'_m(\varrho, v) = (2\pi\varrho v N)\,(4 l_m^{3/2}\pi^{-1/2} v^2 \exp[-l_m v^2]) \qquad (4.206a)$$

where

$$l_m = \frac{1}{2kT}\,\frac{M_m m_H}{M_m + m_H} \qquad (4.206b)$$

with m_H the hydrogen atom mass.

When Eq. (4.206a) is substituted into Eq. (4.205); ξ is reintroduced, and u is substituted for $l_m v^2$, there results:

$$I(x) = \frac{\Gamma}{2\pi x^2}\left\{1 + \frac{4\pi^{1/2}}{\Gamma x^{1/2}}\sum_m \frac{N_m q_m^{3/2}}{\hbar^{3/2}}\gamma_1^{1/2} H_m(x)\right\} \qquad (4.207a)$$

where:

$$H_m(x) = \int_0^\infty e^{-u}\,du \int_0^{(\gamma_1/\gamma_2)u^{1/2}} f_\sigma^2(\xi)\,\frac{d\xi}{\xi} \qquad (4.207b)$$

$$\gamma_1 = \frac{q_m l_m x}{\hbar}; \quad \gamma_2 = \frac{q_m l_m^{1/2}}{\hbar s} \qquad (4.207c)$$

and finally:

$$\sigma^2 = 1 + \frac{\gamma_1^2}{u^2\xi^2} \qquad (4.207d)$$

from Eq. (4.204b). We might note that the quantity $q_m/\hbar$ is the Stark shift which results from placing a charge $Z_m e$ unit distance from a hydrogen atom. We shall have more to say about γ_1 and γ_2 later.

The main problem which Spitzer faced in carrying the theory forward to completion from this point was the evaluation of the integral occurring in Eq. (4.207b). The difficulty arising here is largely due to the fact that $f_\sigma(\xi)$ (as we may recall from Eq. (4.173)) has only been evaluated for integral values of Υ

and hence σ. Spitzer avoids the difficulty presented by this situation by taking the following for x/x_{max} less than 1/16:

$$f_\sigma(\xi) = \begin{cases} f_1(\xi) & 1 < \sigma < 1.5 \\ f_2(\xi) & 1.5 < \sigma < 3 \\ f_4(\xi) & 3 < \sigma < 5 \\ f_\infty(\xi) & 5 < \sigma \end{cases} \tag{4.208a}$$

and for x/x_{max} greater than 1/16:

$$f_\sigma(\xi) = f_\sigma(0) \tag{4.208b}$$

We shall not consider the intervening steps in detail but simply give the solution as:

$$H_m(x) = 4\ln\frac{1}{\gamma_1} - 2.33 + 61\gamma_1 + 6.7\frac{\gamma_1}{\gamma_2} \tag{4.209}$$

which is accurate to within ten percent for γ_1 less than $\gamma_2/10$ and γ_2 less than 0·2. The asymptotic value of $H_m(x)$ for large values of γ_2 is apparent from this equation. For x/x_{max} less than 1/16 this asymptotic result holds for $H_m(x)$. Finally with γ_1 large (γ_2 still assumed small) $H_m(x)$ is given by $H_m'(x)$ where:

$$H_m'(x) = 4\ln\frac{1}{\gamma_1} - 3.02 + 61\gamma_1 \tag{4.210}$$

These then are the values of $H_m(x)$ which are to be utilized in Eq. (4.207a) in order to obtain the intensity distribution for one component of the hydrogen line. Since we are desirous of obtaining the intensity distribution in an actual observed line, we now sum contributions of the form Eq. (4.207a) from the various components weighted by the oscillator strengths of the components concerned. The result is:

$$I(x) = \frac{1}{2\pi x^2}\left\{\sum_j \Gamma_j \varphi_j + 4\frac{\pi^{1/2}}{x^{1/2}\hbar^{3/2}}\sum_j \varphi_i \sum_m N_m q_{mj}^{3/2}\gamma^{1/2}(H_{mj}'(x) - H_{mj}^2(x))\right\} \tag{4.211}$$

where the φ_j are the normalized oscillator strengths:

$$\varphi_j = \frac{f_j}{\sum_j f_j} \tag{4.212}$$

Spitzer's treatment has admittely been primarily concerned with the Lyman-α line. Still it indicates clearly enough the *modus operandi* which may be employed in a pure quantum treatment of some other specific case of ionic Stark broadening.

4.21. Review and prognosis

We have considered: (1) both the simple (Sec. 4.13) and somewhat more complex (Sec. 4.8) forms of the Holtsmark statistical theory of Stark broadening, (2) the ionic and electronic shielding effects on ionic fields (Sec. 4.14), and, lastly,

(3) Quantum theory of ionic Stark broadening. We have not considered Stark broadening by electrons or broadening by combinations of electrons and ions. These latter two will, of course, be treated next.

The Holtsmark theory has proven generally quite adequate for the description of ionic broadening effects, and, of course, its comparative simplicity provides a good deal in the way of its recommendation. This theory, as we have seen, is based on a statistical type probability function for the existence of a particular field at the emitter due to the presence of ions. We may find the spectral line contour from this by simply supposing the original line to be shifted with varying probabilities by the varying fields or to be split and smeared with varying probabilities by these fields.

The shielding of the fields produced by the ions is certainly to be anticipated with the presence of the background of ions and electrons. We have seen how the Debye–Hückel potential has been applied to the treatment of this shielding problem.

The quantum treatment of the ionic Stark broadening consisted, as one would certainly anticipate, principally of mathematical detail. Basically one seeks the square of the state growth coefficient for an arbitrary frequency in the spectral line or an arbitrary photon in the radiation field. In our considerations here we have seen the effect of an adiabatic assumption.

Since we have remarked the applicability of the Holtsmark treatment to ionic Stark broadening, it will be of interest to see the applicability of Interruption theory to electronic Stark broadening. First, however, we shall study the quantum treatment of this phenomenon which, like the ionic case, is replete with specific detail.

Finally, we shall consider the combined Stark broadening effects of ions and electrons. As we would expect, folding of one distribution into another with all the complexities easily imagined is the obvious and oft attempted approach here. There are notable and ingenious exceptions, however. Fortunately, in some cases ionic effects predominate, in others electronic predominate, so that one is not always obliged to appeal to this most complex if complete description.

4.22. Quantum electron broadening of the Lyman alpha line

In this section we shall consider the careful treatment of the broadening of the first Lyman line which was carried out by Kivel, Bloom, and Margenau (KBM)[107]. It may appear that, by beginning with a lengthy considerations of the $2\,p - 1\,s$ transition in hydrogen, one is left with a great deal of ground to cover, but such is not really the case. There has been so little work of such a detailed nature as this that this is hardly a "special case", and, also, there are several physical phenomena encountered in the development which are of general interest of themselves.

If we take R as the coordinate of the free electron and r as the coordinate of the emitting atom, we obtain the by now familiar Hamiltonian:

$$H(\boldsymbol{r}, \boldsymbol{R}) = H_a(\boldsymbol{r}) + H_e(\boldsymbol{R}) + V(\boldsymbol{r}, \boldsymbol{R}) + S(\boldsymbol{r}) \tag{4.213}$$

wherein H_e refers to the perturbing electron, and we have absorbed the unperturbed field Hamiltonian into the unperturbed atom Hamiltonian H_a.

We assume that the solution to the time dependent Schrödinger equation utilizing this Hamiltonian,

$$i\hbar \frac{\partial}{\partial t} \Psi(\boldsymbol{r}, \boldsymbol{R}) = H(\boldsymbol{r}, \boldsymbol{R}) \Psi(\boldsymbol{r}, \boldsymbol{R}) \tag{4.214}$$

is given by:

$$\Psi = \sum_n \sum_\lambda^n a_{n\lambda}(t)\, \psi_n(\boldsymbol{r})\, \varphi_\lambda^n(\boldsymbol{R}) \exp\left[-i(\varepsilon_\lambda^n + E_n)\, t/\hbar\right] \tag{4.215}$$

where $\sum\limits_\lambda^n$ means the sum and/or integral over all perturber states λ while the emitter is in state n. Our atomic wave functions satisfy:

$$H_a(\boldsymbol{r})\, \psi_n(\boldsymbol{r}) = E_n \psi_n(\boldsymbol{r}_n); \quad \int |\psi_n|^2\, \mathrm{d}r = 1 \tag{4.216}$$

The free electron is in the field of an atom in a state n and satisfies the following:

$$[H_e(\boldsymbol{R}) + V_{nn}(\boldsymbol{R})]\, \varphi_\lambda^n(\boldsymbol{R}) = \varepsilon_\lambda^n\, \varphi_\lambda^n(\boldsymbol{R}) \tag{4.217a}$$

$$\int |\varphi_\lambda^n|^2\, \mathrm{d}\boldsymbol{R} = 1; \quad V_{mn}(\boldsymbol{R}) = \int \bar{\psi}_m\, V\, \psi_n\, \mathrm{d}\boldsymbol{r} \tag{4.217b}$$

Now let us substitute Eq. (4.215) into Eq. (4.214) at the same time making use of Eqs. (4.216) and (4.217). The result is immediate:

$$i\hbar\, \dot{a}_{m\mu} \exp\left[-i(\varepsilon_\mu^m + E_m)\, t/\hbar\right] = \sum_n \sum_\lambda^n a_{n\lambda} S_{mn} \int \bar{\varphi}_\mu^m \varphi_\lambda^n\, \mathrm{d}\boldsymbol{R} \exp\left[-i(\varepsilon_\lambda^n + E_n)\, t/\hbar\right]$$

$$+ \sum_{n \neq m} \sum_\lambda^n a_{n\lambda} \int \bar{\varphi}_\mu^m\, V_{mn}\, \varphi_\lambda^n\, \mathrm{d}\boldsymbol{R} \exp\left[-i(\varepsilon_\lambda^n + E_n)\, t/\hbar\right] \tag{4.218}$$

but quite important and informative.

The two terms on the right describe several physically separable effects. The first term on the right refers to what KBM have called "universal broadening". The exponent appearing in this term provides the mathematical means for exchanges of energy between the free electron and the radiations field. The term effectively describes the broadening effects of electrons which are scattered nearly elastically. It ignores first-order Stark effects by failing to include linear combinations of the degenerate states, and it further ignores second-order Stark effects by failing to mix in any of the higher states as would be requisite to this effect.

The second term on the right of Eq. (4.218) describes either quenching or polarization. If "m" and "n" are the initial and final states in the radiating transition in this term, then the collision has effectively quenched the radiation, and the contribution to the broadening is a direct result of this quenching.

This second term may represent polarization of the atom if the states "m" and "n" are degenerate or nearly so. The colliding electron may then either induce a dipole in the atom or reorient the dipole that already exists there. The first polarization effect is accomplished by mixing in other states to effect the induction. The second exchanges two states which effects a re-orientation of the existing dipole. The effects of this second term on the right of Eq. (4.218) will not be considered in the detail we wish to apply to the effects of the first term.

We begin our treatment of the universal effect by supposing the off-diagonal elements of V to disappear. This is most valid as an approximation for an atom carrying out a transition between two isolated levels. For this case our state growth equations take on the form:

$$\dot{a}_\lambda = \sum_r \sum_\mu^0 b_\mu \bar{S}_\nu \exp(i\omega t + i\Omega_{\lambda\mu}t) \int \bar{\varphi}_\lambda{}' \varphi_\mu{}^0 \, \mathrm{d}\boldsymbol{R} \qquad (4.219\text{a})$$

$$\dot{b}_{\nu\mu} = \sum_{\lambda'}^1 a_{\lambda'} S_\nu \exp(-i\omega t - i\Omega_{\lambda'\mu}t) \int \bar{\varphi}_\mu{}^0 \varphi_{\lambda'}{}' \, \mathrm{d}\boldsymbol{R} \qquad (4.219\text{b})$$

where:

$$\omega_r = \omega_{\text{atom}} - \omega; \quad \hbar\,\Omega_{\lambda'\mu} = \varepsilon_{\lambda'}{}' - \varepsilon_\mu{}^0 \qquad (4.219\text{c})$$

KBM have proven and we accept the exponential dependence of a_λ on the time. We let:

$$\varphi_1 \equiv \sum_\lambda{}' a_{\lambda'}{}' \varphi_\lambda{}' \exp[-i\varepsilon_{\lambda'}{}'\hbar]; \quad \varphi_0 \equiv \sum_\mu^0 \varphi_\mu{}^0 \exp(-i\varepsilon_\mu{}^0 t/\hbar) \qquad (4.220)$$

and substitute into Eq. (4.219b) to obtain:

$$\dot{b}_r = \sum_\mu i\hbar\dot{b}_{r\mu} = S_r \,\mathrm{e}^{-(\gamma + i\omega t)} \int \bar{\varphi}_0 \varphi_1 \, \mathrm{d}\boldsymbol{R} \qquad (4.221)$$

The functions φ_0 and φ_1 represent wave packets moving in the field of the atom in state "0" and state "1", respectively. Now suppose the electron is initially in the specific state λ so that $a_{\lambda'}{}' = \delta_{\lambda'\lambda}$. Then from Eq. (4.221) we obtain:

$$b_{r\mu}(\omega) = \frac{\delta_r}{i\hbar} \frac{\int \bar{\varphi}_\mu{}^0 \varphi_\lambda{}' \, \mathrm{d}\boldsymbol{R}}{p + i(\omega + \Omega_{\lambda\mu})} \qquad (4.222)$$

The familiar assumption relating the intensity to the square of the state coefficient after an infinite time is applied with the result:

$$I_U(\omega) = \varrho_r \sum_\mu |b_{r\mu}(\omega)|^2 = \varrho_r \left|\frac{S_r}{\hbar}\right|^2 \frac{\sum_\mu |\int \bar{\varphi}_m{}^0 \varphi_\lambda{}' \, \mathrm{d}\boldsymbol{R}|^2}{\gamma^2 + (\omega + \Omega_{\lambda\mu})^2} \qquad (4.223)$$

A theory related to a result of this type has been discussed by Rudkjobing.[179] When this simplifying assumption as to $a_{\lambda'}{}'$ is not included, the derivation becomes much more complex and occupies the remainder of KBM's considerations of universal broadening. We shall only indicate the approach very generally.

First time development operators* (TDO) are adopted to the state growth coefficients as follows:

$$||T(0)|| = ||U(0)|| = \text{unit matrix}; \quad ||\dot{T}|| = ||p||\,||T||;$$
$$||\dot{U}|| = ||q||\,||U|| \qquad (4.224)$$
$$||a(t)|| = ||T(t)||\,||a(0)||; \quad ||b(t)|| = ||U(t)||\,||b(0)||$$

* We accept these operators as defined by Eq. (4.224).

The $||p||$ and $||q||$ arise from the definitions:

$$\varphi_1 = \sum_\lambda a_\lambda u_\lambda \exp(-i\varepsilon_\lambda t/\hbar) \qquad (4.225\text{a})$$

$$\varphi_0 = \sum_\lambda b_\lambda u_\lambda \exp(-i\varepsilon_\lambda t/\hbar) \qquad (4.225\text{b})$$

$$i\hbar\dot{a}_\lambda = \sum_\mu i\hbar\, p_{\lambda\mu} a_\mu$$

$$H_e u_\lambda = \varepsilon_\lambda u_\lambda \qquad (4.225\text{c})$$

$$i\hbar\dot{b}_\lambda = \sum_\mu i\hbar\, q_{\lambda\mu} b_\mu$$

The u_λ are a set of eigenfunctions independent of the atomic state. Now Eq. (4.221) takes on the form:

$$i\hbar\,||\dot{b}_r|| = ||S_r||\,||b(0)||^\dagger\,||U||^\dagger\,||T||\,||d(0)||\,e^{-(\gamma+i\omega)t} \qquad (4.226)$$

which leads to the intensity expression:

$$I_U = \varrho_r \frac{|J_r|^2}{\hbar^2} Tr\left[||\varrho^0|| \int_0^\infty d\tau\,||T||^\dagger\,||U||\,e^{(\gamma \;\; i\,\omega)\tau} \int d\tau\,||U||^\dagger\,||T||\,e^{-(\gamma+i\,\omega)\tau}\right] \qquad (4.227)$$

where:

$$||\varrho^0|| = ||d(0)||\,||d(0)||^\dagger; \quad ||b(0)||\,||b(0)||^\dagger = \text{unit matrix}$$

Now specific expansions are taken for the two TDO's. These are of the form:

$$||T(t)|| = ||\varepsilon|| + ||T^{(1)}(t)|| + ||T^{(2)}(t)|| + \cdots \qquad (4.228\text{a})$$

where:

$$||T^{(n)}(t)|| = \int_0^t \int_0^{t_n} \cdots \int_0^{t_2} ||p(t_n)|| \cdots ||p(t_1)||\,dt_1 \cdots dt_{n-1}\,dt_n \qquad (4.228\text{b})$$

As it turns out, these expansions limit the validity of the result to broadenings of the order of the natural line width. The expansions are more strongly convergent for small V, and this has the obvious result of restricting what follows to small scattering. What follows is a very involved reduction of Eq. (4.227) after the replacement of the TDO's by their expansions which we shall not carry through. The final result is a long and involved expression which we may write down as:

$$I_U = \frac{\varrho\,|S/\hbar|^2}{\gamma^2+\omega^2}\left\{1 + \frac{R}{\gamma^2+\omega^2}\right\} \qquad (4.229)$$

The authors refer to R as the "redistribution factor". It is given by:

$$R = \sum_\lambda \varrho_{\lambda\lambda}{}^0 \sum_\lambda R_{\lambda\mu} \qquad (4.230\text{a})$$

with:

$$R_{\lambda\mu} = \frac{A_{\lambda\mu}}{\gamma^2+(\omega-\Omega)^2} + \frac{B_{\lambda\mu}}{\gamma^2+(\omega+\Omega)^2} \qquad (4.230\text{b})$$

The various factors appearing in Eqs. (4.230) are defined as:

$$\hbar\Omega = \varepsilon_\lambda - \varepsilon_\mu \qquad (4.231\text{a})$$

$$A_{\lambda\mu} = (P-Q)\,2\,P(\omega^2-\gamma^2-\omega\Omega) \qquad (4.231\text{b})$$

$$B_{\lambda\mu} = (P-Q)\,[P(\gamma^2+\omega^2) - Q(3\omega^2-\gamma^2+2\omega\Omega)] \qquad (4.231\text{c})$$

$$P = (V_{11})_{\lambda\mu}/\hbar; \quad Q = (V_{00})_{\lambda\mu}/\hbar \qquad (4.231\text{d})$$

Eq. (4.229) is as far as the theory can be practically carried in a general form, although we recall that we have already limited it by our expansions of Eqs. (4.228). Eq. (4.229) describes certain physical phenomena which it would behoove us to discuss. In the summation over λ the first term in Eq. (4.230b) has a resonance at $\Omega_{\lambda\mu} = \omega$ which is equivalent to the following:

$$\varepsilon_\lambda - \varepsilon_\mu = \hbar\omega_a - \hbar\omega_r \tag{4.232a}$$

We suppose $\omega_r > \omega_a$. This means that the radiation has gained more energy than the emitting atom gave up. For the equality in the equation to hold, however, $\varepsilon_\mu > \varepsilon_\lambda$, so the electron has also gained energy. Consequently, there was no energy exchange between radiation field and free electron. On the other hand suppose that $\omega_r < \omega_a$. Under these conditions the radiation field did not gain as much energy as the atom emitted. At the same time this requires that $\varepsilon_\lambda > \varepsilon_\mu$ so that the electron also lost energy. As a consequence of all this, it becomes evident that the first term in Eq. (4.230b) represents the case where energy is not conserved. The term is small as we might anticipate under the circumstances. At resonance ($\Omega_{\lambda\mu} = \omega$) its coefficient, $\omega^2 - \gamma^2 - \omega\Omega_{\lambda\mu}$, is $-\gamma^2$. Consequently, it would vanish altogether were it not for the energy uncertainty resulting from the natural line width.

The resonance for the second term in Eq. (4.230b) occurs for $\Omega_{\lambda\mu} = \omega$ with the result that:

$$\varepsilon_\mu - \varepsilon_\lambda = \hbar\omega_a - \hbar\omega_r \tag{4.232b}$$

In this case when the radiation field receives more energy than the atom radiates, that is $\omega_r > \omega_a$, the electron loses energy since $\varepsilon_\lambda > \varepsilon_\mu$. The second term in Eq. (4.230b) therefore represents the case where energy is transferred to the radiation field with the anticipated conservation.

KBM have evaluated Eq. (4.229) for the special case of the first Lyman line where the states involved are $1s$ and $2p$ states of hydrogen. We take $g = 1/N(2\pi)^3$ as the number of states per unit volume in wave number space where N is the perturber density and $n = 1/V$. The summation over final states may then be replaced as follows:

$$\sum_\mu \to \int g\, \mathrm{d}\boldsymbol{k}_\mu$$

wherein $\boldsymbol{k}_\mu$ is the wave-number vector of the free electron which we now suppose to be represented by a plane wave:

$$\mu_\lambda = \frac{1}{\sqrt{V}} e^{i\boldsymbol{k}_\lambda \cdot \boldsymbol{R}} \tag{4.233}$$

The summation over final states in Eq. (4.230a) may now be written as:

$$\sum_\mu R_{\lambda\mu} \to 2\pi g \int_{-1}^{+1} \mathrm{d}(\cos\Theta) \int_0^\infty k_\mu^2\, \mathrm{d}k_\mu\, R_\mu(k_\mu, \cos\Theta) \tag{4.234}$$

In evaluating this KBM used Eq. (4.233), actual hydrogen atom wave functions for the states in question, and the following familiar expression for the electrostatic interaction:

$$V_{nn} = -\frac{e^2}{|\boldsymbol{R}|} + e^2 \int \frac{|\psi_n(\boldsymbol{r})|^2 \mathrm{d}r}{|\boldsymbol{R} - \boldsymbol{r}|} \tag{4.235}$$

The actual evaluation is lengthy and involved, and no particular purpose would be served by following it through. The result is:

$$I_U = \frac{\varrho_r |S_r/\hbar|^2}{\omega^2 + \gamma^2} \left\{1 + f L_U \frac{\omega^2 - \gamma^2}{\omega^2 + \gamma^2}\right\} \tag{4.236a}$$

where:

$$L_U \equiv \int_0^{4a^2 k_\lambda^2} \frac{(F_1 - F_2)^2}{x^2} \mathrm{d}x \tag{4.236b}$$

$$f = \frac{\partial N v_\lambda \sigma_\lambda}{\gamma}; \qquad \sigma_\lambda = \pi \frac{\hbar^2}{(m v_\lambda)^2} \tag{4.236c}$$

where v_λ is the electron velocity in the state λ, so that f is a measure of the relative number of collisions during the lifetime of the excited state; a is the first Bohr radius. More explanation of Eq. (4.236b) is in order. If we define the following:

$$F_{nn} = \int \mathrm{e}^{i\boldsymbol{K}\cdot\boldsymbol{r}} |\psi_n(\boldsymbol{r})| \Psi_n(\boldsymbol{r})|^2 \mathrm{d}\boldsymbol{r}$$

then the symbols in Eq. (4.236b) may be identified as:

$$F_1 = F_{100,\,100}; \; F_2 = \frac{1}{3} \sum_m F_{2pm,\,2pm} \tag{4.236d}$$

Eqs. (4.236) then yield the line profile for the first Lyman line. The factor which is most dependent on the atomic states, is of course, L_U. KBM took electron density as 10^{14}, ε_λ as 1/2 eV, and $\varrho_{\nu\nu}{}^0 = \delta_{\lambda\nu}$. With a value $\gamma = 3{\cdot}12 \times 10^8$/sec they obtained:

$$I_U \doteq \frac{\varrho_r |S_r/\hbar)|^2}{\gamma^2 + \omega^2} \left\{1 - 0{\cdot}13 \left(\frac{\gamma^2 - \omega^2}{\gamma^2 + \omega^2}\right)\right\}$$

which is plotted in Fig. 4.10. The other curves are for other values of the electron density. The dotted curves are meaningless since elements of V beyond the second order are important to these calculations.

From Eq. (4.236) we may obtain the half width for small $f L_U$ as:

$$f = \gamma(1 + f L_U) \tag{4.237a}$$

so that KBM have suggested defining the half width for the universal effect as:

$$\delta_U = \gamma f L_U = N v_\lambda \sigma_\lambda (2 L_U) \tag{4.237b}$$

As a consequence, Eq. (4.236a) can then be shown to be an approximation to first powers in γ_U/γ of:

$$I_U = \frac{\varrho_r |S_r/\hbar|^2 (\gamma + \gamma_U)/\gamma}{\omega^2 + \frac{1}{4}(\gamma + \gamma_U)^2} \tag{4.238}$$

Now we recall that f is a measure of the number of collisions which will occur during the lifetime of an excited state; it is of the order of the quotient of the

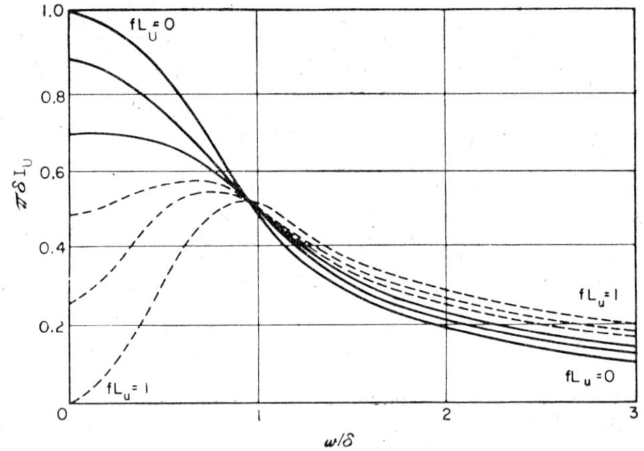

FIG. 4.10. Intensity distributions for universal broadening: $fL_U = 0, 0{\cdot}13, 0{\cdot}5, 0{\cdot}7, 1{\cdot}0$. (After Kivel, Bloom, and Margenau[104])

state lifetime and the average inter-collision time. Thus, if f is much greater than one, there are many collisions during the state lifetime, and some sort of

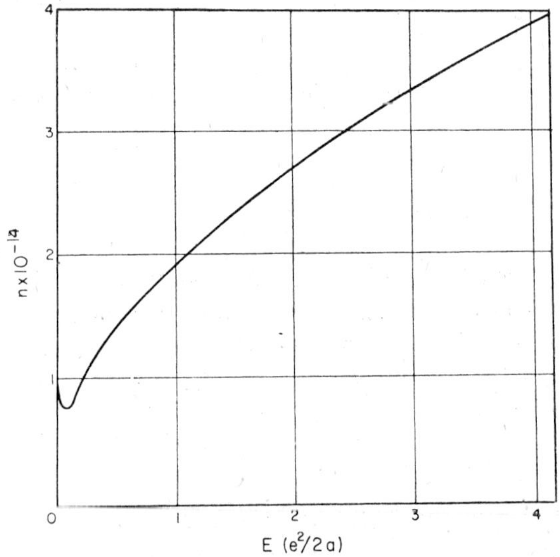

FIG. 4.11. Electron energy (in units of $e^2/2a$) as a function of electron density delimiting region where universal broadening is negligible compared to the natural width. (After Kivel, Bloom, and Margenau[104])

static perturbation treatment should be applicable. For the intermediate value of fL_U we would need the higher powers of the interaction matrix elements which we have neglected. Thus the lower regions of applicability are defined and illustrated on Fig. 4.11. This restriction to small N may be removed by using a free electron wave function distorted by the presence of the emitting atom. KBM felt that the result would be a universal half width not much different from that given by Eq. (4.237b), and Margenau and Kivel[143] later proved this by a derivation in which only the free electron states were considered. This universal effect then is the result of considering the off diagonal elements of the perturbing interactions as negligible.

Our considerations of the second term on the right of Eq. (4.218) and the physical phenomena implied will be comparatively brief. In their polarization study of the first Lyman line KBM began by considering only three states, namely, $\psi_+ = 1/\sqrt{2}\,(\psi_{2s0} + \psi_{2p0})$, $\psi_- = 1/\sqrt{2}\,(\psi_{2s0} - \psi_{2p0})$, and ψ_{1s0}. They later considered the requisite additional states $\psi_{2p\pm}$, whose application to certain of the interaction matrix element evaluations led to nuisancesome divergences. The "polarization by orientation" is illustrated by the matrix element linking the states ψ_+ and ψ_-. The "polarization by induction" is illustrated by the interaction element mixing in $\psi_{2p\pm}$, with one of the other three wave functions. The quenching effects which are also to be found in Eq. (4.218) have been discussed.

The polarization effect is much more important to the Lyman α line than is the universal effect, so it might seem reasonable for us to consider it at somewhat greater length. However, Kolb has shown that his classical path theory yields the same result, so we shall not concern ourselves with the specific development. It might be pointed out here that the Kolb treatment does not yield the polarization by reorientation effect, but this is comparatively small. The result of the complete KBM calculation is given in Table 4.2.

TABLE 4.2

BROADENING OF THE $L\alpha$ LINE BY ELECTRON IMPACTS AT AN ELECTRON DENSITY 10^{14}, MEAN ENERGY 1/2 eV AND NATURAL LINE WIDTH $3{\cdot}12 \times 10^8$ sec^{-1}

"Universal" broadening	0·13
Polarization by reorientation	0·027
Induction	2·7
Quenching	0·0013
Stark broadening by ions	80

It is apparent that in the case of the line treated by these three authors the electron effects are trivial compared to the ionic effects. Such is not always the case as Kivel[103] pointed out when he applied the same theory to the $3d \rightarrow 2p$ line in the helium spectrum. In this latter case the electronic effects predominate over the ionic effects.

4.23. Quantum electron broadening of the Balmer lines

We begin with assumptions as to the Hamiltonian and wave function base similar to those of the last section. The important and basic approximation which is subsequently introduced supposes that the perturbing Hamiltonian possesses no matrix elements not involving the initial state. This perturbing Hamiltonian includes the emitter field as well as a sum of terms referring to collision interactions between the emitter and the various broadening electrons. Now this means that only the initial state can radiate a photon and collision induced transitions can only proceed into or out of the initial state. This further allows the restriction of the state growth equations for no photons in the field to elements of the collision interaction and for a photon of frequency ν in the field to elements of the particle-field interaction. A long range cut-off is introduced into a collision interaction differing somewhat from Eq. (4.235):

$$V_{rR} = \begin{cases} -\dfrac{e^2}{\boldsymbol{R}} + \dfrac{e^2}{|\boldsymbol{R} - \boldsymbol{r}|} & \text{for } 0 \leq R \leq \mathrm{d} \\ 0 & \text{for } \quad R > \mathrm{d} \end{cases} \tag{4.239}$$

As is hardly surprising the initial state growth coefficient decays exponentially when:

$$\frac{3}{2}\frac{kT}{h} \gg \Gamma_v + \Gamma_s; \qquad \omega_0 \gg \Gamma_v + \Gamma_s$$

In this equation ω_0 is the frequency of the undisplaced line; Γ_v is the collisional decay constant of the initial state, and Γ_s is the natural decay constant. One may obtain the intensity as:

$$I = \frac{(\Gamma_v + \Gamma_s)/\pi}{(\omega - \omega_0)^2 + (\Gamma_v + \Gamma_s)^2} \tag{4.240}$$

where:

$$\Gamma_s = \sum_n \Gamma_{0n}{}^s; \; \Gamma_v = \sum_n \Gamma_{0n}{}^v \tag{4.241}$$

For Γ_s the summation is over all states into which the atom can radiate from the initial state while Γ_v is the summation over all states into which the atom can be knocked by a collision.

Landwehr[118] has calculated these collisional decay factors as required by the development of line contours for the first three Balmer lines of hydrogen. The half half-width Γ_v which arises from collision broadening of the atomic state is:

$$\Gamma_{v(n_0 l_0 m_0)} = n_0{}^2 \sqrt{\frac{2\pi m}{kT}}\, e^4 N_e n_0{}^2$$
$$\times \{A - B \ln (\hbar^2/\sqrt{N_e}\, 2\, m\, k\, T\, n_0\, a_0) + C\,\varepsilon\,(\hbar^2/2\, m\, k\, T\, n_0{}^2\, a_0{}^2)\} \tag{4.242}$$

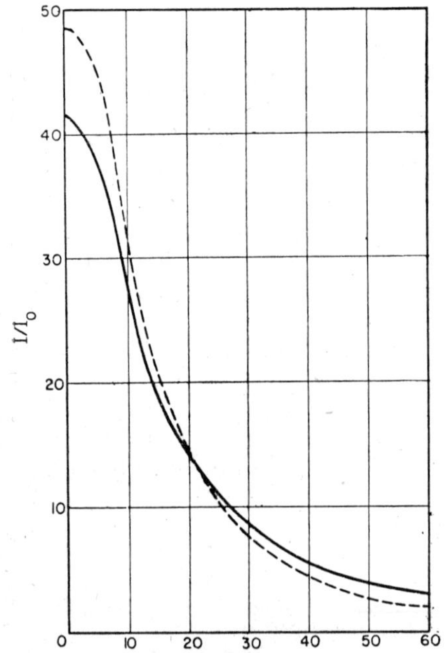

FIG. 4.12. A comparison of the Landwehr with the Interruption result. (Margenau and Lewis[217])

where A, B, and C are constants which depend on (n_0, l_0, m_0). $\varepsilon(x)$ is the exponential integral of x.

This is quite similar to the classical Lindholm result, and, indeed, Griem[65] has shown that the classical Lindholm theory yields half-widths differing from the Landwehr result by no more than the errors inherent in the development. There is some difference between the contours predicted by the two theories, however. The Interruption result is symmetric as is the Landwehr result; however, the latter is lower at line center and higher in the wings than is the former. This is illustrated in Fig. 4.12.

4.24. A classical path consideration to include degeneracy

We shall now consider the theory developed by Kolb[106] and discussed in some detail by Margenau and Lewis[217]. It is not as precisely quantum mechanical as the theories of the last few sections; we have shown and shall illustrate somewhat more conclusively, however, that the classical analogs are generally effective. The study will include the effects of adiabatic and diabatic* hypotheses and a rather direct comparison between them. The development will go about as follows:

We begin with an expression for the intensity in the line based on the quantum average of the square of the dipole moment as given by Eq. (2.151). The matrix elements of the dipole moment which appear in this expression are evaluated in terms of wave functions satisfying a time-dependent Schrödinger equation in the Hamiltonian of which is included the collisional perturbation of the radiator. These eigenfunctions in turn are developed from a base consisting of the unperturbed emitter functions. Under the weak collision assumption the matrix elements of the dipole moment are evaluated in terms of the state growth coefficients relating the perturbed to the unperturbed base. Then for the adiabatic case all elements save the diagonal ones for the dipole moment are supposed zero. Subsequent to this, the intensity expression — in terms of these elements, of course — is expressed in Correlation Function form. The term in this Correlation Function which involves the interactions is supposed separable into electron and ion contributions. When it is assumed that the ionic field produces Statistical Broadening, its contribution ceases to be a problem. The electrons are treated by the additive phase shift approximation, and there is some consideration of the form of the phase shift and the method of the averaging to be carried out. The diabatic approximation is then treated; the result is only applicable in the line wings. This is the general form of the treatment which we shall describe in somewhat greater detail, but no general panacea is presented by the result, and specific consideration for specific circumstances is still required.

We may write down the expression for the difference between the absorption and the induced emission or net absorption as:

$$I_n(\omega) \propto \left\langle \sum_n [\varrho_m(0) - \varrho_n(0)] \lim_{T\to\infty} \frac{1}{T} \left| \int_0^T \mathrm{d}t\, \mu_{nm}^V(t)\, \mathrm{e}^{-i\omega t} \right|^2 \right\rangle \quad (4.243)$$

* Or non-adiabatic for those with pleonastic tendencies such as the author's.

where the ϱ_n and ϱ_m are Boltzmann factors for the states involved in the absorbing transition. The dipole matrix element, $\mu_{nm}{}^V(t)$, is defined by:

$$\mu_{nm}{}^V(t) = \int \bar{\chi}_n(t)\,\mu\,\chi_m(t)\,d\tau \tag{4.244}$$

where:

$$i\hbar\,\dot{\chi}_n(t) = [H_0 + V(t)]\,\chi_n(t) \tag{4.245}$$

Now we suppose the states m and n degenerate in the absence of perturbations and label the corresponding sets of states α_i and ε_f. We now rewrite Eq. (4.243) to include these degeneracies:

$$I(\omega) = \sum_{\varepsilon_f} I_{\varepsilon_f}(\omega) \propto \sum_i \left\{ \sum_{\varepsilon_f \alpha_i} [\varrho_{\alpha_i}(0) - \varrho_{\varepsilon_f}(0)]\, I_{\varepsilon_f \alpha_i} \right\} \tag{4.246a}$$

where:

$$T_{\varepsilon_f \alpha_i} \equiv \frac{1}{2\pi} \left\langle \lim_{T\to\infty} \frac{1}{T} \left| \int_0^T dt\, \mu_{\varepsilon_f \alpha_i}{}^V(t)\, e^{-i\omega t} \right|^2 \right\rangle \tag{4.246b}$$

In our introductory remarks above we mentioned the weak collision hypothesis but did not define it specifically. Let us recall our remarks on directional adiabaticity in Section 2.11. The close collisions that turn the atom around we consider as strong collisions. The distant collisions which do not turn the atom around are the weak collisions of our interest here. We now suppose:

$$\chi_\alpha(t) = \sum_{\alpha'} a_{\alpha'\alpha}(t)\,\psi_{\alpha'}{}^0 \exp\left[-\frac{i}{\hbar} \int_0^t \{E_{\alpha'}{}^0 + V_{\alpha'\alpha'}\}\,dt' \right] \tag{4.247a}$$

where:

$$V_{\alpha'\alpha'} \equiv \int \bar{\psi}_{\alpha'}{}^0\, V\, \psi_{\alpha'}{}^0\, d\tau \tag{4.247b}$$

and, in combination with Eq. (4.245) these equations yield the state growth equations:

$$\dot{a}_{\alpha''\alpha}(t) = \frac{i}{\hbar} \sum_{\alpha'\neq\alpha''}{}' a_{\alpha'\alpha}(t)\, V_{\alpha''\alpha'} \exp[-i\{\omega_{\alpha'\alpha''}{}^0 t + P_{\alpha'\alpha''}\}] \tag{4.248a}$$

where:

$$\omega_{\alpha'\alpha''}{}^0 = \omega_{\alpha'}{}^0 - \omega_{\alpha''}{}^0; \quad \omega_{\alpha'}{}^0 = E_{\alpha'}{}^0/\hbar$$
$$P_{\alpha'\alpha''} = P_{\alpha'} - P_{\alpha''}; \quad P_{\alpha'} = \frac{1}{\hbar}\int_0^t V_{\alpha'\alpha'}\,dt \tag{4.248b}$$

The familiar initial condition assigning the atom to the state at time zero is assumed. Then the weak collision idea allows us to hold down the other state growth coefficients to negligible values for a long enough period of time to allow the following solution to Eq. (4.248a):

$$a_{\alpha''\alpha}(t) = -\frac{i}{\hbar}\int_0^t V_{\alpha''\alpha} \exp[-i(\omega_{\alpha\alpha''}{}^0 t + P_{\alpha\alpha''})]\,dt' \tag{4.249}$$

and Eq. (4.247a) becomes:

$$\chi_\alpha(t) = \psi_\alpha{}^0 \exp[-i(\omega_\alpha{}^0 t + P_\alpha)] + \sum_{\alpha'\neq\alpha} a_{\alpha'\alpha}(t)\,\psi_{\alpha'}{}^0 \exp[-i(\omega_{\alpha'}{}^0 t + P_{\alpha'})] \tag{4.250}$$

Now we may evaluate the matrix elements of the dipole moment as:

$$\begin{aligned}\mu_{\varepsilon\alpha}{}^V(t) = {} & \mu_{\varepsilon\alpha}{}^0 \exp[i(\omega_{\varepsilon\alpha}{}^0 t + P_{\varepsilon\alpha})] + \sum_{\varepsilon\neq\alpha} \mu_{\varepsilon'\alpha}{}^0\, a_{\varepsilon'\varepsilon}(t) \exp[i(\omega_{\varepsilon'\alpha} t \\ & + P_{\varepsilon'\alpha})] + \sum_{\alpha\neq\varepsilon} \mu_{\varepsilon\alpha'}{}^0\, a_{\alpha\alpha'}(t) \exp[i(\omega_{\varepsilon\alpha'}{}^0 t + P_{\varepsilon\alpha'})]\end{aligned} \tag{4.251}$$

The treatment now diverges into an adiabatic or a diabatic one. For the adiabatic treatment we suppose there to be only diagonal elements of the dipole moment and neglect the second and third terms in Eq. (4.251). We now substitute Eq. (4.251) into Eq. (4.246b) to obtain:

$$I_{\varepsilon\alpha} = \frac{|\mu_{\varepsilon\alpha}{}^0|^2}{\delta\pi} \left\langle \lim_{T\to\infty} \frac{1}{T} \left| \int_0^T dt \exp\left[-i(\Delta\omega_{\varepsilon\alpha}{}^0 - P_{\varepsilon\alpha})\right] \right|^2 \right\rangle \tag{4.252}$$

with $\Delta\omega_{\varepsilon\alpha}{}^0 = \omega - \omega_{\varepsilon\alpha}{}^0$; Eq. (4.252) is then expressed in the familiar Correlation Function form:

$$I_{\varepsilon\alpha} = \frac{|\mu_{\varepsilon\alpha}{}^0|^2}{\pi} \Re \int_0^\infty d\tau \exp(i\,\Delta\omega_{\varepsilon\alpha}{}^0\tau) \langle \exp[-i\,P_{\varepsilon\alpha}(\tau)]\rangle \tag{4.253}$$

In $P_{\varepsilon\alpha}$ we have interactions involving both ions and electrons. A splittability is presumed, $P_{\varepsilon\alpha} = P_{\varepsilon\alpha}{}^i + P_{\varepsilon\alpha}{}^e$, which contemplates the separate calculation of the averages over the two particle types. The ions are treated by Statistical Theory so that:

$$P_{\varepsilon\alpha}{}^i = \frac{\Omega_2{}^{\varepsilon\alpha}}{e} E\tau \tag{4.254}$$

where E is the ionic electric field and $\Omega_2{}^{\varepsilon\alpha}$ arises from Eq. (4.134a).

The electrons are treated by the phase shift or Interruption Theory, and their correlation contribution is:

$$\langle \exp[-i\,P_{\varepsilon\alpha}{}^e(\tau)]\rangle = \left\langle \exp\left[-\frac{i}{\hbar}\int_0^\tau \{V_{\varepsilon\varepsilon}{}^e - V_{\alpha\alpha}{}^e\}\,dt\right]\right\rangle \tag{4.255}$$

wherein V^e is the interaction of the emitter (absorber) with all electrons contributing to the broadening. We suppose the perturbations to be additive as:

$$V^e = \sum_p V^p = \sum_{i,v,l} N(i,v,l)\,V(i,v,l) \tag{4.256}$$

$N(i, v, l)$ is the number of electrons having particular values of the three parameters i, v, l while $V(i, v, l)$ is the operator for such a set of parameters. We do not specify the parameters, for during the averaging process it is most convenient to shift to a different set. We have seen several different forms of the averaging process in Chapter 2, and we do not propose to investigate another here. The result of such a process is:

$$\langle \exp[-i\,P_{\varepsilon\alpha}{}^e(\tau)]\rangle = \exp[-(u_1{}^{\varepsilon\alpha} + i\,u_2{}^{\varepsilon\alpha})\,\tau] \tag{4.257a}$$

$$\begin{pmatrix} -u_1{}^{\varepsilon\alpha} \\ -u_2{}^{\alpha\varepsilon} \end{pmatrix} = \begin{pmatrix} \Re \\ \Im \end{pmatrix} 2\pi N_e \int_0^\infty v\,W(v)\,dv \int_0^R \varrho\,d\varrho \int_0^\Sigma \frac{d\sigma}{\Sigma} \{\exp[-i\,\varphi_{\varepsilon\alpha}(\infty)] - 1\} \tag{4.257b}$$

$$\varphi_{v,l}(\tau) \equiv \varphi_{\varepsilon\alpha}(\tau) \equiv \hbar^{-1} \int_0^\tau [(V(v,l))_{\varepsilon\varepsilon} - (V(v,l))_{\alpha\alpha}]\,dt \tag{4.257c}$$

These results allow us to rewrite Eq. (4.253) as:

$$\begin{aligned} I_{\varepsilon\alpha} &= \frac{|\mu_{\varepsilon\alpha}{}^0|^2}{\pi} \Re \left\langle \int_0^\infty \exp\left[i\left(\Delta\omega_{\varepsilon\alpha}{}^0 - u_2{}^{\varepsilon\alpha} - \frac{\Omega_2{}^{\varepsilon\alpha}}{e} E\right)\tau - u_1{}^{\varepsilon\alpha}\tau\right] d\tau \right\rangle_{\text{ions}} \\ &= \frac{|\mu_{\varepsilon\alpha}{}^0|^2}{\pi} \left\langle \frac{u_1{}^{\varepsilon\alpha}}{(\Delta\omega_{\varepsilon\alpha}{}^0 - \Omega_2{}^{\varepsilon\alpha} E/e - u_2{}^{\varepsilon\alpha})^2 + (u_1{}^{\varepsilon\alpha})^2} \right\rangle_{\text{ions}} \end{aligned} \tag{4.258}$$

In Eq. (4.257b) $W(v)$ is the Maxwell–Boltzmann velocity distribution; v is the electron velocity; ϱ is the distance of closest approach of the broadener, and σ is an open parameter.

From this point on one must treat specific spectral lines, so we shall not follow matters quite so closely. A specific consideration has been carried out by Kolb to determine (1) the error which is introduced by those collisions which do not satisfy the Interruption criterion and (2) the effect of close collisions. If we suppose the broadening due solely to electrons and

$$\varphi_{\varepsilon\alpha}(\infty) = \frac{2\,\Omega_2^{\varepsilon\alpha}}{\varrho\, v} \cos\vartheta$$

where ϑ is the additional parameter, σ, which was mentioned previously. Then from Eq. (4.134a) we obtain:

$$\Omega_2^{\varepsilon\alpha} = \frac{3e^2 a_0}{2\,\hbar}\{[n^\varepsilon(k_1^\varepsilon - k_2^\varepsilon)] - [n^\alpha(k_1^\alpha - k_2^\alpha)]\}$$

and Eq. (4.257b) becomes:

$$u_2^{\varepsilon\alpha} = 0 \tag{4.259a}$$

$$u_1^{\varepsilon\alpha} = 2\pi N_e \int_0^\infty v\, W(v)\, dv \int_0^{\varrho_m} \varrho\, d\varrho \left[1 - \frac{\sin(2\,\Omega_2^{\varepsilon\alpha}/\varrho v)}{(2\,\Omega_2^{\varepsilon\alpha}/\varrho v)}\right] \tag{4.259b}$$

In Eq. (4.259b) ϱ_m is a cut-off distance for which a reasonable choice appears to be the Debye length. If someone can think of something else to choose, one simply has a justification argument. We shall not follow through the specific evaluation any further, but, rather, we turn to the two points mentioned above.

Now we may recall the criterion for an Interruption assumption as:

$$\Delta\omega \ll v\left(\frac{\hbar\, v}{C_n}\right)^{1/(n-1)}$$

When we substitute the Stark constant into this equation, recalling that $n = 2$ for the linear Stark effect, there results:

$$\Delta\omega \ll v\left(\frac{v}{\Omega_2^{\varepsilon\alpha}}\right) \iff v \gg \sqrt{\Omega_2^{\varepsilon\alpha} \cdot \Delta\omega} \tag{4.260}$$

With Eq. (4.260) and the result of a somewhat more specific evaluation of Eq. (4.259b) one can estimate the error involved in the inclusion of all velocities for a specific case. For 10000 °K, for $H\beta$, and for a wave length separation of thirty angstroms from line center, one finds that ninety percent of the velocity contribution to the result come from velocities which satisfy Eq. (4.260).

Insofar as close collisions are concerned, they may be treated by simple Interruption Theory. If one divides the contribution from Eq. (4.259b) into a weak collision contribution ($\varrho > \varrho_c$) as treated above and a strong collision contribution ($\varrho < \varrho_c$, $\varrho_c = 2\,\Omega_2^{\varepsilon\alpha}/\langle v\rangle$) as treated by simple Interruption Theory and calculate the result, one obtains the following criterion: the contributions from close collisions may be neglected if,

$$(u_1^{\varepsilon\alpha})_{\varrho<\varrho_c}/(u_1^{\varepsilon\alpha})_{\varrho>\varrho_c} \ll 1$$

or:

$$\ln(\varphi_n \langle v\rangle/2\,\Omega_2^{\varepsilon\alpha}) \gg 1.5 \tag{4.261}$$

As an example, if the electron density is 10^{15}, the temperature 20,000 °K and $\Omega_2^{\varepsilon\alpha}/A = 1.1$ (where $A \equiv 3/2\, e^2 a_0 \hbar$) then $\ln(\varrho_m \langle v \rangle / 2\Omega_2^{\varepsilon\alpha}) \doteq 6$. Now recall that this has been an adiabatic consideration of the broadening based on the assumption that only the first term in Eq. (4.251) was non-zero.

We neglect the ion contributions, so that the states involved are degenerate without question. Further, we still maintain the weak collision approximation so that $|P_{\varepsilon\alpha}^e| \ll 1$. The application of this approximation to Eq. (4.251) results in:

$$\mu_{\varepsilon\alpha}^V(t) = \mu_{\varepsilon\alpha}^0 \exp(i\,\omega_{\varepsilon\alpha}^0 t) \left\{ 1 + i\,P_{\varepsilon\alpha}^e + \sum_{\varepsilon' \neq \varepsilon} \frac{\mu_{\varepsilon'\alpha}^0}{\mu_{\varepsilon\alpha}^0} \bar{a}_{\varepsilon'\varepsilon} + \sum_{\alpha' \neq \alpha} \frac{\mu_{\varepsilon\alpha'}^0}{\mu_{\varepsilon\alpha}^0} a_{\alpha'\alpha} \right\} \tag{4.262}$$

The last three terms on the right of Eq. (4.262) are small enough so that the bracketed expression can be considered the expansion of the exponential:

$$\mu_{\varepsilon\alpha}^V(t) = \mu_{\varepsilon\alpha}^0 \exp(i\,\omega_{\varepsilon\alpha}^0\, t) \exp(i\,\Phi_{\varepsilon\alpha}) \tag{4.263a}$$

so that:

$$\Phi_{\varepsilon\alpha} \equiv P_{\varepsilon\alpha}^e - i \left[\sum_{\varepsilon' \neq \varepsilon} \frac{\mu_{\varepsilon'\alpha}^0}{\mu_{\varepsilon\alpha}^0} \bar{a}_{\varepsilon'\varepsilon} + \sum_{\alpha' \neq \alpha} \frac{\mu_{\varepsilon\alpha'}^0}{\mu_{\varepsilon\alpha}^0} a_{\alpha'\alpha} \right] \tag{4.263b}$$

Now the fact that the states are truly degenerate here zeroes out the $\omega_{\alpha\alpha''}^0$ in Eq. (4.249). The weak collision approximation then renders the exponent very nearly zero so that:

$$a_{\alpha'\alpha} = -\frac{i}{\hbar} \int_0^t V_{\alpha'\alpha}^e \exp(-i\,P_{\alpha\alpha'}^e)\, dt' \doteq -\frac{i}{\hbar} \int_0^t V_{\alpha'\alpha}^e\, dt' \tag{4.264}$$

As in the adiabatic case, scalarly additive perturbations are supposed so that:

$$\Phi_{\varepsilon\alpha} \equiv \sum_{i,v,l} N(i,v,l)\, \varkappa_{ivl} \tag{4.265a}$$

$$\varkappa_{ivl}(t) = \frac{1}{\hbar} \int_0^t dt' \left\{ \sum_{\varepsilon' \neq \varepsilon} \frac{\mu_{\varepsilon'\alpha}^0}{\mu_{\varepsilon\alpha}^0} [V(i,v,l)]_{\varepsilon'\varepsilon} - \sum_{\alpha' \neq \alpha} \frac{\mu_{\varepsilon\alpha'}^0}{\mu_{\varepsilon\alpha}^0} [V(i,v,l)]_{\alpha'\alpha} \right\} \tag{4.265b}$$

In further analogy to the adiabatic case we obtain:

$$I_{\varepsilon\alpha} = \frac{|\mu_{\varepsilon\alpha}^0|^2}{\pi} \frac{u_1^{\varepsilon\alpha}}{(\Delta\omega_{\varepsilon\alpha}^0 - u_2^{\varepsilon\alpha})^2 + (u_1^{\varepsilon\alpha})^2} \tag{4.266a}$$

$$\begin{pmatrix} -u_1^{\varepsilon\alpha} \\ -u_2^{\varepsilon\alpha} \end{pmatrix} = \begin{pmatrix} \mathfrak{R} \\ \mathfrak{J} \end{pmatrix} 2\pi N_e \int_0^\infty v\, W(v)\, dv \int_0^R \varrho\, d\varrho \int_0^\Sigma \frac{d\sigma}{\Sigma} \{\exp[-i\,\varkappa_{\varepsilon\alpha}(\infty)] - 1\} \tag{4.266b}$$

Eqs. (4.266) are really valid only in the wings of the line since $|P_{\varepsilon\alpha}| \ll 1$ is certainly violated after a long enough period of time. In the line wings Eqs. (4.266) become:

$$I_{\varepsilon\alpha} = \frac{u_1^{\varepsilon\alpha}}{\pi(\Delta\omega_{\varepsilon\alpha}^0)^2} \tag{4.267a}$$

$$u_1^{\varepsilon\alpha} = 2\pi N_e \int_0^\infty W(v)\, v\, dv \int_0^R \varrho\, d\varrho \int_0^\Sigma \frac{d\sigma}{\Sigma} \frac{1}{2\hbar^2} \cdot \left[\int_{-\infty}^{+\infty} dt \left\{ \sum_{\varepsilon' \neq \varepsilon} \mu_{\varepsilon'\alpha}^0 V_{\varepsilon\varepsilon'} - \sum_{\alpha' \neq \alpha} \mu_{\alpha'\alpha}^0 V_{\alpha'\alpha} \right\} \right] \tag{4.267b}$$

We have again arrived at the departure point of special cases. Margenau and Lewis considered the broadening of the Lyman lines by both the adiabatic and diabatic theories discussed above. We shall not detail their considerations, but we do point out the important conclusion to be drawn from them. The adiabatic case wherein the off-diagonal elements are neglected yields only one third the broadening of the diabatic case. The idea now is to use the knowledge acquired from this specific case to estimate the diabatic contributions to other related situations. For example, Kolb, in accounting for the combined effects of ions and electrons on the Lyman α, applies an adiabatic treatment and then supposes that the diabatic effects are approximately the same as they are in the case of electrons alone.

4.25. The applicability of the classical treatment

The success of the Holtsmark Theory over the years for those cases where the dominant broadening agency is ions should be sufficient balm to the consciences of those who wish to use classical treatments for ionic Stark broadening. The size and lumbering nature of the particles involved makes this almost intuitive. Electrons are something else again, however, One would expect that these particles would almost always require quantum treatments. We shall now consider why this is not the case. First, we shall develop a general criterion established by Margenau and Lewis[217] for the classical path and, subsequently, we shall consider the specific comparison established by Meyerott and Margenau[154] between the spectrum half widths prognosticated by the KBM universal broadening and the classical impact theory.

Let us begin by recalling, for example, Jablonski's appeal to correspondences (Section 3.7) to equate classical and quantum angular momenta for high angular momentum values. If one divides the equivalence assumption, $mrv \doteq l\hbar$, by one form of the uncertainty principle, $m\,\Delta r\,\Delta v = \hbar$, one obtains $(\Delta r/r)\,(\Delta v/v) = 1/l$. Then the criterion yielded here by large angular momentum (large l) for the classical path assumption is:

$$\frac{\Delta r}{r}\frac{\Delta v}{v} \ll 1$$

Now if we consider the perturber as a quantum wave packet of mean spatial extent a, then a minimum criterion for classical description will surely be that $a \doteq r$ where r is the separation of emitter and packet center. Now let that region of space where the interaction is sufficient to warrant our consideration be of dimension d. Then, if the perturber diffuses at a velocity $\hbar/ma$, the time of collision or traversal of the collision region must be much greater than the time of diffusion through the emitter perturber separation r. This must also hold for $a = r$, and if we recall the de Broglie wavelength, λ, there results:

$$a \gg \sqrt{d\lambda} \tag{4.268}$$

We shall develop a somewhat more specific criterion from this.

We recall the interaction energy of collision as generally expressible in the form:

$$V(r) = c_j/r^j = c_j/(\varrho^2 + v^2 t^2)^{-j/2}$$

Then the momentum exchange and the phase shift may be evaluated as:

$$\Delta p = - \int_{-d/2}^{+d/2} \Delta V(t)\,\mathrm{d}t = - c_j \int_{-d/2}^{+d/2} (\varrho^2 + v^2 t^2)^{-j/2} \frac{\mathrm{d}x}{v}$$
$$= j \frac{C_j \sqrt{\pi}\, \Gamma[(j+1)/2]}{v \varrho^j\, \Gamma[(j+2)/2]} \doteq \frac{V(\varrho)}{v} \tag{4.269a}$$

$$\eta = \frac{1}{\hbar} \int_{-d/2}^{+d/z} V(t)\,\mathrm{d}t = \frac{c_j}{\hbar v \varrho^{j-1}} \sqrt{\pi} \frac{\Gamma[(j-1)/2]}{\Gamma(j/2)} \tag{4.269b}$$

We obtain C_j from Eq. (4.269b) and substitute it in Eq. (4.269a) to obtain:

$$\Delta p = \left(\frac{\varrho_c}{\varrho}\right)^{j-1} \left(\frac{j-1}{\varrho}\right) \hbar \tag{4.270}$$

where ϱ_c is the distance of closest approach corresponding to unit phase shift, critical to the simple Interruption Theory. We may assure the correctness of the classical picture if we suppose the momentum exchange to be greater than the momentum uncertainty:

$$\Delta p \doteq \frac{V(\varrho)}{v} \gg \frac{\hbar}{a} \Rightarrow \varrho \ll \alpha_j \varrho_c \left(\frac{a}{\varrho}\right)^{1/(j-1)} \tag{4.271}$$

from Eq. (4.270) and by letting $\alpha_j = (j-1)^{1/(j-1)}$. Now $a \leq r$ during the collision and r may become equal to ϱ so that the following inequality should hold:

$$\varrho \ll \alpha_j \varrho_c \tag{4.272}$$

Since Eq. (4.272) is essentially the condition $\eta < 1$, this means that collisions within the optical collision diameter may be treated classically. Next, from Eqs. (4.268) and (4.272) we find:

$$\sqrt{d\lambda} \ll \varrho_c \tag{4.273}$$

The spatial extent of the perturbing interaction may be taken as the Debye radius, $6{\cdot}90 \sqrt{T/N}$, with fully the justification of our shielding considerations. When we substitute for λ and ϱ_c, Eq. (4.273) becomes our specific criterion for the classical path:

$$\frac{T^{1/(j-1)}}{\sqrt{N}\, M^{3/2/(j-1)}} \ll \frac{[\gamma_j (c_j/\hbar)]^{2/(j-1)}}{6.90\, \hbar (3k)^{1/2(j-1)}} \tag{4.274}$$

where we take v as the *rms* velocity of the perturber $\sqrt{3kT/M}$, and γ_j are of the order of unity.

Some illustrative values of the limiting densities and temperatures for the application of classical path treatments to electrons and protons are given in the following two tables. These were developed by Margenau and Lewis[217].

Now we shall consider the close relationship which proves to exist between the half widths predicted by the universal broadening considerations of Kivel, Bloom, and Margenau[104] and those predicted by classical Interruption Theory. Although the result applies specifically to the hydrogen atom case due to the selection of the potential, the justification for the application of the classical theory to an even wider domain is indicated.

One may express L_U (see Eq. (4.236b)) approximately as:

$$L_U = 4.31(1 - e^{-4.1x}) \tag{4.275}$$

TABLE 4.3

BROADENING OF $H\alpha$ BY FIRST-ORDER STARK EFFECT. THE TEMPERATURES ARE THOSE BELOW WHICH THE CLASSICAL PATH HOLDS FOR ELECTRONS AND FOR PROTONS FOR THE VALUES OF THE DENSITY INDICATED

N cm^{-3}	°K protons	°K electrons
10^{12}	4×10^6	50
10^{14}	4×10^7	500
10^{16}	4×10^8	5000
10^{18}	4×10^9	50000

TABLE 4.4

BROADENING BY SECOND-ORDER STARK EFFECT. THE TEMPERATURES ARE THOSE BELOW WHICH THE CLASSICAL PATH HOLDS

N cm^{-3}	°K protons	°K electrons
10^{12}	10^7	0·1
10^{14}	10^{10}	100
10^{16}	10^{15}	10^5
10^{18}	10^{16}	10^8

The collision frequency ($\nu_c = \gamma_U$) obtainable through Eq. (4.237b) from Eq. (4.275) is plotted in Fig. 4.13. The classical result with which we shall compare this may be taken from, say, Eq. (2.26a) as:

$$\gamma_c = Nv \int_0^\infty 4\pi \varrho \, d\varrho \sin^2 \frac{2\eta}{2} \tag{4.276}$$

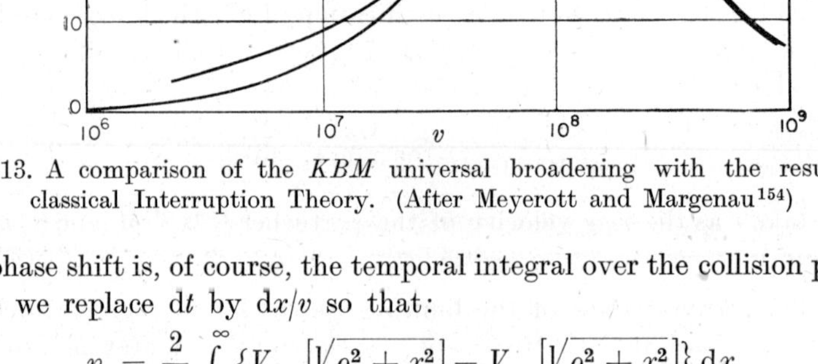

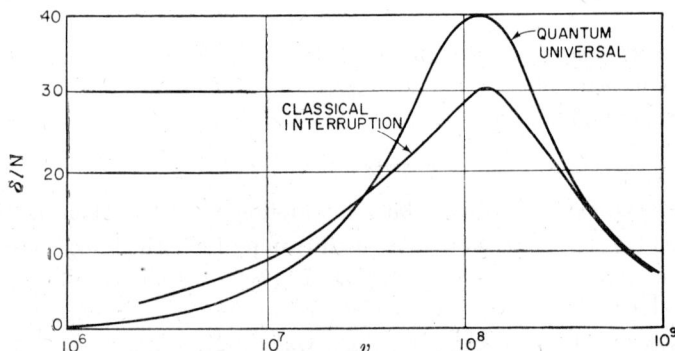

FIG. 4.13. A comparison of the *KBM* universal broadening with the results of classical Interruption Theory. (After Meyerott and Margenau[154])

The phase shift is, of course, the temporal integral over the collision potential, and here we replace dt by dx/v so that:

$$\eta = \frac{2}{v} \int_0^\infty \{V_{2p}[\sqrt{\varrho^2 + x^2}] - V_{1s}[\sqrt{\varrho^2 + x^2}]\} \, dx \tag{4.277}$$

The potentials in question are those acting on the free electron when the hydrogen atom is in its upper state:

$$V_{2p}(r) = \frac{e}{a}\left[\frac{1}{r} + \frac{3}{4} + \frac{r}{4} + \frac{r^2}{24}\right] e^{-r} \tag{4.278a}$$

and those acting when the atom is in its lower state:

$$V_{1s}(r) = \frac{e}{a}\left[\frac{1}{r} + 1\right] e^{-2r} \tag{4.278b}$$

Angular dependence has been averaged out. The results of numerical integration of Eq. (4.277) are shown in Fig. 4.13.

This calculation tends to indicate that electrons are amenable to at least reasonably approximate classical treatment over a wide range. Let us re-emphasize two facets of this universal effect. Unfortunately, it neglects first-order Stark effects by its ignoration of degeneracy, and secondly, it ignores second-order Stark effects by failing to consider matrix elements linking higher states. At any rate, Eq. (4.274) and Tables 4.3 and 4.4 indicate rather clearly that there will seldom be a necessity for quantum considerations of ionic broadening and that one will often be able to apply classical treatments to electrons.

4.26. Synthesis of electronic and ionic Stark broadening

We have considered the Stark broadening by ions alone and the broadening by electrons alone. To a limited extent in Section 4.24 we considered the simultaneous broadening by ions and electrons; however, we dropped the dualism before the development was completed. Surely though, it has been apparent that some synthesis of the two effects is called for and we now turn our attention to this synthesis.

By applying Correlation Function treatment we were able to show the limiting positions which The Statistical and Interruption Theories occupy with respect to it. Before this development, however, it was felt that some synthesis of Interruption and Statistical effects was called for, and it was well known that one could not simply add half widths or do something equally inane with regard to the line shapes themselves. In this regard it was suggested by Margenau and others that this synthesis be accomplished by folding one distribution into the other. Suppose we have the simple Interruption Theory expression for a particular frequency. In order to get the actual spectral line intensity corresponding to that frequency we average the Interruption expression over the Statistical distribution. Now this approach, as we have remarked, has for obvious reasons lost favor in broadening situations such as, say, straight van der Waals, but it has been applied recently to the combination of ionic and electronic Stark Broadening.

Let us discuss one way in which we might carry this out. Consider Eq. (4.77). $I(E, \nu)$ is a function which describes the frequency dependent intensity between the outermost Stark components as determined by the electric field strength E.

Now we took this intensity distribution as a rectangle which, of course, we knew it was not. Let us describe it as follows:

The field strength E takes however many Stark components there are and places them here and there between the outermost ones. This field is provided by the ionic distribution. The electrons broaden each component into a symmetrical Interruption Shape. $I(E, \nu)$ now describes this distribution of symmetrical Interruption shapes corresponding to E. To complete the synthesis we now carry out the integration over all field strengths and corresponding broadened component distributions. This calculation can be carried out on an IBM 704 without a great deal of trouble. Although it has not been carried out, similar smearing distributions have been so calculated.

As an example, Griem[65] has chosen the following smeared distribution for a Balmer line:

$$I(\lambda) = \frac{\Delta}{\pi} \frac{I_0(\lambda)}{(\lambda - \lambda_0)^2 + (\delta/2)^2} + \frac{\delta/2}{\pi E_0} \int_{-\infty}^{+\infty} \frac{\langle W(x/E_0) \rangle \, dx}{(\delta/2 - x)^2 + (\delta/2)^2} \tag{4.279}$$

Here λ_0 is the wavelength of the unperturbed line. The first term yields the effect of the unshifted component which one encounters in the case of these lines. The function $\langle W(x/E_0) \rangle$ includes averaging over different Stark components. The reader may refer to Margenau and Lewis for a discussion of the improvement of this treatment made in the agreement with experiment. This

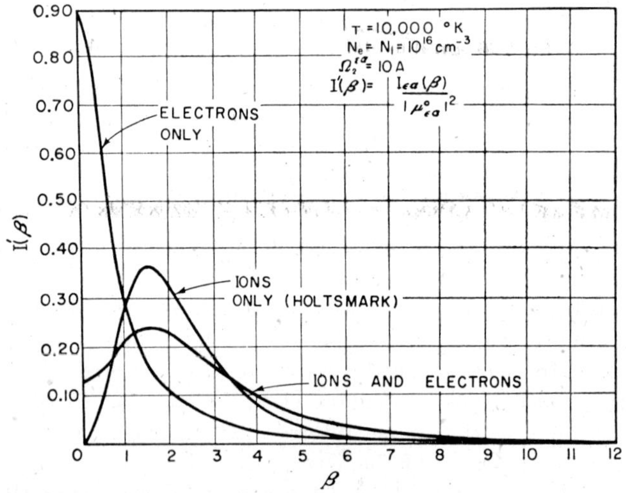

FIG. 4.14. Line contours for electrons and ions individually as well as for ions and electrons in combination. (After Holb[106])

would hardly seem surprising if the synthesis and its parts are justified. Actually the assumption of a classical electron distribution here is not justified for a first-order Stark Effect since this phenomenon contributes to the broadening through small phase shifts. For the second order Stark Effect it apparently is a justifiable treatment.

The method of this classical synthesis should be apparent. The actual application of it will, in general, have to be considered for specific cases. In the previous paragraph we remarked that the classical consideration of electron contributions was not often justified. This means we appeal to the quantum treatment of them which was given by Kolb[106] and we now consider briefly.

The equation which we apply is Eq. (4.258), and we further assume the adiabatic approximation, so that Eqs. (4.259) also apply. The result is:

$$I_{\varepsilon\alpha} = \frac{|\mu_{\varepsilon\alpha}{}^0|^2}{\pi} \int_0^\infty W(\Omega)\, d\Omega \left\{ \frac{u_1{}^{\varepsilon\alpha}}{(\Delta\omega_{\varepsilon\alpha}{}^0 - \Omega_2{}^{\varepsilon\alpha}\Omega)^2 + (u_1{}^{\varepsilon\alpha})^2} \right\} \tag{4.280}$$

In this equation $u_1{}^{\varepsilon\alpha}$ is independent of Ω. Here $W(\Omega)$ is the simple Holtsmarkian distribution. After some mathematical manipulation culminating in a contour integration Kolb obtains:

$$I_{\varepsilon\alpha} = \frac{2}{\pi} \frac{|\mu_{\varepsilon\alpha}{}^0|^2}{(\Omega_2{}^{\varepsilon\alpha})^2} \int_0^\infty \xi\, d\xi \exp\left[(-4{\cdot}21\, \Omega_2{}^{\varepsilon\alpha} n_i\, \xi^{3/2} - u_1{}^{\varepsilon\alpha}\xi)/\Omega_2{}^{\varepsilon\alpha}\right]$$
$$\cdot \left[\Delta\omega_{\varepsilon\alpha}{}^0 \sin\left(\frac{\Delta\omega_{\varepsilon\alpha}{}^0\,\xi}{\Omega_2{}^{\varepsilon\alpha}}\right) + u_1{}^{\varepsilon\alpha}\cos\left(\frac{\Delta\omega_{\varepsilon\alpha}\,\xi}{\Omega_2{}^{\varepsilon\alpha}}\right)\right] \tag{4.281}$$

The reduction to the Holtsmark distribution for zero electron density or the Interruption distribution for zero ion density is pointed out by Kolb. Should the reader be interested in applying the Kolb theory to a particular spectral line, he will face the necessity of numerical integration of Eq. (4.281). Kolb did carry out evaluations of Eq. (4.281) for the asymptotic cases of large $\Delta\omega_{\varepsilon\alpha}{}^0$ and small $\Delta\omega_{\varepsilon\alpha}{}^0$. These we shall not detail. We do remark Fig. 4.14, however, which is an example worked out by Kolb of the individual and composite effects of ions and electrons.

4.27. The Baranger–Mozer treatment of electron and ion broadening

Of quite recent vintage is the work of Baranger and Mozer[6,7] who have divided the contributions of the ions and electrons into low and high frequency components of the field, respectively. Their picture of the physical phenomena involved in Stark Broadening goes about as follows:

The electric field is made up of the two components, generated by ions and electrons. The ions contributing to the low frequency component have their field shielded by the cloud of electrons which surrounds them and accompanies their relatively slow motion or low frequency. In addition, these authors consider the correlations which take place between pairs of these particles, and these ion-ion correlations must account for any shielding of ionic fields by ions. On the other hand the electrons whose rapid motions produce the high frequency field component are not treated by a shielded field of, say, the Debye–Hückel type, but are considered as acting against a uniform neutralizing background. The shielding effect of other electrons may be considered as entering here, however, through the correlations introduced into their field.

In deriving the cluster expansion we begin by considering that the electric field E at a point is made up of a contribution E_1 from a particle at x_1, E_2 from a particle at x_2 and so on, with the result:

$$\boldsymbol{E} = \boldsymbol{E}_1 + \boldsymbol{E}_2 + \boldsymbol{E}_3 + \cdots + \boldsymbol{E}_n \tag{4.282}$$

Although the field strength probability function $W(E)$ is desired, it is found to be more convenient to compute its Fourier transform:

$$F(\boldsymbol{k}) = \int \exp(i\,\boldsymbol{k}\cdot\boldsymbol{E})\,W(\boldsymbol{E})\,\mathrm{d}\boldsymbol{E} \tag{4.283}$$

which may be re-written in terms of the probability of finding the field producing particle distribution corresponding to the field:

$$F(\boldsymbol{k}) = \int \exp[i\,\boldsymbol{k}\cdot(\boldsymbol{E}_1 + \cdots + \boldsymbol{E}_n)]\,P(x_1, \ldots, x_n)\,\mathrm{d}x_1 \ldots \mathrm{d}x_n \tag{4.284}$$

We recall the familiar artifice

$$\exp(i\,\boldsymbol{k}\cdot\boldsymbol{E}_i) = 1 + [\exp(i\,\boldsymbol{k}\cdot\boldsymbol{E}_i) - 1] = 1 + \varphi_i$$

and expand the product of n such factors as:

$$1 + \sum_1 \varphi_i + \sum_2 \varphi_i\varphi_j + \cdots \tag{4.285}$$

In Eq. (4.285) the second term is the sum over all particles, the third term is the sum over all pairs of particles, and so on. Eq. (4.284) is now replaced by a sum with terms corresponding to those in Eq. (4.285). For a given term, say the M-th, one can surely integrate over all coordinates save those referring to the M particles in question. Thus, we may replace the configurational probability function of Eq. (4.284) by a corresponding function referring only to the M particles, Eq. (4.284), therefore, is replaced by:

$$F(\boldsymbol{k}) = \sum F_M(\boldsymbol{k}) \tag{4.286a}$$

$$F_M(\boldsymbol{k}) = \sum_M \int \varphi_i \ldots \varphi_s\,P_M(x_i, \ldots, x_s)\,\mathrm{d}x_1 \ldots \mathrm{d}x_s \tag{4.286b}$$

The sum in Eq. (4.286b) is over all possible combinations of the n particles taken M at a time.

If we suppose two body correlations to be less important than no correlations, three body less so than two, and so on, an expansion whose successive terms involve successive orders of correlations is indicated. Let us consider it for the M particles in Eq. (4.286b). If there is no correlation, we obtain a probability function which is simply the product of M single-particle probability functions. The second term of series of terms is concerned with two-body interactions or correlations and so on. The probability expression may then be written as:

$$\begin{aligned} V^M P_M(x_i, \ldots, x_s) &= \Pi\, g_1(x_i) + \sum_2 g_2(x_j, x_k)\,\Pi\, g_1(x_i) \\ &+ \sum_{22} g_2(x_i, x_k)\,g_2(x_l, x_m)\,\Pi\, g_1(x_i) + \sum_{222}\cdots + \sum_3 g_3(x_j, x_k, x_l)\,\Pi\, g_1(x_i) \\ &+ \sum_{33} g_3(x_j, x_k, x_l)\,g_3(x_m, x_n, x_p)\,\Pi\, g_1(x_i) \\ &+ \sum_{333}\cdots + \sum_{32} g_3(x_j, x_k, x_l)\,g_2(x_m, x_n)\,\Pi\, g_1(x_i) + \cdots \end{aligned} \tag{4.287}$$

In Eq. (4.287) $\sum_2$ stands for the sum over all pairs among the particles; $\sum_{22}$ is the summation over all pairs of pairs; $\sum_{32}$ is the sum over all combinations of a triplet and a pair, and, again, so on. Remark that the volume has been removed from the probability functions so that the $g_1(x_i)$ are unity for no charge at the origin. Eq. (4.287) is substituted into Eq. (4.286b) for the limit of very large V and n: the result is:

$$F(\boldsymbol{k}) = G_1(\boldsymbol{k})\, G_2(\boldsymbol{k})\, G_3(\boldsymbol{k}) \tag{4.288a}$$

$$\begin{aligned} G_p(\boldsymbol{k}) &= 1 + V^{-p} \sum_p \int \varphi_i \ldots \varphi_0\, g_p(x_i, \ldots, x_s)\, \mathrm{d}x_1 \ldots \mathrm{d}x_s \\ &+ V^{-2p} \sum_{pp} \int \varphi_i \ldots \varphi_v\, g_p(x_i, \ldots, x_s)\, g_p\,(x_t, \ldots, x_v)\, \mathrm{d}x_i \ldots \mathrm{d}r_v + \cdots \end{aligned} \tag{4.288b}$$

Here, for example, $\sum_p$ runs over all combinations of particles of the total n. The symbol $\sum_{pp}$ refers to all possible distinct combinations of two clusters containing P particle each. Now n is allowed to become infinite with the result:

$$G_p(\boldsymbol{k}) = \exp\,[(n^p/p!)\, h_p(\boldsymbol{k})] \tag{4.289a}$$

$$h_p(\boldsymbol{k}) = \int \varphi_i \ldots \varphi_p\, g_p(x_i, \ldots, x_p)\, \mathrm{d}x_i \ldots \mathrm{d}x_p \tag{4.289b}$$

so that Eq. (4.284) becomes:

$$F(\boldsymbol{k}) = \exp\left[\sum_{p=1}^{\infty} (n^p/p!)\, h_p\,(\boldsymbol{k})\right] \tag{4.290}$$

Each term in the series involves larger and larger clusters, the first term alone reducing, as is to be expected, to a Holtsmark distribution. The field strength probability function is, of course:

$$W(\boldsymbol{E}) = \frac{1}{(2\pi)^3} \int \exp\,(-i\boldsymbol{k} \cdot \boldsymbol{E})\, F(\boldsymbol{k})\, \mathrm{d}\boldsymbol{k} \tag{4.291}$$

In treating the high frequency component of the field due to the electrons, Baranger and Mozer choose the unshielded Coulomb field:

$$E_i = \frac{e\,\boldsymbol{x}_i}{r_i^{\,3}} \tag{4.292}$$

Since the origin is neutral, g, is unity and when Eq. (4.292) is substituted into Eq. (4.289b) the result is:

$$h_1(k) = -\frac{4}{15}\,(2\,\pi\, e\, k)^{3/2} \tag{4.293}$$

The pair correlation function is taken from Debye–Hückel Theory as:

$$\zeta\,(x_1, x_2) = \frac{1}{V^2} \exp\,(-e^2\, \Phi_{12}/kT) \tag{4.294a}$$

$$\Phi_{12} = \frac{\exp\,(-|\boldsymbol{x}_1 - \boldsymbol{x}_2|/D)}{|\boldsymbol{x}_1 - \boldsymbol{x}_2|}\,; \qquad D = \sqrt{\frac{kT}{4\pi\, N\, e^2}} \tag{4.294b}$$

From Eq. (4.287)

$$g_2(x_1, x_2) = \exp(-e^2 \Phi_{12}/kT) - 1$$

or since the exponent here is assumed small:

$$g_2(x_1, x_2) = -\frac{e^2 \Phi_{12}}{kT} \tag{4.294c}$$

so that

$$h_2(\boldsymbol{k}) = -\frac{e^2}{kT} \int\int \varphi_1 \varphi_2 \Phi_{12} \, dx_1 \, dx_2 \tag{4.294d}$$

Now all factors in the integrand are expanded in spherical harmonics:

$$\varphi_i = \sum_l i^l \sqrt{4\pi(2l+1)} \left[j_l \left(\frac{e k}{r_i^2} \right) - \delta_{lD} \right] Y_{l\vartheta}(\vartheta_i, \omega_i) \tag{4.295a}$$

$$\begin{aligned} \Phi_{12} &= \sum_l \sqrt{\frac{1}{4\pi}(2l+1)}\, f_l(r_1 r_2 D)\, Y_{l\vartheta}(\vartheta_{12}, \omega_{12}) \\ &= \sum_{lm} f_l(r_1 r_2 D)\, \overline{Y}_{lm}(\vartheta_1, \omega_1)\, Y_{lm}(\vartheta_2, \omega_2) \end{aligned} \tag{4.295b}$$

Where the j_l are spherical Bessel functions. The angular integration in Eq. (4.294d) can now be carried out; and the resulting series is alternating, convergent, and well approximated by the first three terms.

When a unit length is defined as

$$\frac{4}{15}(2\pi)^{3/2} r_3{}^0 N = 1$$

A unit field strength, $E_0 = e/r_0^2$, and the Holtsmarkian dimensionless field strength $\beta = E/E_0$ may be defined, and $x = kE_0$ then

$$H(\beta) = 4\pi^2 W(\beta) = \frac{2\beta}{\pi} \int_0^\infty \sin(\beta x) F(x)\, x \, dx \tag{4.296}$$

The first and second terms in Eq. (4.290) may be written as:

$$N h_1(\boldsymbol{k}) = -x^{3/2} \tag{4.297a}$$

$$\frac{1}{2} N^2 h(\boldsymbol{k}) = x^{3/2} \psi(x, y); \quad y = \frac{r_0}{D} \tag{4.297b}$$

The function ψ is dependent on $\sqrt{x}\, y$ and plotted in Fig. 4.15. As a consequence Eq. (4.290) becomes:

$$F(x) = \exp\left\{-x^{3/2}\left[1 - \psi\left(\sqrt{x}\, y\right)\right]\right\} \tag{4.298}$$

This is good when the two body correction is small compared to the Holtsmark term and this is true for $0 \leq r_0/D \leq 1$. Although the authors developed a series expansion for Eq. (4.296) from Eq. (4.298) for large β, for most cases of interest the former was numerically evaluated.

We recall that this derivation has been concerned with the high frequency component of the field which is due to the unshielded electrons and given by

Eq. (4.292). Now we consider the electron shielded fields of the ions, that is, the low frequency field component. The shielded field due to an ion is (again adopting the Debye–Hückel potential):

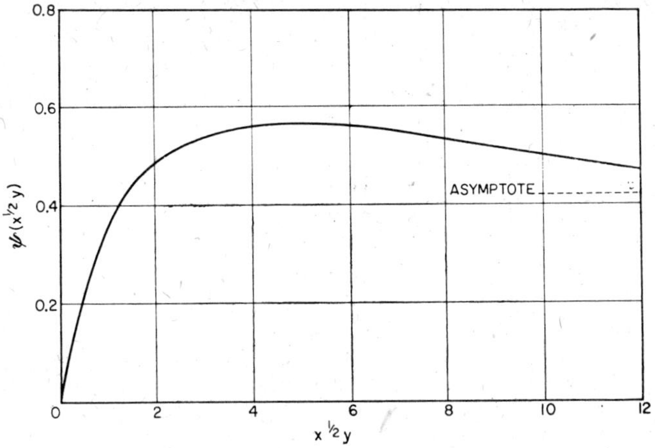

FIG. 4.15. The Baranger Mozer pair correlation function $4(x\,1/2\,y)$. (After Baranger and Mozer[6])

$$E = -\frac{e}{r^3}\left(1+\frac{r}{D}\right)e^{-r/D} \tag{4.299}$$

with D still given by Eq. (4.294b). From Eqs. (4.284) and (4.289b):

$$N\,h_1(\boldsymbol{k}) = -x^{3/2}\,\chi'\left(\sqrt{x}\,y\right) \tag{4.300}$$

Where $\chi'\left(\sqrt{x}y\right)$ is plotted in Fig. 4.15. The pair correlations are still given by Eqs.(4.300) except that now we replace Φ by Φ' which differs since now $D' = D/\sqrt{2}$. This is true because now both ions and electrons are involved in the shielding so that N is replaced by $2N$ in the Debye length. We now obtain:

$$\frac{1}{2}N^2\,h_2(\boldsymbol{k}) = x^{3/2}\,\psi'\left(\sqrt{x}\,y\right) \tag{4.301}$$

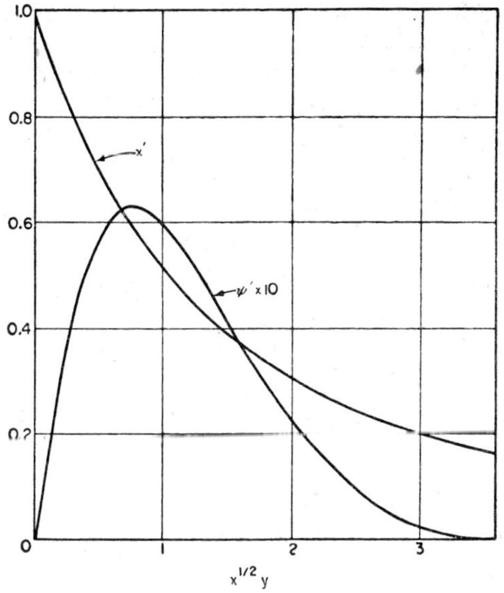

FIG. 4.16. The functions x' and ψ' appearing in the Baranger-Mozer theory of Stark Broadening. (After Baranger and Mozer[7])

from which

$$F(x) = \exp\left\{-x^{3/2}\left[\chi'\left(\sqrt{x}\,y\right) - \psi'\left(\sqrt{x}\,y\right)\right]\right\} \tag{4.302}$$

where χ' and ψ' are plotted in Fig. 4.16. From Eq. (4.29) one may obtain numerical value for the field strength probability function.

Baranger and Mozer also consider the case where the emitter is charged.* Here it is an *hf* component at the ion or an *lf* component at an electron. The function $g_1(x)$ is no longer unity, rather it is the pair correlation function for electrons or ions with one member at the origin. Hence:

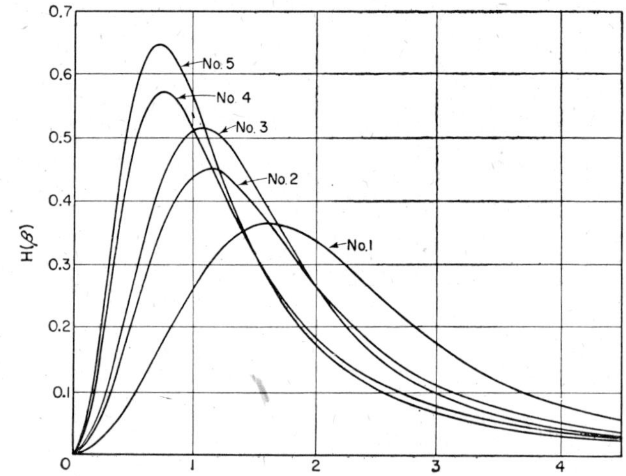

FIG. 4.17. Distributions for $r_0/D = 0{\cdot}8$. No. 1: Holtsmark. No. 2: High frequency component at a neutral point. No. 3: High frequency component at an electron. No. 4: Low frequency component at a neutral point. No. 5: Low frequency component at an ion. (Baranger and Mozer[7])

$$g_1(x) = \frac{\exp\left[-e^2 (r\,k\,T)^{-1} \exp(-r/D)\right]_{HF}}{\exp\left[-e^2 (r\,k\,T)^{-1} \exp(-r/D')\right]_{LF}} \qquad (4.303)$$

The two body cluster now effectively becomes a three body cluster with the result that it may be neglected.

The results of the various numerical calculations are indicated in Fig. 4.17.

4.28. Review and summary

In a plasma containing ions and electrons we have Stark broadening effects due to the fields of the slowly moving ions and the rapidly moving electrons. The fields of the ions are shielded by the fields of the surrounding ions and electrons, while the electrons may be considered as acting against a uniform background. Further there are effects due to thse interaction of perturbers of the same kind.

In approaching the Stark broadening problem one would first consider whether or not the classical path treatment was justified. In particular Tables 4.3 and 4.4 tell us whether the classical path treatment is applicable in the cases

* Mayer[151], Broyles[18] and Margenau and Lewis[217] have also treated the case when the radiator is an ion.

of first- and second-order Stark effect, respectively. It is certainly going to be seldom, if ever, when the classical path treatment is inapplicable to the ions.

If the classical treatment is applicable to the ions we turn to the Holtsmark Theory which we have described in some detail. Then we would apply Eq. (4.77) for the line intensity or, perhaps, the simpler Holtsmark version as given by Eqs. (4.128) or (4.129).

If the classical treatment is applicable to the electrons according to the two tables mentioned above, we apply the Interruption Theory to the electrons with the proper value of n, of course, corresponding to linear or quadratic effects. Now suppose the classical treatment is sufficiently poor for the electrons that no connection with reality can be claimed. Then one turns to something specific in the way of evaluation for the particular emitter involved. We have given an example of such a treatment in the case of the Kivel, Bloom, and Margenau treatment of the first Lyman line. In our treatment of this we have only treated the Universal Broadening effect with any sort of attention to detail. Now, the specifics of the emitter enter this effect through Eq. (4.236b) via the form factors contained therein, and it would be there that the vineyard laborer would be required to use his good atomic and free electron wave functions to obtain his specific results. But, of course, this would not be the only place that he would have to do so. He would, in addition, be required to use these functions in calculation of quenching and polarization effects. It would, indeed, behoove this laborer, if he is in any hurry at all, to use the classical Interruption treatment if it is possible. Finally, we have given other quantum treatments of the problem, but there is one basic common denominator among them all. Specific, detailed, and voluminous calculations are required for each particular emitter.

Now we have the effects of the ion in the form of the Holtsmark theory possibly shielded after some fashion which we have discussed and the effects of the electrons in a classical or quantum form. What is desired is the combined effects of the two types of particles. Here the folding of one distribution into the other is required. Eq. (4.279) gives as good an example of this as any. We smear the Interruption Broadened Stark components over a statistical distribution which tells us how the ionic fields shift around the components themselves. In Eq. (4.280) we smear the quantum electron broadening description over the same thing.

In toto, then, one may say that for a given broadening situation there are many guides as to how to proceed but the proceeder must apply these guides to specific calculation for his particular Stark Broadening problem.

CHAPTER 5

RESONANCE BROADENING

WE HAVE considered the broadening of spectral lines resulting from the presence of other molecules in the neighborhood of the emitter primarily for those cases which arise when the broadening molecules are molecules of a different type than the emitter. As we have seen, this broadening by foreign gases arises from an interaction of, say, the Stark or van der Waals type between the emitter and broadener and results in the distortion of the energy levels of the former with the consequent spreading and shifting of the spectral line. We now devote our attention to the case of resonance or self broadening in which the width of the spectral line arises from the presence of molecules of the same type as the emitter. As was early recognized, this type of broadening is much stronger than that caused by foreign molecules. The reason for this greater broadening can be qualitatively understood either by considering the situation from a classical or quantum viewpoint.

5.1. The qualitative basis of self broadening

We have noted repeatedly that, excepting Doppler and natural line width, broadening is caused by an interaction between the emitting molecule and a neighbor. In the classical sense, the similar molecules of a radiating gas are made up of a collection of like oscillators of the same natural frequency. Due to the sameness of this natural frequency, a strong coupling of these oscillators may be expected to occur in the usual classical manner given some basic, say electrostatic, coupling force. This in turn will cause a spread of the coupled oscillator frequencies about the natural frequency of a single oscillator.

From a quantum mechanical sense, the same strong broadening can be inferred. Let us hypothesize a two molecule system in which the level degeneracy resulting from the sameness of the molecules results in an energy perturbation dependent on the inverse cube of the molecular separation. We recall that under the same interaction force two unlike molecules result in an inverse sixth power dependence on the molecular separation. Thus, obviously greater broadening will result from the presence of like molecules. We shall later discuss the reduction in state lifetime for the case of resonance broadening which results in a broadening of the spectral line, but the state degeneracy which we have mentioned above we shall allow to suffice as an introductory consideration.

5.2. The Holtsmark theory of coupled oscillators

The first attack on the problem of resonance broadening was made by Holtsmark [82] in 1925.* This author considered the problem from a classical point of view which depicted the molecules of the absorbing, self-broadening gas as classical oscillators.

Now we suppose a coupling force to be present which is dependent on the electric dipole moment of the molecules. Let us begin by considering two of these classical oscillators each consisting primarily of a "quasi-elastically bound" electron. If x_1 is the vibrational coordinate of the first oscillator and x_2 that of the second, then the Lagrangian for the problem is surely:

$$L = T - V = \left(\frac{1}{2} m \dot{x}_1^2 + \frac{1}{2} m \dot{x}_2^2\right) - \left(\frac{1}{2} k_1 x_1^2 + \frac{1}{2} k_2 x_2^2 + k x_1 x_2\right) \qquad (5.1)$$

where, as usual, $k_1 = m\omega_1^2$ and $k_2 = m\omega_2^2$, and ω_1 and ω_2 are the natural frequencies of the two oscillators which have now been coupled through k_3. We have assumed a quadratic potential energy expression, thus inferring harmonic oscillation. If we now apply Lagrange's equations we obtain:

$$m \ddot{x}_1 + m \omega_1^2 x_1 + k_3 x_2 = 0 \qquad (5.2a)$$

$$m \ddot{x}_2 + m \omega_2^2 x_2 + k_3 x_1 = 0 \qquad (5.2b)$$

Utilizing the standard assumptions $x_1 = c_1 e^{i\omega t}$ and $x_2 = c_2 e^{i\omega t}$ we obtain Lagrange's secular determinant:

$$\begin{vmatrix} \omega^2 - \omega_1^2 & -\dfrac{k_3}{m} \\ -\dfrac{k_3}{m} & \omega^2 - \omega_2^2 \end{vmatrix} = 0$$

which yields for the two frequencies of the now coupled oscillators:

$$\omega^2 = \frac{1}{2}\left(\omega_1^2 + \omega_2^2 \pm \sqrt{(\omega_1^2 - \omega_2^2) + 4\left(\frac{k_3}{m}\right)^2}\right) \qquad (5.3)$$

If the two atoms are unlike, then $|\omega_1^2 - \omega_2^2| \gg k_3/m$ and we obtain:

$$\omega^2 - \omega_{1,2}^2 = \pm \frac{k_3}{m(\omega_1^2 - \omega_2^2)} \qquad (5.4)$$

whereas, in the case $\omega_1 = \omega_2 = \omega_0$ (like molecules), we obtain:

$$\omega^2 - \omega_0^2 = \pm \frac{k_3}{m} \qquad (5.5)$$

A consideration of Eqs. (5.4) and (5.5) serves to illustrate the much larger frequency shift accompanying the coupling of like oscillators since k_3 is always

* At about the same time Mensing [153] attacked the problem on much the same general basis, utilizing the "old" quantum theory. Since little is to be gained by carrying through the problem using the "old" and then again using the "new" quantum theory, we shall begin our quantum theoretical considerations of the problem with the work of Frenkel [49] in 1930.

small. This essentially forms the basis for Holtsmark's considerations. Let us first determine, with Holtsmark, k_3, and then proceed with the problem.

We have assumed our coupling force to arise from the electric dipole moments of the oscillators, and we now specify these moments by $\mathbf{p}_k(t) = e\,\boldsymbol{x}_k$. Eq. (2.19) may be rewritten slightly if we let the unit vector in the direction of $\mathbf{r}$ be given by $\mathbf{r}_0 = \mathbf{r}/|\mathbf{r}|$:

$$\mathbf{E}_i = \frac{1}{r^3}\left[\mathbf{p}_i - 3\,\mathbf{r}_0\,(\mathbf{r}_0 \cdot \mathbf{p}_i)\right] \tag{2.19}$$

which is the electric field at distance r (in the direction of $\mathbf{r}_0$) due to the electric dipole $\mathbf{p}_i$. Then the potential due to the interaction of a dipole $\mathbf{p}_k$ at distance r is:

$$V = \mathbf{E}_i \cdot \mathbf{p}_k = \frac{1}{r^3}\left[\mathbf{p}_i \cdot \mathbf{p}_k - 3\,(\mathbf{r}_0 \cdot \mathbf{p}_i)\,(\mathbf{r}_0 \cdot \mathbf{p}_k)\right] \tag{5.6a}$$

If we utilize the coordinates of Fig. 3.4 here, we obtain Eq. (3.1) from this. Eq. (5.6a) may be rewritten as:

$$V = \frac{e^2}{r^3}\left[r_i\, r_k \cos\gamma_{ik} - 3 r_i\, r_k \cos\gamma_i \cos\gamma_k\right] \tag{5.6b}$$

In Eq. (5.6b) r_i and r_k are the axes of the dipoles of molecule one and molecule two, respectively; γ_{ik} is the angle between the two dipoles, and γ_i and γ_k are the angles between $\mathbf{p}_i$ and $\mathbf{r}$ and $\mathbf{p}_k$ and $\mathbf{r}$ respectively. Holtsmark assumed the special case $\mathbf{p}_i || \mathbf{p}_k$ so that he obtained:

$$V = \frac{e^2}{r^3}\left[1 - 3\cos^2\gamma_{ik}\right] x_i\, x_k \tag{5.6c}$$

where γ_{ik} is now the angle between either $\mathbf{p}_i$ or $\mathbf{p}_k$ and $\mathbf{r}$.

Then the force on molecule i due to the field produced by molecule k is, since $\mathbf{F} = -\nabla V$:

$$F = \frac{e\, w_{ik}\, p_k(t)}{r_{ik}^{\,3}} = \frac{e^2\, w_{ik}}{r_{ik}^{\,3}}\, x_k$$

so that:

$$k_3 = \frac{e^2\, w_{12}}{r_{12}^{\,3}}$$

We may let:

$$a_{ik} = \frac{e^2\, w_{ik}}{m\, r_{ik}^{\,3}} \tag{5.7}$$

Now if we consider a system of N_1 atoms instead of our original two atomic system we may obtain, after the same fashion as we obtained Eq. (5.2), N_1 equations of the form:

$$\ddot{x}_i + \omega_0^2\, x_i + \sum a_{ik}\, x_k = 0 \qquad \begin{array}{l} k \lessgtr i \\ i = 1, 2, \ldots, N_1 \end{array} \tag{5.8}$$

In arriving at Eq. (5.8) we must, of course, assume that all the atoms of our system are identical and, as a result, possess the same natural frequency.

From Eq. (5.8) we may obtain the secular determinant for the problem:

$$\begin{vmatrix} \Lambda & a_{12} & a_{13} & \dots & a_{1N_1} \\ a_{21} & \Lambda & a_{23} & \dots & a_{2N_1} \\ a_{31} & a_{32} & \Lambda & \dots & a_{3N_1} \\ \dots & \dots & \dots & \dots & \dots \\ a_{N_1 1} & a_{N_1 2} & a_{N_1 3} & \dots & \Lambda \end{vmatrix} = 0 \tag{5.9}$$

where the abbreviation $\Lambda = \omega^2 - \omega_0^2$ has been utilized.

To this point the development has been on quite firm ground, but Holtsmark's continuation of the problem has been seriously disputed by, among others, Weisskopf[222], Lenz[121], and Margenau and Watson[218]. The difficulty arises out of Holtsmark's attempts on the rather appalling determinant of Eq. (5.9).

Eq. (5.9) will have N_1 roots, the determinant resulting in an N_1-th order equation. The N_1 roots Λ_τ will be of the form:

$$\Lambda_\tau = \omega_\tau^2 - \omega_0^2 \tag{5.10}$$

Thus, the absorbing atoms are now capable of absorbing the N_1 frequencies ω_τ instead of simply the natural frequency ω_0 so that we obtain an absorption line of a definite width.

The equation resulting from the determinant of Eq. (5.9) will have no Λ^{N_1-1} term. Now in an N_1-th order equation in which the coefficient of the N_1-th power term is unity the coefficient of the $(N_1 - 1)$-st power term is equivalent to the sum of the roots of the equation. This fact leads us to the conclusion that in this case:

$$\sum_{\tau=1}^{N_1} \Lambda_\tau = 0$$

In addition, the relations between the roots and coefficients of a polynomial are such that the coefficient of the $(N_1 - 2)$ nd power term is given by:

$$A_2 = -\sum_{\substack{i=1 \\ i<k}}^{N_1} \sum_{k=2}^{N_1} \Lambda_i \Lambda_k = -\sum_{\substack{i=1 \\ i<k}}^{N_1} \sum_{k=2}^{N_1} a_{ik} a_{ki} \tag{5.11}$$

The term on the extreme right of Eq. (5.11) is the coefficient of the $(N_1 - 2)$nd power term resulting from the expansion of the determinant.

A secular determinant arising from a vibrational problem of this nature is automatically symmetric since a_{ik} and a_{ki} are each equal to one half of the constant for some individual coordinate cross product term in the quadratic potential energy expression for the problem. Thus $a_{ik} = a_{ki}$.

Holtsmark now makes the assumption that the summation over $a_{ik}a_{ki}$ in Eq. (5.11) may be replaced by an integration. When one considers the large number of molecules involved and the large distribution of a_{ik} as given by Eq. (5.7) this would appear to be a reasonable assumption. Let us take the probability that a_{ik} lies between a_{ik} and $a_{ik} + da_{ik}$ in value as:

$$w(a_{ik})\, da_{ik}$$

Then if a random spatial distribution of the molecules is assumed:

$$w(a_{ik})\,\mathrm{d}a_{ik} = \frac{2\pi}{V} r^2\,\mathrm{d}r \sin\gamma_{ik}\,\mathrm{d}\gamma_{ik} \tag{5.12}$$

where γ_i and γ_k are the angles between the axes of the i-th and k-th molecules and the line joining these molecules, respectively. We thus may take γ_{ik} as the polar angle in our volume element. Since $a_{ik} = a_{ki}$, we may replace $a_{ik}a_{ki}$ by a_{ik}^2, and, when we integrate a_{ik} and Eq. (5.12) over all space we will obtain the mean value of a_{ik}^2. Now there are $N_1(N_1 - 1)/2 \doteq N_1^2/2$ of the a_{ik}^2 terms in Eq. (5.11) so that the results of this integration must be multiplied by $N_1^2/2$ to obtain A_2.

Thus:

$$A_2 \doteq -\frac{N_1^2}{2}\int w(a_{ik})\,a^2{}_{ik}\,\mathrm{d}a_{ik} = -\frac{N_1^2}{V}\frac{e^4\pi}{m^2}\int_{\gamma_0}^{\infty}\int_0^{\pi}\frac{(1-\cos^2\gamma_{ik})}{r_{ik}^4}\sin\gamma_{ik}\,\mathrm{d}\gamma_{ik}\,\mathrm{d}r_{ik}$$

$$= -\frac{8\pi}{15}\frac{e^4}{m^2}\frac{NN_1}{\varrho_0^3} \tag{5.13}$$

since $V = N_1/N$. ϱ_0 is here taken as the closest distance of approach of two molecular centers.

Since — as we have deduced from the fact that the equation resulting from the determinant has no Λ^{N_1-1} term — the sum of the Λ_r is zero, it is true that:

$$\sum_{\substack{i=1\\ i<k}}^{N_1}\sum_{k=2}^{N_1}\Lambda_i\Lambda_k = -\frac{1}{2}\sum_{\tau=1}^{N_1}\Lambda_\tau^2$$

Thus, according to Eq. (5.11) we may now write for Eq. (5.13):

$$\Sigma\Lambda_\tau^2 = \frac{16\pi}{15}\frac{e^4}{m^2}\frac{NN_1}{\varrho_0^3} \tag{5.14}$$

According to Eq. (5.10) this becomes:

$$\frac{\Sigma\Lambda^2{}_\tau}{N_1} = \langle(w_\tau^2 - \omega_0^2)^2\rangle = \frac{16\pi}{15}\frac{e^4}{m_2}\frac{N}{\varrho_0^3} \tag{5.15}$$

If we let $\omega - \omega_0 = \omega_\tau'$ then

$$\omega_\tau^2 - \omega_0^2 = \omega_\tau\,\omega_\tau' + \omega_0\,\omega_\tau' \doteq 2\omega_0\,\omega_\tau'$$

since we have assumed $(\omega_\tau - \omega_0) \ll \omega_0$. We now obtain from Eq. (5.15):

$$\sqrt{\langle\omega_\tau'^2\rangle} = \sqrt{\frac{16\pi}{15}\frac{e^2}{m}\frac{1}{\varrho_0^{3/2}}}\sqrt{N}\,\frac{1}{2\,\omega_0} \tag{5.16}$$

At this point Holtsmark makes the assumption that the distribution of ω_τ' and hence the resulting intensity distribution in the spectral line is a Gaussian error curve. It is particularly with this assumption that Weisskopf[222] and later Margenau and Watson[218] find themselves in disagreement. In the development to this point there is no more basis for the choice of a Gaussian error curve than there is for any other reasonably conceivable curve. In addition, the experimental evidence does not provide much support for the assumption. Be this as it may, however, when this assumption is made there results:

$$W(\omega')\,d\omega' = \frac{1}{\sqrt{2\pi\langle\omega'^2\rangle}}\,e^{-\omega'^2/\langle\omega'^2\rangle}\,d\omega' \tag{5.17}$$

and for the half-width:

$$2\,\omega' = 2.36\sqrt{\langle\omega'^2\rangle} = 103.5\times 10^7\,\frac{\sqrt{N}}{\varrho_0^{3/2}}\,\frac{1}{2\,\omega_0} \tag{5.18}$$

In essence, then, the theory yields a half-width which is directly proportional to the root of the density and inversely proportional to a distance ϱ_0 which is rather arbitrary and may be likened to the optical collision diameter of the Lorentz–Lenz–Weisskopf-etc. theory. This half-width is predicated on a line shape which could just as justifiably have been assumed at the beginning of the formulation as at the end. The theory is certainly important, however, in that it provided the entree into the field of resonance broadening. A few years after the presentation of this theory by Holtsmark, Frenkel[49] attacked the problem on the basis of the quantum theory, although the physical considerations were essentially the same.

5.3. Quantum resonance in binary interactions

Let us begin, with Frenkel, by considering a system of two molecules. The interaction energy when the two molecules are unlike has already been given by Eq. (3.15). Now for the case of two like molecules, let us again use Holtsmark's potential $k_3 x_1 x_2$ for comparison. We find ourselves dealing with a case of quantum mechanical degeneracy.

Let us assume that one member of our two particle system will proceed from a state of energy E_1 to the ground state of energy E_0 with accompanying radiation. The two molecules of the system we designate as 1 and 2, and the eigenfunctions for the first molecule alone in the ground and first excited states as $\psi_0(1)$ and $\psi_1(1)$ respectively. Then, the unperturbed eigenfunction for the system in the ground state is:

$$\Psi_0 = \psi_0(1)\,\psi_0(2) \tag{5.19}$$

Ψ_0 is the eigenfunction going with the state having energy E_0. On the other hand, if the system is possessed of energy E_1, the eigenfunction may be:

$$\psi_{11} = \psi_1(1)\,\psi_0(2) \quad \text{or} \quad \psi_{21} = \psi_0(1)\,\psi_1(2)$$

Since either of these two eigenfunctions satisfies the Schrödinger equation for energy E_1, the eigenfunction for the state is a linear combination of the two possibilities:

$$\Psi_{1j} = c_1\,\psi_{11} + c_2\,\psi_{21} = \sum_{j'} c_{1j'}\,\psi_{1j'} \tag{5.20}$$

In addition, the two states ψ_{11} and ψ_{21} of equal energy indicate that a quantum degeneracy problem* has arisen, and Eq. (4.141) yields two equations for the $c_{1j'}$. These equations have a solution, according to Cramer's rule, only if:

$$\begin{vmatrix} E_1^{(1)} - H_{11}' & -H_{12}' \\ -H_{12}' & E_1^{(1)} - H_{22}' \end{vmatrix} = 0 \qquad (5.21)$$

$H_{11} = H_{22} = 0$ so that:†

$$E_1^{(1)} = \pm H_{12}' \qquad (5.22)$$

We substitute Eq. (5.22) into an equation of the form Eq. (4.141) to obtain:

$$c_{11} = \pm c_{12}$$

so that Eq. (5.20) becomes:

$$\Psi_{1s} = c_{11}\,\psi_{11} + c_{11}\,\psi_{21} \qquad (5.23\text{a})$$

$$\Psi_{1a} = c_{11}\,\psi_{11} - c_{11}\,\psi_{21} \qquad (5.23\text{b})$$

To find c_{11} we normalize Eq. (5.23a):

$$c_{11}^2 \int |\psi_{11}|^2\,d\tau + c_{11}^2 \int |\psi_{21}|^2\,d\tau + 2\,c_{11}^2 \int \psi_{11}\,\psi_{21}\,d\tau = c_{11}^2 + c_{11}^2 + 0 = 1$$

due to the orthonormality of the original eigenfunctions. From this, it is apparent that $c_{11} = 1/\sqrt{2}$.

An inspection of Eq. (5.23a) shows it to be symmetric with respect to exchange of the two molecules while Eq. (5.23b) is antisymmetric. In the case of electric dipole radiation in which the transition proceeds from state one to be ground state the matrix element is:

$$M_{10} = \int \overline{\Psi}_{1s}\,\mathbf{M}\,\Psi_0\,d\tau \qquad (5.24)$$

Ψ_0 has been given by Eq. (5.19) and is symmetrical with respect to an exchange of the two molecules, as is also the electric dipole moment in this case. Since an integrand is nothing more nor less than a number after integration, it must surely be symmetrical. On this basis we throw out Eq. (5.23b) and merely retain (5.23a).

We have thus found our first excited state eigenfunction for the system. The first order energy is given by:

$$\begin{aligned} E_1^{(1)} &= \int \overline{\Psi}_{1s}\,H'\,\Psi_{1s}\,d\tau \\ &= k_3 \int x_1\,x_2\,\overline{\psi_1(1)\,\psi_0(2)}\,\psi_0(1)\,\psi_1(2)\,d\tau_1\,d\tau_2 \qquad (5.25) \\ &= +\,k_3\,|x_{01}|^2 \end{aligned}$$

Eq. (5.25) tells us that the presence of one molecule shifts the spectral line by the amount:

$$\nu' = +\frac{k_3}{m}\,|x_{01}|^2 \qquad (5.26\text{a})$$

* See *supra*, Sec. 4.16.

† Cf. *supra*, Eq. (5.1).

Let us rewrite Eq. (5.5) slightly for comparison:

$$\omega - \omega_0 = \pm \frac{k_3}{(\omega + \omega_0)\, m} \doteq \pm \frac{k_3}{2\,\omega_0\, m}$$

or

$$\nu' \doteq \pm \frac{k_3}{8\pi^2 m \nu_0} = \pm \frac{e^2}{r^3} \frac{(1 - \cos^2\gamma)}{8\pi^2 m \nu_0} \tag{5.26b}$$

The quantum mechanical equation (5.26a) apparently yields one shifted frequency, but we shall see that when the different possible values of the magnetic quantum number are considered two frequencies will result.

Under the assumption in k_3 that:

$$\langle(1 - 3\cos^2\gamma)^2\rangle = 1 - \langle 6\cos^2\gamma\rangle + \langle 9\cos^4\gamma\rangle = 1 - 2 + \frac{9}{5} = \frac{4}{5}$$

Frenkel obtains for the root mean square value of ν':

$$\sqrt{\langle\nu'^2\rangle} = \frac{2}{\sqrt{5}} \frac{e^2}{r^3} \frac{|x_{10}|^2}{h} \tag{5.27}$$

In addition, Frenkel utilizes the more exact expression Eq. (5.6a) instead of Eq. (5.6c) for V to obtain:

$$\sqrt{\langle\nu'^2\rangle} = \sqrt{\frac{2}{3}} \frac{1}{r^3} \frac{|\mathbf{p}_{10}|^2}{h} \tag{5.28}$$

As Margenau and Watson[218] have done, we may also use Eq. (3.1) for V. If we take hydrogen-like wave functions and use precisely the method of calculation which we have utilized in obtaining Eq. (5.25), we obtain for E_1:

$$E_1^{(1)} = h\nu' = \left\{ \begin{array}{ll} -\dfrac{2}{3}\dfrac{e^2}{r^3}|r_{12}|^2 & \text{for } m = 0 \\[2ex] +\dfrac{1}{3}\dfrac{e^2}{r^3}|r_{12}|^2 & \text{for } m = \pm 1 \end{array} \right. \tag{5.29a}$$

where, as usual, m is the magnetic quantum number, and where:

$$|r_{12}|^2 = \frac{3h}{8\pi^2 m \nu_0} f_{12}$$

so that:

$$\nu' = \left\{ \begin{array}{ll} -\dfrac{e^2}{r^3}\dfrac{2}{8\pi^2 m \nu_0} f_{12} & \text{for } m = 0 \\[2ex] +\dfrac{e^2}{r^3}\dfrac{1}{8\pi^2 m \nu_0} f_{12} & \text{for } m = \pm 1 \end{array} \right. \tag{5.29b}$$

which compares at least reasonably well with Eq. (5.26b), although the symmetry is obviously lacking.

5.4. The statistical resonance result

Eq. (5.29b) may be rewritten as:

$$E^{(1)} = h\nu' = \gamma\left(\frac{e^2 h f_{12}}{8\pi^2 m \nu_0}\right)\frac{1}{r^3} \tag{5.30}$$

where γ is the statistical weight factor associated with m. After the fashion of Margenau and Watson, let us equate γ to unity. We consider this an averaging process over the possible molecular orientations.* The result is:

$$V = E^{(1)} = c\,r^{-m} = \left(\frac{e^2 h f_{12}}{8\pi^2 m \nu_0}\right) r^{-3} \tag{5.31a}$$

so that:

$$c = \frac{e^2 h f_{12}}{8\pi^2 m \nu_0} \tag{5.31b}$$

From Eq. (3.126b) the half width of the resonance broadened spectral line is:

$$\delta = \frac{2V_0}{h} = \frac{4}{3}\frac{\pi^2}{h} N c$$

so that:

$$\delta = \frac{e^2 f_{12}}{6 m \nu_0} N \tag{5.32}$$

Eq. (5.32) was originally obtained by Margenau and Watson[218] in a slightly different manner.

5.5. Resonance broadening by many molecules (Frenkel)

Let us now turn our attention, with Frenkel, to a system of N_1 like molecules. Again we are only interested in the ground state for the system in which none of the N_1 molecules is excited and the first excited state for the system in which one of the N_1 molecules is excited. Our first excited state eigenfunction is given by Eq. (5.20), and the c_{jn} are given by N_1 equations of the form Eq. (4.141). In Eq. (4.141) — again utilizing Eq. (5.6c) for V — the H'_{lj} are given by:

$$H_{lj}' = \sum_{l'}\sum_{\substack{j' \\ l'<j'}} k_{l'j'} \int \ldots \int x_{l'}\, x_{j'}\, \psi_{l'}\, \psi_{j'}\, d\tau_1 \ldots d\tau_{N_1} \tag{5.33a}$$

where here $H_{lj}' = \sum_{l<j} k_{lj}\, x_l\, x_j$.

It would appear to be worthwhile from the point of view of clarity to illustrate this equation with an example.

Let us suppose that we have a three molecular system, and let the eigenfunctions for the three molecules in the ground state by given by $\psi_1(0)$, $\psi_2(0)$, and $\psi_3(0)$. In the first excited state we shall replace 0 by 1. Then the first excited state for the system must be a linear combination of the following functions:

$$\left.\begin{aligned} \psi_1 &= \psi_1(1)\,\psi_2(0)\,\psi_3(0) \\ \psi_2 &= \psi_1(0)\,\psi_2(1)\,\psi_3(0) \\ \psi_3 &= \psi_1(0)\,\psi_2(0)\,\psi_3(1) \end{aligned}\right\} \tag{5.34}$$

* Cf. *supra*, Eqs. (3.14) and subsequent.

From Eq. (5.34) we may obtain Eq. (5.33a) for this case as, for example:

$$\left.\begin{aligned} H_{12}' &= \int (k_{12} x_1 x_2 + k_{13} x_1 x_3 + k_{23} x_2 x_3)\, \psi_1 \psi_2 \, d\tau_1 \, d\tau_2 \, d\tau_3 \\ &= k_{12} \int x_1 \psi_1(1)\, \psi_1(0)\, d\tau_1 \int x_2 \psi_2(0)\, \psi_2(1)\, d\tau_2 \int \psi_3(0)\, \psi_3(0)\, d\tau_3 \\ &\quad + k_{13} \int x_1 \psi_1(1)\, \psi_1(0)\, d\tau_1 \int \psi_2(0)\, \psi_2(1)\, d\tau_2 \int x_3 \psi_3(0)\, \psi_3(0)\, d\tau_3 \\ &\quad + k_{23} \int \psi_1(1)\, \psi_1(0)\, d\tau_1 \int x_2 \psi_2(0)\, \psi_2(1)\, d\tau_2 \int x_3 \psi_3(0)\, \psi_3(0)\, d\tau_3 \\ &= k_{12} \, |x_{10}|^2 + 0 + 0 = k_{12} \, |x_{10}|^2 \end{aligned}\right\} \quad (5.35)$$

In general, Eq. (5.33a) may be written:

$$H_{lj}' = k_{lj} \, |x_{10}|^2 \tag{5.33b}$$

Now let us rewrite Eq. (5.8) under the assumption that $x_i = x_i \, e^{i\omega t}$ — the assumption which leads to the secular determinant.

$$(\omega^2 - \omega_0^2)\, x_l = \frac{1}{m} \Sigma\, k_{lj}\, x_j$$

or:

$$(\nu - \nu_0)\, x_l \doteq \frac{1}{8\pi^2 m \nu_0} \Sigma\, k_{lj}\, x_j \tag{5.36}$$

Eq. (5.33) is the classical analog of Eq. (4.141) in the same fashion as Eq. (5.26b) is the analog of Eq. (5.26a). Frenkel used this analogy as his basis for replacing $1/8\pi^2 m\nu_0$ by $1/h\ |x_{10}|^2$ in Eq. (5.16) to obtain the quantum mechanical equivalent of this equation:

$$\sqrt{\langle \nu'^2 \rangle} = 4 \sqrt{\frac{\pi}{15} \frac{\sqrt{N}}{\varrho_0^{3/2}}} \frac{e^2 |x_{10}|^2}{h} \tag{5.37a}$$

Frenkel also uses Eq. (5.6a) instead of Eq. (5.6c) to obtain:

$$\sqrt{\langle \nu'^2 \rangle} = \frac{2}{3} \sqrt{\frac{2\pi N}{\varrho_0^3}} \frac{1}{h} |\mathfrak{p}_{10}|^2 \tag{5.37b}$$

Even if Eq. (5.37b) holds without question we still have no information about the half-width, since the assumption of the Gauss error curve is not very well justified. Eq. (5.37b) does not hold without question, however, as we may now show.

We have N_1 eigenfunctions of the form Eq. (5.20) for the first excited state of the system. We have one eigenfunction for the ground state. The probability of a transition from a state going with one of these N_1 eigenfunctions and the ground state is given by the square of the matrix element of the electric dipole moment. Thus, in order to find the root mean square frequency shift, the proper weights must be given to each of these shifts. The intensity distribution also would result from the calculating of these weighted eigenvalues which, as in Eq. (5.29a), would each have three different values according as $m = 0, \pm 1$. Frenkel did launch an attack on this problem, but with no notable success.

The question of resonance broadening is again confronted by the obstacle presented to progress by Eq. (5.9).

Whether or not the methods used by Holtsmark and Frenkel to obtain their results are correct, the results themselves as given by Eqs. (5.15) and (5.34) are open to serious question in that these equations claim a dependence of the half-width on the root of the density. A consideration of these equations will probably indicate to the reader, however, that ϱ_0 offers a bonnet out of which we might possibly produce a different pressure dependence. This is admittedly a posteriori reasoning, since Schutz–Mensing[186] (nee Mensing) has done just this.

5.6. A linear relation between half-width and density

From Eq. (5.11) and the fact that $2\omega_0\omega_\tau' \doteq \omega_\tau^2 - \omega_0^2$ we may write:

$$\frac{1}{2}\sum 4\,\omega_0^2\,\omega_\tau'^2 = \left(\frac{e^2}{m}\right)^2 \sum_{i\neq k} b_{ik}^2 \tag{5.38a}$$

where $b_{ik} = (1 - 3\cos^2\gamma_{ik})/r^3_{ik}$. Eq. (5.38a) may be rewritten as:

$$\sum \omega_\tau^{2\prime} = \frac{2}{4\,\omega_0^2}\left(\frac{e^2}{m}\right)^2 \sum_{i\neq k} b_{ik}^2$$

or:

$$\langle \nu'^2\rangle = \frac{\sum \nu_\tau'^2}{N} = \frac{2}{4(2\pi)^4\,\nu_0^2}\left(\frac{e^2}{m}\right)^2 \frac{1}{N}\sum_{i\neq k} b_{ik}^2 \tag{5.38b)*$$

Now we recall that Holtsmark replaced the sum in Eq. (5.38b) by a weighted integration over b_{ik}^2. Schutz–Mensing considers only the i-th molecule initially and replaces the summation over k by an integration and multiplication by N_1, there now being $(N_1 - 1)$ instead of $N_1(N_1 - 1)/2$ terms.

As in Eq. (5.13):

$$\sum_k a^2_{ik} \doteq 2\pi \frac{N_1}{V}\int_{\varrho_i}^{\infty}\int_0^{\pi}\frac{1}{r_i^4}(1 - 3\cos^2\gamma_i)^2 \sin\gamma_i\, d\gamma_i\, dr_i = \frac{16\pi}{15}\frac{N}{\varrho^2_i} \tag{5.39}$$

We may note that the lower limit has been designated by ϱ_i. ϱ_i is taken as the separation of the i-th molecule and the closest adjoining molecule and is, hence, a variable. Finally, Schutz–Mensing takes the average ϱ_i to obtain the complete summation in Eq. (5.38b).

$$\sum_{ik} b_{ik}^2 \doteq \left(\frac{1}{2}\right)\frac{16\pi}{15}\frac{NN_1}{\langle\varrho\rangle^3} \tag{5.40}$$

In her earlier work, Schutz–Mensing had calculated a value $0{\cdot}55\,N^{-1/3}$ for $\langle\varrho\rangle$. Thus, Eq. (5.38b) becomes:

$$\langle \nu'^2\rangle = \frac{16\pi}{15}\,6\,\frac{1}{4(2\pi)^4\nu_0^2}\left(\frac{e^2}{m}\right)^2 N^2 \tag{5.38c}$$

* A minor point to the effect that Schutz–Mensing left out the 2 in the numerator of this equation might be mentioned.

† Schutz–Mensing does not obtain the factor 1/2, but since there are only a total of $N_1^2/2$ terms arising out of the summation over b, it would seem paradoxical to end by obtaining a total of N_1. This appears true even though, when only the latter portion of Eq. (5.39) is considered, there do remain $(N_1 - 1)$ to occupy the space point a distance ϱ_i from the i-th molecule. It should finally be noted that the two in the numerator of Eq. (5.38b) and the one-half in Eq. (5.40) do combine to yield Eq. (5.18) with ϱ_0 replaced by $\langle\varrho\rangle$.

Thus, it is apparent that, should we now use Eq. (5.18) for the half-width, we would obtain a linear dependence on the density.

She also carried through a calculation in which an integration is carried out over the probability that ϱ lies between ϱ and $\varrho + d\varrho$ to obtain:

$$\Sigma b_{ik}^2 = -\frac{64}{15}\pi^2 N^2 N_1 \left[-c - \log\left(\frac{4\pi}{3}\varrho_0^3 N\right)\right] \tag{5.41}$$

where the bracketed term is the first term in the expansion of the exponential integral $Ei(-4(\pi/3)\varrho_0^3 N)$.

After the Frenkel manner, Schutz–Mensing next considered the replacement of $1/8\pi^2\nu_0 m$ by the square of the matrix element divided by h to obtain the "corresponding" quantum mechanical cases.

The lack of success attendant on the method of attack utilized by Holtsmark[82] and Frenkel[49] was perhaps the reason that Weisskopf[209] considered the problem from a different viewpoint. As has been mentioned previously, Weisskopf objected to Holtsmark's use of a Gaussian error curve with the resulting relation between the half-width and the mean square of the deviation $\langle\Delta\omega^2\rangle$. He pointed out that in the case of an intensity distribution of the form $(\gamma/\pi)\,1/[(\Delta\omega)^2 + \gamma^2]$ the half-width is γ but $\langle\Delta\omega^2\rangle$ is infinite.

5.7. The Weisskopf resonance broadening theory

Let us first obtain the Lorentz–Lorenz relation. From Maxwell's equations we obtain the following relation between the electric displacement $\boldsymbol{D}$, the electric field $\boldsymbol{E}$, and the polarization $\boldsymbol{I}$:

$$\boldsymbol{D} = \varepsilon\boldsymbol{E} = \boldsymbol{E} + 4\pi\boldsymbol{I} \tag{5.42a}$$

where ε is the dielectric constant for the medium. Now the force on a unit charge in a polarizable medium may be shown to be:

$$\boldsymbol{F} = \boldsymbol{E} + 3\pi\boldsymbol{I} \tag{5.42b}$$

If the polarizability of a molecule is α, then the induced electric dipole moment of the molecule is:

$$\boldsymbol{p} = \alpha\boldsymbol{F} \tag{5.42c}$$

From Eq. (5.42c) the polarization or the electric moment per unit of volume is:

$$\boldsymbol{I} = N\boldsymbol{p} = N\alpha\boldsymbol{F} = N\alpha\left(\boldsymbol{E} + \frac{4\pi}{3}\boldsymbol{I}\right) \tag{5.43}$$

where N is the number of molecules per unit of volume. If we eliminate $\boldsymbol{I}$ between Eqs. (5.42a) and (5.43) we obtain the Lorentz–Lorenz relation:

$$\frac{\varepsilon - 1}{\varepsilon + 2} = \frac{4\pi}{3} N\alpha \tag{5.44a}$$

or, if we recall that the dielectric constant for a medium is the square of the refractive index:

$$\frac{n_c^2 - 1}{n_c^2 + 2} = \frac{4}{3}\pi N\alpha \tag{5.44b}$$

where $n_c = n(1 + i\varkappa)$.

Weisskopf asserted that the "... width is given completely by the course of the absorption coefficient n ..."[209] in Eq. (5.47b), and he based this assertion on the manner of the development of this equation by Oseen[164], Ewald[43], and Bothe[17]. These authors had treated the molecules of a gas as classical oscillators and had obtained the equation in question after a computation in which the "reciprocal effect" of the oscillators on one another had been included. Thus, Weisskopf felt that Eq. (5.44b) included the coupling effect.

Now let us make the usual approximation to the effect that in our gas $n_c \doteq 1$ and $n_c^2 - 1 \doteq 2(n_c - 1)$ so that Eq. (5.44b) becomes:

$$n_c - 1 \doteq 2\pi N\alpha \tag{5.44c}$$

Next, we must needs find $N\alpha$, and, since $\alpha \boldsymbol{E} = \Sigma e \boldsymbol{x}_k$, $N\alpha$ is given by:

$$N\alpha = \frac{N}{\boldsymbol{E}} \Sigma e \boldsymbol{x}_k \tag{5.45}$$

We set up Newton's equation for the amplitude $\boldsymbol{x}_k$ under the assumption that the electron of mass m possesses the natural frequency ω_0, is subject to a velocity proportional damping force (the constant of proportionality being m), and is acted upon by a force $e\boldsymbol{E}$ due to the presence of the electric field $\boldsymbol{E}$. The resulting equation is:

$$m\ddot{\boldsymbol{x}}_k = -m\gamma\dot{\boldsymbol{x}}_k - m\omega_0^2\boldsymbol{x}_k + e\boldsymbol{E} \tag{5.46}$$

If we suppose the time dependent portion of x_k to be the same as that of E, the solution of Eq. (5.46) for the time independent portion of the amplitude is:

$$\boldsymbol{x}_k = \frac{e}{m}\frac{\boldsymbol{E}_0}{\omega_k^2 - \omega^2 + i\omega\gamma} \doteq \frac{e}{m\omega_k}\frac{\boldsymbol{E}_0}{\omega_k - \omega + i\gamma}$$

so that:

$$N\alpha = \frac{Ne^2}{m\omega_0}\frac{1}{\omega_0 - \omega + i\gamma} f_{nm} \tag{5.47}$$

Here f_{mn} is the so-called oscillator strength for the transition from level m to level n. In this case f_{mn} will give the number of dispersion electrons belonging to the transition under consideration. From Eqs. (5.44c) and (5.47c) we obtain:

$$n_c = 1 + \frac{2\pi Ne^2}{m\omega_0}\frac{1}{\omega_0 - \omega + 2\gamma} f_{mn} \tag{5.48}$$

n_c has proviously been identified as $n(1 + i\varkappa)$, and $n\varkappa$ is the imaginary part of Eq. (5.48) so that:

$$n\varkappa = -\frac{2\pi Ne^2\gamma}{m\omega_0}\frac{1}{(\omega_0 - \omega)^2 + \gamma^2} f_{mn} \tag{5.49}$$

γ may be considered as $\gamma_1 + \gamma_2$ where γ_1 and γ_2 are the radiation and collision damping respectively. We may neglect γ_2 in the case of very low pressures to obtain the effects of the natural line width. We again consider Eq. (5.44). This may be written as:

$$n_c^2 - 1 = \frac{4\pi N\alpha}{\left(1 - \frac{4}{3}\pi N\alpha\right)} \tag{5.50}$$

where we have not assumed $n \doteq 1$. If we substituted the value of $N\alpha$ from Eq. (5.47) into Eq. (5.50) and let $A = (Ne^2/m\omega_0) f_{mn}$, we obtain:

$$n_c^2 - 1 = \frac{4\pi A}{\omega_0 - \omega - \frac{4}{3}\pi A + i\gamma} \tag{5.51}$$

If:

$$B = 4\pi A; \quad \overline{\omega_0} = \omega_0 - \frac{4}{3}\pi A; \quad \overline{\Delta\omega} = \omega - \omega_0 \tag{5.52a}$$

Eq. (5.51) becomes:

$$n_c^2 = x + iy = \frac{(-\overline{\Delta\omega} + i\gamma + B)(\overline{\Delta\omega} + i\gamma)}{(\overline{\Delta\omega})^2 + \gamma^2} \tag{5.53)*$$

If the transformation:

$$b = \frac{B}{\gamma}; \quad \Delta = \frac{\overline{\Delta\omega}}{\gamma} \tag{5.52b}$$

is utilized in Eq. (5.53) the result is:

$$n_c^2 = x + iy = -\frac{(-\Delta + b + i)(\Delta + i)}{\Delta^2 + 1} \tag{5.54}$$

We recall that the imaginary part of the root of n_c^2 is $n\varkappa$, which is desired. It may be recalled that:

$$\sqrt{n_c^2} = r^{1/2}\, e^{i\left(\frac{\vartheta + 2k\pi}{2}\right)}$$

where:

$$r = \sqrt{x^2 + y^2}, \quad \vartheta = \text{arc tan}\,\frac{y}{x}$$

Thus, we may obtain from Eq. (5.54):

$$n\varkappa = \Im\sqrt{n_c^2} = \sqrt[4]{\frac{(\Delta - b)^2 + 1}{\Delta^2 + 1}}\,\sin\left\{\frac{1}{2}\arctan\left(\frac{b}{(\Delta - b)\Delta + 1}\right)\right\} \tag{5.55)†$$

for the absorption coefficient.

For the case $\Delta \gg b$, that is, in the wings of the spectral line, Eq. (5.55) reduces to Eq. (5.49).

* Weisskopf gave $(\overline{\Delta\omega} + i\gamma + B)$ instead of $(-\overline{\Delta\omega} + i\gamma + B)$ in his first paper[209] but corrected it in his second[210].

† Weisskopf obtained $(\Delta + b)^2 + 1$ under the root instead of $(\Delta - b)^2 + 1$ in his first paper[209].

It might be mentioned at this point that our half-width as given by $\gamma_1 + \gamma_2$ will be directly proportional to the pressure — above those pressures at which γ_1 noticeably affects the sum — through γ_2. In this theory, which is directly connected to Weisskopf's foreign gas broadening theory,* the optical collision diameter must be taken a good bit larger than in the case of foreign gas broadening. As an example Weisskopf estimated a value of 44 *A* for the optical collision diameter in the case of the Na-D_1 line.

Weisskopf himself was not very happy about the results of his considerations for in the final analysis he was forced to ignore the coupling of the atoms. As we have mentioned, Weisskopf's main conclusion is his assumption that "... the collisions cause an enlargement in the damping of the individual oscillators ..."[209] an enlargement of greater magnitude than in the case of foreign gas broadening.**

We may write out our half width for purposes of consideration if we let the mean free path as given by Eq. (1.13) be l:

$$\delta = \gamma = \frac{2}{\tau} = \frac{2\langle v\rangle}{l} = 2\pi\varrho^2\langle v\rangle N \tag{5.56}$$

where $\langle v\rangle$ is taken by Weisskopf as the mean relative velocity. We have here the optical collision diameter as an essentially unknown and conveniently adjustable parameter. We may calculate a value for it, however, if we again assume† that a phase change of unity defines an optical collision. Eq. (4.7) may be written for this case as:

$$\Delta(r) = \Delta\left(\sqrt{v^2t^2 + \varrho^2}\right) = \frac{K}{r^3}; \quad \Theta = \int_{-\infty}^{+\infty} \frac{K\,dt}{(v^2t^2+\varrho^2)^{3/2}} \sim 1 \tag{5.57a}$$

Now if we let $dx = v\,dt/\varrho$ we obtain from Eq. (5.56):

$$K\int_{-\infty}^{+\infty} \frac{\varrho\,dx}{\varrho^3 v(x^2+1)^{3/2}} = \frac{K}{v\varrho^2}\int_{-\infty}^{+\infty}\frac{dx}{(x^2+1)^{3/2}} = 1$$

or:

$$\varrho = J\sqrt{\frac{K}{v}} \tag{5.51b)††}$$

* See *supra*, Chap. 2.

** Weisskopf also included Doppler broadening in one stage of the development. To do this one simply replaces ω in Eq. (5.47) by $\omega + \Upsilon - \Upsilon$ is the frequency displacement due to the Doppler effect), — multiplies by the probability of the displacement, namely, the Maxwell-Boltzmann probability $\frac{1}{\Upsilon_0\sqrt{\pi}}\exp(-\Upsilon^2/\Upsilon_0^2)\,d\Upsilon$ to obtain, in place of Eq. (5.47):

$$N\alpha = \frac{A}{\Upsilon_0\sqrt{\pi}}\int_{-\infty}^{+\infty}\frac{\exp[-\Upsilon^2/\Upsilon_0^2]\,d\Upsilon}{(\omega_0-\omega-\Upsilon)+i\gamma}$$

† Cf. *supra*, Sec. 4.2.

†† In frequency units ν: $\varrho = \sqrt{\frac{4\pi\varkappa}{v}}$ (5.57b′)

where:

$$J = \left(\int_{-\infty}^{+\infty} \frac{dx}{(1 + x^2)^{3/2}} \right)^{1/2} = \sqrt{2} \tag{5.58}$$

A consideration of Eq. (5.31a), for example, serves to evaluate K.

$$\Delta\omega = \left(\frac{e^2 f_{nm}}{4\pi m \nu_0}\right) r^{-3}$$

$$= \frac{e^2}{2 m \omega_0 r^3} f_{nm} = \frac{K}{r^3} \tag{5.59}$$

where we have averaged over the angle function and multiplied by the oscillator strength of the transition giving rise to the line under consideration. Utilization of Eqs. (5.57) and (5.58) in Eq. (5.56) yields:

$$\delta = 2\pi \frac{e^2}{m \omega_0} f_{mn} N \tag{5.60}$$

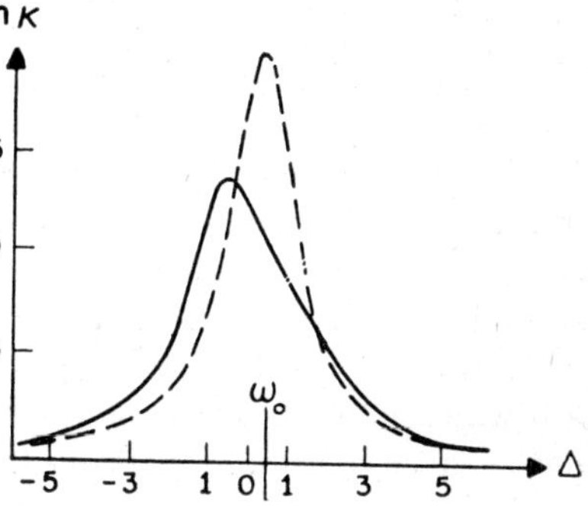

FIG. 5.1. Line shape according to the more sophisticated Weisskopf theory. (After Weisskopf[209])

In a short paper on the subject Weisskopf[210], plotted Eq. (5.55) for the case $b = 4$ and the collision width much greater than the natural width. Weisskopf's figure is reproduced in Fig. 5.1, and a consideration of the figure serves to illustrate the shift and asymmetry provided by the theory.

5.8. Review of self-broadening to this point

Thus far then we have considered two theories as to the origin of the resonance interactings resulting in self broadening and four theories propounded for the purpose of explaining this self broadening as such. The first of these interaction theories as advanced by Holtsmark considered the self broadening gas as a collection of coupled oscillators whose coupling resulted in a splitting of the frequency about the unperturbed line position. Unfortunately, the calculational difficulties involved in this theory led to results which were questionable. In particular we refer to Eq. (5.9). Without too much justification Holtsmark used his interaction results within the assumption of a Gaussian line shape to obtain Eq. (5.18) for the line half width.

The quantum mechanical analog of this coupled-oscillator interaction theory provided the other resonance interaction. This quantum mechanical analog was essentially predicated on the degeneracy resulting from the presence of N_1 particles any one of which may be excited. It too faced extreme difficulties of computation. As a consequence, although the basic theory appeard sound, the possibility of obtaining results from this theory seemed remote. At any rate, on an appeal to correspondences Frenkel obtained, in analogy to Eq. (5.18), Eqs. (5.37) for the line half width.

Interaction results are obtainable for the case of two molecules, however, as for example, Eqs. (5.29). Margenau and Watson used this binary interaction, within the framework of the Statistical Theory, to obtain Eq. (5.32) for the half width. This result should be a reasonable one for those situations in which the Statistical Theory may be expected to hold and for pressure sufficiently low that we need not consider the simultaneous interaction of more than two molecules.

The fourth broadening theory which was advanced was nothing more nor less than the application of the Michelson–Lorentz theory of Interruption broadening to the case of self broadening. Here a parameter — the optical collision diameter — is conveniently present which, if it is taken a good bit larger than in the case of foreign gas broadening, does yield results. If we wish to calculate this parameter under the assumptions utilized in obtaining Eq. (5.60) we could still adjust the final result by adjusting the phase change which defines a collision. Weisskopf mentions the fact that in some cases the theory only yields a ϱ which is of the order of magnitude of the correct result and in other cases not even this accuracy is attained. These other cases can, of course, be explained as "due to other causes".

5.9. Qualitative consideration of the energy transfer theory

Some four years after the publication of Weisskopf's theory Furssow and Wlassow[51] entered the field with what did amount to a new approach. Their entry was marked by the intriguing statement that the Weisskopf theory — and the Lenz theory in passing — was completely incorrect.* Although one obvious reason for their statement may have been to pave the way for the introduction of their theory, it would appear reasonable for us to briefly consider the reasons which they advanced for this conclusion.†

For greatest accuracy of consideration let us quote the Furssow and Wlassow statement verbatim before attempting to consider their reasoning.

"To be sure Weisskopf's considerations of the collisions between the like atoms is not entirely correct. Weisskopf without foundation applies the correct concept of the mechanism of the collision damping in the case of non-extinguishing atoms of different types to the case of like atoms. As is known the broadening of the lines through collisions result not only when the wave train emitted by the atom is propagated after the collision (extinguishing gases) but also at the time when at the collision a change of the vibration phase of the excited atom sets in (non-extinguishing gases). In order to compute the phase shift during the collision one must take the change of the frequency $\Delta\omega$ of the emitting atom which occurs through the interaction with foreign atoms and integrate over the collision time. If one makes this integral equal to one then the magnitude of the optical collision diameter can be evaluated. The colli-

* Houston took mild exception to this statement to the effect that "... the criticism of Weisskopf's work contained in this paper does not appear to be justified ..."[89] but did not reply to the Furssow and Wlassow criticism specifically.

† One is reminded of King Tarquin lopping the heads off the poppies.

sions of the like atoms Weisskopf takes instead of $\Delta\omega$ arbitrarily the difference between the frequency of one of the normal vibrations of a system of two dipole linked like linear oscillators and the frequency of the isolated atom. It is clear that this computation is based on a misunderstanding. $\Delta\omega$ is in terms of its nature the change of the frequency of the emitting atom, the light of which is analyzed according to Fourier. That however which Weisskopf substitutes instead of it has nothing at all to do with the matter because of the degeneracy ... the concept that in the collision of two like atoms one of which is excited and the other of which is unexcited the vibration phase of the excited atom is changed does not all correspond to reality."[51]

It might be remarked first that the $\Delta\omega$ which they believe to have been improperly chosen seems to be the very one whose choice they advocate for the following reason "... take the change of the frequency $\Delta\omega$ ... and integrate over the collision time".[57] Now it is true that integration is not extended over these limits in Eq. (5.57a), but let us recall that this latter equation is an approximation which has its roots in the more nearly correct Eq. (4.6). Further, the remarks which they make about coupled oscillators seems less applicable to Weisskopf than to almost any other resonance author. If they base their argument on the "degeneracy" — they do not define this degeneracy but we can assume that it is that degeneracy arising from the indistinguishability of the atoms of our ensemble — then the question of which approach is more nearly correct has to be answered, and no one seems to have done this to date, at least not to the satisfaction of more than a few. Surely though the adjustable-parameter dependent Weisskopf theory has not been so devastated by the above quoted argument that it cannot be considered as remaining a reasonable approximation.

The Furssow and Wlassow theory** really has its basis in a conception advanced by Kallmann and London[94] in 1929 to the effect that like atoms may simply exchange energy between each other without accompanying radiation. Even a cursory consideration of this conception gives an inkling of an application to the theory of self-broadening. We may examine it from either a quantum or a classical viewpoint.

First the classical viewpoint. To begin with we consider our molecules as classical oscillators, and again we have the dipole interaction between two like oscillators, one of which is excited and the other of which is unexcited. As a result of this interaction the amplitude of oscillation of the excited molecule decreases while that of the unexcited molecule increases, that is, an energy transfer occurs. This is effectively the same as damping the oscillations of the excited molecule, and, as in the case of the Lorentz damping*, will result in a broadening of the spectral line. Let us now consider qualitatively the quantum mechanical explanation of this effect.

** Although the physical theory here is different, we shall see that it leads to precisely the mathematical results of Secs. 5.3 and 5.5.

* Cf. *supra*, Chap. 1.

Energy is assumed to be transferred as a result again of the dipole interaction from an excited to an unexcited molecule of the same kind. Whatever the lifetime of the excited state would be under conditions of no transfer, this lifetime will be greatly shortened by a high probability of transfer before radiation. Now we may recall that one form of the Heisenberg uncertainty principle states that $\Delta E \Delta t > h$. Thus, if, as is the case in the ground state, the state lifetime is infinite, the ground state will be infinitely well defined or virtually infinitely narrow. As soon as we consider a state with a finite lifetime, however, the situation changes. A certain lifetime Δt will give us a certain indefiniteness ΔE in the state energy or a certain level width. This level width in turn will mean that a spectral line arising from the combination of this level with another will be broadened as a consequence. Thus, when we increase our state lifetime by this energy transfer, quantum mechanics decrees that we indirectly broaden our spectral line by widening the energy level.

These then are the classical and quantum forms of the theory as advanced by Furssow and Wlassow. It now remains only to rewrite our qualitative conjectures after a quantitative fashion. Let us consider the classical approach first.

5.10. The classical energy transfer (low pressure)

To begin with it is, of course, necessary to make a few simple assumptions regarding our system. In this consideration we are assuming our molecule to be a classical harmonic oscillator. Now we shall suppose that the electronic transition which gives rise to our broadened spectral line proceeds to the ground state. We will assume that only one valence electron is responsible for the spectral line under consideration. Finally let us suppose that the excited molecule, of which our system contains one for our purposes here, moves in the neighborhood of the remainder of the unexcited molecules. The excited molecule we then consider as moving rectilinearly with velocity v. We might possibly bring up the same type of objection to the utilization of a linear velocity here as was brought up by Jablonski in objection to Weisskopf's utilization of a linear velocity in a central force problem. However this is probably minor at this stage.†

In considering our interaction which leads to the self broadening of the line, we shall assume this interaction to take place between our emitter and one of the unexcited broadening molecules. It has already been assumed that the broadening molecules are stationary so we shall place our broadener at the origin of the inertial reference frame. The coordinates of the valence electron of this unexcited molecule let us designate as $x_1 y_1 z_1$. We next establish a moving reference frame at the excited molecule. The coordinates of the valence electron are taken to be $x_2 y_2 z_2$ which coordinates are, of course, referred to the moving frame. Finally, let us designate by R the separation of the two reference frames.

As has been mentioned earlier, it is here desired to use as the potential of the interaction the electric dipole potential. Thus, we are essentially faced with the

† Most authors appear to consider it minor at any stage, although this is not conclusive.

same task with which we were presented in deriving Eq. (3.1). We make the same assumptions, that is, we first assume that R is very large compared to the size of the molecules and hence compared to the separation of the electron from the nucleus. Now we assume that the moving reference frame is so oriented with respect to the inertial frame that one axis moves parallel to the corresponding axis in the inertial frame and the other corresponding axes remain parallel to each other. Let us further suppose that the angles made by the radius vector R with the coordinate axes are α, β, γ. Under these assumptions the Coulomb interaction may be expanded in terms of α, β, γ.* The first term in this expansion will, as we have seen, be the electric dipole potential, which potential we shall assume to be responsible for the interaction between the excited and unexcited molecules. The electric dipole potential is given by:

$$\left.\begin{aligned} V = \frac{e^2}{R^3}[&x_1x_2(1-3\cos^2\alpha) + y_1y_2(1-3\cos^2\beta) + z_1z_2(1-3\cos^2\gamma) \\ &-3(x_1y_1\cos\alpha\cos\beta \\ &+x_1z_2\cos\alpha\cos\gamma + y_1x_2\cos\beta\cos\alpha + y_1z_2\cos\beta\cos\gamma \\ &+z_1x_2\cos\gamma\cos\alpha + z_1y_2\cos\gamma\cos\beta)] \end{aligned}\right\} \quad (5.61)$$

If our molecule was a one dimensional harmonic oscillator which was undergoing no interaction with any other oscillators, Newton's equation of motion would be $\ddot{x} + \omega_0^2 x = 0$, assuming, of course, that motion is along the x-axis. Here ω_0 is the natural frequency of the molecular oscillator. Now, however, we must modify this equation to include the interaction with the second oscillator and the effects of three dimensional oscillation. The modification to include the electric dipole interaction may be taken directly from Eq. (5.61) to yield the following system of equations:

$$\left.\begin{aligned} \ddot{x}_1 + \omega_0^2 x_1 + \Lambda[(1-3\cos^2\alpha)x_2 - 3\cos\alpha\cos\beta\, y_2 - 3\cos\alpha\cos\gamma\, z_2] &= 0 \\ \ddot{y}_1 + \omega_0^2 y_1 + \Lambda[(1-3\cos^2\beta)y_2 - 3\cos\beta\cos\gamma\, z_2 - 3\cos\beta\cos\alpha\, x_2] &= 0 \\ \ddot{z}_1 + \omega_0^2 z_1 + \Lambda[(1-3\cos^2\gamma)z_2 - 3\cos\gamma\cos\alpha\, x_2 - 3\cos\gamma\cos\beta\, y_2] &= 0 \\ \ddot{x}_2 + \omega_0^2 x_2 + \Lambda[(1-3\cos^2\alpha)x_1 - 3\cos\alpha\cos\beta\, y_1 - 3\cos\alpha\cos\gamma\, z_1] &= 0 \\ \ddot{y}_2 + \omega_0^2 y_2 + \Lambda[(1-3\cos^2\beta)y_1 - 3\cos\beta\cos\gamma\, z_1 - 3\cos\beta\cos\alpha\, x_1] &= 0 \\ \ddot{z}_2 + \omega_0^2 z_2 + \Lambda[(1-3\cos^2\gamma)z_1 - 3\cos\gamma\cos\alpha\, x_1 - 3\cos\gamma\cos\beta\, y_1] &= 0 \end{aligned}\right\} \quad (5.62)$$

where:

$$\Lambda = \frac{e^2}{m}\frac{1}{R^3}$$

Furssow and Wlassow solved the system of equations given by Eq. (5.62) by using a method of successive approximation. First, let us assume that the bracketed terms in this equational system are all zero. The solution of the system of equations resulting from this approximation we will consider our zeroth-order solution. Initially, the unexcited molecule whose electronic coordinates are

* Cf. *supra*, Eq. (3.1).

$x_1y_1z_1$ is undergoing no electronic vibrations. For simplicity of consideration, we shall assume that the excited molecule is carrying out vibrations along its z-axis. When these approximations are taken into account the zeroth-order solution to the problem is as follows:

$$x_1^{(0)} = y_1^{(0)} = z_1^{(0)} = x_2^{(0)} = y_2^{(0)} = 0 \tag{5.63a}$$

$$z_2^{(0)} = A \cos \omega_0 t \tag{5.63b}$$

We may now substitute the values for $x_1, y_1, \ldots, z_2$ back into Eq. (5.62). The solution of the resulting system of equations yields the first approximate solution.

Let us consider briefly one example of the method used in obtaining this first approximate solution. To begin with after we have substituted our zeroth-order approximation Eqs. (5.62), we will obtain equations of the type:

$$\ddot{\eta} + \omega_0^2 \eta = N(t) \tag{5.64a}$$

The solution to Eq. (5.64a) under the assumption that $\eta(0) = \eta(0) = 0$, is the following:*

$$\eta(t) = \frac{1}{\omega_0} \int_0^t N(t) \sin \omega_0 (t - \tau)\, d\tau \tag{5.64b}$$

Now let $N(t) = \alpha(t) \cos \omega_0 t$. This $N(t)$, in general, represents the function which will arise out of the bracketed terms in Eq. (5.62). ω_0 is, of course, of the order of magnitude of the frequency of the emitted radiation. $\alpha(t)$ on the other hand is a function of the angles between the radius vector and the coordinate systems, and, as a result of this, its time change will depend on the heat motion of the molecules. Thus $\alpha(t)$ will vary quite slowly in comparison to $\cos \omega_0 t$. Now let us substitute the expression for $N(t)$ into Eq. (5.64b) and expand the resulting cos sin integrand to obtain the following:

$$\eta(t) = \frac{1}{2\omega_0} \left\{ \sin \omega_0 t \int_0^t \alpha(t)(1 + \cos 2\omega_0 \tau)\, d\tau - \cos \omega_0 t \int_0^t \alpha(t) \sin 2\omega_0 \tau\, d\tau \right\} \tag{5.65a}$$

If we let $T = 2\pi/\omega_0$, that is, T identical to the period of the electronic oscillation, Eq. (5.65a) becomes:

$$\left.\begin{aligned} \eta(t) = \frac{1}{2\omega_0} \sum_k \Bigg\{ & \sin \omega_0 t\, \alpha(\tau_k) \left[\int_{kT}^{(k+1)T} d\tau + \int_{kT}^{(k+1)T} \cos 2\omega_0 \tau\, d\tau \right] \\ & - \cos \omega_0 t\, \alpha(\tau \omega_k') \int_{kT}^{(k+1)T} \sin 2\omega_0 \tau\, d\tau \Bigg\} \end{aligned}\right\} \tag{5.65b}$$

We have taken $\alpha(\tau_k)$ outside the integral sign due to the fact that it varies so slowly during one period of electronic oscillation. The summation over k suffices to give us any over-all time period which we desire. We may note that

* See page 282 of Reference 225a.

the integral of the cosine and the integral of the sine over a period vanishes. Thus Eq. (5.65b) becomes:

$$\eta(t) = \frac{1}{2\,\omega_0} \sin \omega_0 t \sum_k \alpha(\tau'_k)\, \Delta\tau_k = \frac{1}{2\,\omega_0} \int_0^t \alpha(\tau)\, d\tau \sin \omega_0 t \tag{5.66}$$

Eq. (5.66) furnishes the general form for the first approximate solution to the equational system given by Eq. (5.62) as modified by the zeroth-order solution. Eq. (5.66) yields, for the specific cases under consideration, the following solutions:

$$\left.\begin{aligned} x_1^{(1)} &= \frac{3A}{2\,\omega_0} \int^t A \cos\alpha \cos\gamma\, dt \sin \omega_0 t \\ y_1^{(1)} &= \frac{3A}{2\,\omega_0} \int^t A \cos\beta \cos\gamma\, dt \sin \omega_0 t \\ z_1^{(1)} &= -\frac{A}{2\,\omega_0} \int^t A (1 - 3\cos^2\gamma)\, dt \sin \omega_0 t \\ x_2^{(1)} &= y_2^{(1)} = 0; \quad z_2^{(1)} = A \cos \omega_0 t \end{aligned}\right\} \tag{5.67}$$

Eqs. (5.67) are very enlightening in that they serve to show the manner in which the energy is transferred from the excited molecule to an unexcited molecule during the course of an optical collision. We see that z_2 has not been changed by the collision. In addition we see how vibrations along all three axes in the system of the initially unexcited molecule have been excited.

Furssow and Wlassow carried forward the problem to higher orders of approximation, and they reported generally on the results which they had attained. According to their statement, they found that the second-order approximation yielded no changes in the vibrational amplitudes $x_1 y_1 z_1$. They found that the third-order approximate solution did yield a small change in the vibrational amplitudes for the initially unexcited molecule. In like manner the fifth-order approximation and higher order odd approximations yielded a rapidly converging series of corrections to the amplitude. Since this series did converge rapidly they felt that Eqs. (5.67) should suffice to describe the energy transfer as a result of an optical collision at large separation. We must now introduce, after the manner of Furssow and Wlassow, a few additional quantities which will prove useful in the continuation of the calculation. First of all, we may recall that we have considered the initially excited molecule as moving rectilinearly among an ensemble of stationary, initially unexcited molecules. We again assume an interaction between one unexcited molecule and our excited molecule. Let us take the vector $\boldsymbol{\varrho}$ as the perpendicular from the unexcited molecule to the velocity vector of the excited molecule. Let us designate by $\alpha_1 \beta_1 \gamma_1$ the angles which the velocity vector $\mathbf{v}$ makes with the coordinate axes. By the symbols $\alpha_2 \beta_2 \gamma_2$ we will denote the angles which the vector $\boldsymbol{\varrho}$ makes with the coordinate axes. No ambiguity is introduced here by simply referring to coordinate axes as we may recall that the frame of the excited molecule remains parallel to the frame of the unexcited molecule.

Let us assume that the optical collision, which serves to transfer energy from the excited to the unexcited molecule, lasts a time interval τ. If this is the case, and if we assume the collision to begin at time $t = 0$, then the collision must be initiated at a distance $-v\tau/2$ from ϱ.

Then after a collision of duration τ, the amplitude we may obtain from Eq. (5.61):

$$X_1^{(1)} = \frac{3A}{2\omega_0}\int_0^{\tau} \Lambda \cos\alpha \cos\gamma \, dt \tag{5.68}$$

It is apparent as we have mentioned previously that $\alpha\beta\gamma$ are time dependent through the heat motion of the atoms. We would like then to find some expression in terms of time for these three angles. Straightforward geometrical considerations yield the following expressions for these three angle functions:

$$\left.\begin{aligned}
\cos\alpha &= \varrho\frac{\cos\alpha_2 + v\left(t-\frac{1}{2}\tau\right)\cos\alpha_1}{R}\\
\cos\beta &= \varrho\frac{\cos\beta_2 + v\left(t-\frac{1}{2}\tau\right)\cos\beta_1}{R}\\
\cos\gamma &= \varrho\frac{\cos\gamma_2 + v\left(t-\frac{1}{2}\tau\right)\cos\gamma_1}{R}\\
R^2 &= \varrho^2 + v^2\left(t-\frac{1}{2}\tau\right)^2
\end{aligned}\right\} \tag{5.69}$$

The relations among the three previously defined vectors **R**, **v**, and ϱ as given by Eq. (5.69) are quite obvious ones.

We may substitute Eq. (5.69) into Eq. (5.68). When the variable $t' = t - 1/2\,\tau$ is substituted for t in the resulting expression we obtain:

$$X_1^{(1)} = \frac{3A}{2\omega_0}\frac{e^2}{m}\int_{-\tau/2}^{+\tau/2}\frac{(\varrho\cos\alpha_2 + vt'\cos\alpha_1)(\varrho\cos\gamma_2 + vt'\cos\gamma_1)}{\sqrt{(\varrho^2 + v^2t'^2)^5}}\,dt' \tag{5.70}$$

It is apparent that we could also obtain analogous expressions for $Y_1^{(1)}$ and $Z_1^{(1)}$. We may make a simplification for the purpose of evaluating the integral in Eq. (5.70) by letting τ go to infinity and thus extending the limits on the integral to $-\infty$ and $+\infty$. Again this may be justified, as has been done in similar cases, in that only that portion of time immediately before and after the time of optical collision is of any import. Thus, the extension of the limits to minus and plus infinity serves merely to encompass regions which contribute nothing to the integral in question.

We may, by elementary means, evaluate the integrals of the form Eq. (5.70):

$$\left.\begin{aligned} X_1^{(1)} &= \frac{e^2}{m\,\omega_0}\,\frac{A}{\varrho^2\,v}\,(2\cos\alpha_2\cos\gamma_2+\cos\alpha_1\cos\gamma_1)\\ Y_1^{(1)} &= \frac{e^2}{m\,\omega_0}\,\frac{A}{\varrho^2\,v}\,(2\cos\beta_2\cos\gamma_2+\cos\beta_1\cos\gamma_1)\\ Z_1^{(1)} &= \frac{e^2}{m\,\omega_0}\,\frac{A}{\varrho^2\,v}\,(2\cos\gamma_2+\cos^2\gamma_1-1) \end{aligned}\right\} \quad (5.71)$$

We are now desirous of obtaining the amount of energy which is transferred during the collision from the initially excited molecule to the initially unexcited one. In doing so, let us first recall that the energy of a classical oscillator is given by $E = (1/2)\,m\dot{x}^2 + (1/2)\,kx^2$ where we have here assumed a linear harmonic oscillator which is vibrating in the x-direction. Since $\dot{x}$ is given by $\omega_0 A\cos\omega_0 t$ and k is given by $m\omega_0^2$, E becomes $(1/2)\,m\,\omega_0^2A^2\cos\omega_0 t + (1/2)\,m\,\omega_0^2A^2\sin\omega_0 t$. Thus, the energy is $(1/2)\,m\omega_0^2A^2$ or the square of the vibrational amplitude multiplied by $(1/2)\,m\omega_0^2$. Hence, in the present case the energy E which has been transferred to the initially unexcited molecule will be given by the sum of the squares of the amplitude components as given by Eq. (5.71) multiplied by $(1/2)\,m\omega_0^2$. We then obtain for the transferred energy:

$$\varepsilon = \frac{e^4}{m^2\,\omega_0^2}\,\frac{1}{\varrho^4\,v^2}\sin^2\gamma_1\,E \quad (5.72)$$

where E, the initial energy of the excited molecule, is, of course, $(1/2)\,\omega_0^2A^2$.

Eq. (5.72) is predicated on the first approximate solutions as given by Eq. (5.67) to Eq. (5.72). The validity of these Eqs. (5.64) is assumed only for large transit distances. Furssow and Wlassow defined the minimum distance of closest approach — transit distance — as that distance at which the transferred energy is equal to the energy at time zero of the initially excited oscillator, that is, $\varepsilon = E$. Thus if ϱ_0 is taken as the minimum transit distance Eq. (5.72) leads us to the following criteria for large distances:

$$\frac{1}{\varrho_0^2\,v} = \frac{m\,\omega_0}{e^2} \quad (5.73)$$

In Eq. (5.73a) $\sin\gamma_1$ has been replaced by its maximum value.

We have thus determined the energy which will be transferred during an optical collision between an initially unexcited molecule and an initially excited one. Now in order to determine the total energy which our initially excited molecule will lose in the course of time it will be necessary to ascertain the total loss of energy to all the molecules in the neighborhood. Let us assume that our unexcited molecules are uniformly distributed, there being N of them per unit of volume. Let us consider a layer of thickness Δs lying perpendicular to the direction of motion of the excited molecule. Then during the time when the excited molecule is in this layer of thickness Δs, those molecules will be capable of obtaining energy from it which lie at distances of ϱ_0 or at greater distances. Thus, the increment of energy loss by the excited molecule during its passage through this layer will be given by:

$$\mathrm{d}E = -2\pi N\Delta s \int_{\varrho_0}^{\infty} \varepsilon \varrho \,\mathrm{d}\varrho = -\pi N\Delta s \frac{e^4}{m^2 \omega_0^2} \sin^2 \gamma_1 \frac{1}{\varrho_0^2 v^2} E \qquad (5.74\text{a})$$

Now Δs may also be written as $v\,\mathrm{d}t$. When this is done and when the value for $1/\varrho_0^2 v$ is substituted from Eq. (5.73):

$$\mathrm{d}E = -\pi N \frac{e^2}{m\omega_0} \sin^2 \gamma_1 E \,\mathrm{d}t \qquad (5.74\text{b})$$

From Eq. (5.74b) Furssow and Wlassow obtained:

$$E = E_0 \,\mathrm{e}^{-\Upsilon_0' t} \qquad (5.74\text{c})$$

where:

$$\Upsilon_0' = \frac{2\pi}{3} \frac{Ne^2}{m\omega_0}$$

In obtaining Eq. (5.74c) Furssow and Wlassow averaged Eq. (5.74c) over γ_1 and integrated the separable differential equation which resulted. In a slightly more refined treatment in which they assumed motion on the part of the initially unexcited molecules and where averaging was carried out over all orientations for this motion, they obtained instead of the Υ_0' given above:

$$\Upsilon = \frac{\pi}{2} \frac{e^2}{m\omega_0} N \left(1 + \frac{1}{2} \sin^2 \alpha\right)$$

Here α is the angle between the electric moment and the direction of motion of the emitter.

We have thus attained in Eq. (5.74c) an expression for the attenuation of the energy of the initially excited molecule with time. This is, of course, not the desired objective. We are desirous of obtaining the broadening of the emitted spectral line as a result of this energy attenuation. In order to arrive at this we shall consider the analogous case of damping by some other means such as a radiation damping of the molecular oscillations.

First consider the equation for a molecular oscillation of natural frequency ω_0 and subjected to a damping of damping constant γ:

$$m\ddot{x} + m\omega_0^2 x + m\gamma\dot{x} = 0 \qquad (5.75)$$

For the case $\gamma \ll \omega_0$ — which we shall find to be an excellent approximation — the solution of Eq. (5.75) can be shown to be:

$$x = x_0 \,\mathrm{e}^{-\gamma t/2} \,\mathrm{e}^{i\omega_0 t} \qquad (5.76)$$

It may be seen, as we have specifically mentioned earlier,* that the energy of the oscillator whose amplitude is governed by Eq. (5.76) is $E_0 \mathrm{e}^{-\gamma t}$. This γ corresponds to the Υ_0' in Eq. (5.74c). Here we know that the electric vector of

* See *supra*, Eq. (5.72) and preceding.

the radiation emitted by our molecular oscillator will also contain the exponential of Eq. (5.76).* The electric vector of the radiation field will be of the form:

$$\mathbf{E} = \mathbf{E}_0 \, e^{-\gamma t/2} \, e^{i\omega_0 t}$$

In the usual manner the electric vector may be expanded in a Fourier integral in order to obtain the amplitude of this vector as a function of frequency:

$$E(\nu) = \frac{1}{2\pi} \frac{1}{i(\omega - \omega_0) - \gamma/2}$$

We may thus obtain the intensity of the emitted spectral line as a function of frequency by simply taking the absolute square of the amplitude of the electric vector:

$$I(\nu) \doteq |E(\nu)|^2 \propto \text{const} \frac{\gamma/2}{(\omega - \omega_0)^2 + (\gamma/2)^2} \tag{5.77}$$

It can then be seen from the familiar Eq. (5.74) that γ or Υ is the half-width of the emitted spectral line. Thus, from Eq. (7.71c) the half-width of the emitted spectral line is found to be:

$$\delta_r' = \frac{2\pi}{3} \frac{e^2}{m\omega_0} N \tag{5.78}$$

It is then apparent that the effect of this energy transfer at large transit distances is a damping of the molecular oscillations resulting in a broadening of the spectral line whose half-width is given by Eq. (5.78). Finally the half-width δ_r' should be multiplied by the oscillator strength of the transition under consideration.

We thus have treated collisions at transit distances of greater than ϱ_0 as defined by Eq. (5.73). Now with Furssow and Wlassow we may essentially define ϱ_0 in analogy to the optical collision diameter of Weisskopf. There the optical collisions, which have already been treated, resulted in only a change in the amplitude of the electronic oscillations. Furssow and Wlassow considered collisions at less than this optical collision diameter as changing not only the amplitude but also the orientation of the electronic oscillation. For these latter cases they simply took the half-width as given by the Lorentz–Lenz–Weisskopf-etc. collision theory:†

$$\delta_s' = \pi \varrho_0^2 N \langle v \rangle = 2\pi \frac{e^2}{m\omega_0} N \tag{5.79}$$

where again δ_s' is to be multiplied by the oscillator strength for the collision.

Finally, the line broadening as a result of near and distant collisions may be assumed as the sum of the widths due to the two types. Thus, we eventually arrive at the line half-widths as obtained from the classical energy transfer theory:

$$\delta = f(\delta_r' + \delta_s') = \frac{8\pi}{3} \frac{e^2}{m\omega_0} N f \tag{5.80}$$

* Cf. *supra*, Sec. 1.8.

† All of which would appear rather paradoxical.

A comparison of Eq. (5.80) with Eq. (5.60) shows that the Furssow–Wlassow theory yields line widths 4/3 times as great as the Weisskopf theory.

The manner in which the dipole interaction between two like molecules causes a transfer of the oscillatory energy from the initially excited molecules to the remaining initially unexcited molecules has been shown. This energy transfer then acts as a damping force on the oscillatory motion of the molecule and, as a result, the emitted radiation is broadened into a spectral line of finite width. We shall see that a quantum consideration under certain specific assumptions leads to approximately the same results.

5.11. Quantum treatment of low pressure self-broadening

We have sketched qualitatively the quantum mechanical theory of resonance broadening by energy transfer, and our first task in a quantitative consideration will be to ascertain the time change of eigenfunctions of the excited molecule due to this energy transfer.

Again the system is initially taken as two like molecules one of which is excited and one of which is unexcited. The motion of the molecule will be considered classically* as is normally done in problems of this kind with quite reasonable justification. The potential of interaction between the two atoms is still the dipole potential of Eq. (5.61). The symbols such as ϱ, α_1, etc. which were utilized in the classical consideration will again appear with the same connotation.

Since specific assumptions regarding the states of the two molecules must be made, the simplest ones possible are utilized, namely, the unexcited molecule is assumed to be initially in a ground state where $\psi_{nlm}(1) = \psi_{n00}(1)$, and the excited molecule is assumed in a p state, with eigenfunction $a\psi_{n10}(2) + c\psi_{n00}(2)$. This eigenfunction assumes that the emitted radiation is polarized in the z direction since $\Delta m = 0$ — the only transition possible from this eigenfunction — corresponds to such a polarization. We might look ahead a bit at this point by a consideration of the eigenfunction of the excited molecule. The probability that the emitter is in the state $m10$† at time t is given by $|a(t)|^2$. Thus the behavior of $a(t)$ will tell us the probability that the emitter will transfer its energy to the initially unexcited molecule in the course of time. Qualitatively at least it is apparent that this determines the state lifetime change and the resulting line broadening.

In the considerations which led to Eq. (5.20) no spatial degeneracy was considered so that only a two-fold degeneracy resulted for the system. When we consider the three values which m may assume when either of our molecules is in the first excited state, a six fold degeneracy for the first excited state of the system results. The ground state of the system remains, of course, non-

* This appears intuitively justifiable, but for a mathematical justification one may consult Reference 155.

† The "m" in m 10 refers to a particular principal quantum number. Concentration by the reader will allay any confusion of this symbol with the identical symbol referring to the magnetic quantum number.

degenerate. On the basis of these considerations we choose as eigenfunctions to describe the system during the time of transit:

$$\left.\begin{aligned}\psi(1,2,t) = &\{a_1(t)\,\psi_{n00}(1)\,\psi_{m10}(2) + a_2(t)\,\psi_{n00}(1)\,\psi_{m11}(2)\\ &+ a_3(t)\,\psi_{n00}(1)\,\psi_{m1-1}(2) + b_1(t)\,\psi_{m10}(1)\,\psi_{n00}(2)\\ &+ b_2(t)\,\psi_{m11}(1)\,\psi_{n00}(2) + b_3(t)\,\psi_{m1-1}(1)\,\psi_{n00}(2)\}\\ &\cdot e^{-i\frac{E_n+E_m}{\hbar}t} + c_1(t)\,\psi_{n00}(1)\,\psi_{n00}(2)\,e^{-i2\frac{E_n}{\hbar}t}\end{aligned}\right\} \quad (5.81)$$

The Schrödinger equation for our system — neglecting the radiation field — is:

$$[H(1) + H(2) + V(1,2,t)]\,\psi(1,2,t) = i\hbar\frac{\partial}{\partial t}\psi(1,2,t) \qquad (5.82\text{a})$$

In Eq. (5.82a) $H(1)$ is the unperturbed Hamiltonian for the initially unexcited molecule; $H(2)$ is the Hamiltonian for the initially excited molecule, and $V(1, 2, t)$ is the perturbing Hamiltonian introduced by the dipole interaction as given by Eq. (5.61). Let us now substitute Eq. (5.81) into Eq. (5.82a) to obtain:

$$\left.\begin{aligned}&i\hbar\sum_{i=1}^{3}\dot{a}_i\psi_i + i\hbar\sum_{i=1}^{3}\dot{b}_i\psi_i + i\hbar\,\dot{c}_1\,\psi_{n00}(1)\,\psi_{n00}(2)\,e^{i2\frac{E_n}{\hbar}t}\\ &\quad+\sum_{i=1}^{3}a_i E_l\psi_i + \sum_{i=1}^{3}b_i E_l\psi_i + 2c_1E_n\psi_{n00}(1)\,\psi_{n00}(2)\\ &\quad=\sum_{i=1}^{3}a_i[H(1)+H(2)]\,\psi_i + \sum_{i=1}^{3}b_i[H(1)+H(2)]\,\psi_i\\ &\quad+c_1[H(1)+H(2)]\,\psi_{n00}(1)\,\psi_{n00}(2) + V(1,2,t)\,\psi(1,2,t)\end{aligned}\right\} \quad (5.82\text{b})$$

In Eq. (5.82b) $E_l = E_n + E_m$ and ψ_i represents the $\psi_{nlm}(1)\,\psi_{nlm}(2)$ going with the appropriate a_i or b_i. Let us find $\dot{a}_1$ an example.

Eq. (5.82b) is first multiplied through on the left by $\overline{\psi_{n00}(1)\,\psi_{m10}(2)}$ and the resulting expression is integrated over all space. The result is:

$$i\hbar\,\dot{a}_1 + a_1E_l = a_1E_n + a_1E_m + \int\overline{\psi_{n00}(1)\,\psi_{m10}(2)}\,V(1,2,t)\,\psi(1,2,t)\,d\tau \qquad (5.83)$$

The results obtained here are entirely due to the orthonormality of the unperturbed eigenfunctions of the two molecules. In the integral on the right all terms containing $\int\overline{\psi_{n00}(1)}\,V\psi_{n00}\,d\tau$ will disappear since $V(1, 2, t)$ averaged over the spherically symmetrical ground state is zero.* As a consequence, if we adopt the convention:

$$V^{n00;m10}_{m10;n00} = \int\overline{\psi_{n00}(1)\,\psi_{m10}(2)}\,V(1,2,t)\,\psi_{m10}(1)\,\psi_{n00}(2)\,d\tau \qquad (5.84)$$

we obtain for a:

$$i\hbar\,\dot{a}_1 = V^{n00;m10}_{m10;n00}\,b_1 + V^{n00;m10}_{m11;n00}\,b_2 + V^{n00;m10}_{m1-1;n00}\,b_3 \qquad (5.85)$$

* Cf. *supra*, Eq. (3.1).

In this manner we may obtain the equation for the time rate of change of the coefficients in Eq. (5.81) as:

$$\left.\begin{aligned}
i\hbar\,\dot{a}_1 &= V^{n00;m10}_{m10;n00}\,b_1 + V^{n00;m10}_{m11;n00}\,b_2 + V^{n00;m10}_{m1-1;n00}\,b_3\\
i\hbar\,\dot{a}_2 &= V^{n00;m11}_{m10;n00}\,b_1 + V^{n00;m11}_{m11;n00}\,b_2 + V^{n00;m11}_{m1-1;n00}\,b_3\\
i\hbar\,\dot{a}_3 &= V^{n00;m1-1}_{m10;n00}\,b_1 + V^{n00;m1-1}_{m11;n00}\,b_2 + V^{n00;m1-1}_{m1-1;n00}\,b_3\\
i\hbar\,\dot{b}_1 &= V^{m10;n00}_{n00;m10}\,a_1 + V^{m10;n00}_{n00;m11}\,a_2 + V^{m10;n00}_{n00;m1-1}\,a_3\\
i\hbar\,\dot{b}_2 &= V^{m11;n00}_{n00;m10}\,a_1 + V^{m11;n00}_{n00;m11}\,a_2 + V^{m11;n00}_{n00;m1-1}\,a_3\\
i\hbar\,\dot{b}_3 &= V^{m1-1;n00}_{n00;m10}\,a_1 + V^{m1-1;n00}_{n00;m11}\,a_2 + V^{m1-1;n00}_{n00;m1-1}\,a_3\\
i\hbar\,\dot{c}_1 &= 0
\end{aligned}\right\}\qquad(5.86)$$

It is reasonably obvious that the selection of two higher and more degenerate states would have complicated the problem a bit.

Now it may again be assumed that the interaction between the two molecules begins at time $t = 0$ when the emitter is a distance $v\tau/2$ from ϱ. On this basis we may solve the quantum mechanical problem by a series of approximations.

We have assumed that previous to the introduction of the dipole interaction the upper state is a p state with $m = 0$. Thus at time $t = 0$ there is a finite probability of $\psi_{n00}(1)\,\psi_{m10}(2)$ or $\psi_{n00}(1)\,\psi_{n00}(2)$ — there will always be a finite probability for the existence of the ground state — but no probability for the existence of any other state. These conditions lead us to the zero order solution:

$$\left.\begin{aligned}
a_1(0) &= a; \quad a_2(0) = a_3(0) = 0\\
b_1(0) &= b_2(0) = b_3(0); \quad c_1(0) = c
\end{aligned}\right\}\qquad(5.87)$$

For the first approximation it may be assumed that a_1 remains a constant during time τ. This would mean that no probability for $\psi_{m11}(2)$ or $\psi_{m1-1}(2)$ exists during this interval. This yields:

$$\left.\begin{aligned}
a_1 &= a; \quad a_2 = a_3 = 0\\
b_1 &= \frac{a}{i\hbar}\int_0^\tau V^{m10;n00}_{n00;m10}\,\mathrm{d}t; \quad b_2 = \frac{a}{i\hbar}\int_0^\tau V^{m11;n00}_{n00;m10}\,\mathrm{d}t\\
b_3 &= \frac{a}{i\hbar}\int_0^\tau V^{m1-1;n00}_{n00;m10}\,\mathrm{d}t
\end{aligned}\right\}\qquad(5.88)$$

In order to determine the matrix elements of V, it is apparent from Eq. (5.61) that the matrix elements of x, y, and z must first be determined. Hydrogen-like wave functions of the form of Eq. (3.7) are utilized for this purpose. As an example:

$$\begin{aligned}
(n00\,|x|\,m11) &= \int \psi_{n00}\,x\,\psi_{m11}\,\mathrm{d}\tau = \frac{\sqrt{6}}{8\pi}\int_0^\infty\int_0^{2\pi}\int_0^\pi\\
&\times [R_{n0}(r)]\,[r\sin\vartheta\cos\varphi]\,[R_{m1}(r)\sin\vartheta\,\mathrm{e}^{i\varphi}]\,r\sin\vartheta\,\mathrm{d}\vartheta\,\mathrm{d}\varphi\,\mathrm{d}r\\
&= \frac{1}{2}\sqrt{\frac{2}{3}}\,r_{nm} = (m11\,|x|\,n00)
\end{aligned}$$

where r_{nm} is the radial matrix element $\int_0^\infty r^3 R_{n0}(r)\,R_{m1}(r)\,\mathrm{d}r$.

In like manner the non-vanishing matrix elements may be found as:

$$\left.\begin{aligned}
(n\,00\,|x|\,m\,11) &= (m\,11\,|x|\,n\,00) = \frac{1}{2}\sqrt{\frac{2}{3}}\,r_{n\,m}\\
(n\,00\,|x|\,m\,1-1) &= (m\,1-1\,|x|\,n\,00) = -\frac{1}{2}\sqrt{\frac{2}{3}}\,r_{n\,m}\\
(n\,00\,|y|\,m\,11) &= (m\,11\,|y|\,n\,00) = \frac{i}{2}\sqrt{\frac{2}{3}}\,r_{n\,m}\\
(n\,00\,|y|\,m\,1-1) &= (m\,1-1\,|y|\,n\,00) = -\frac{i}{2}\sqrt{\frac{2}{3}}\,r_{n\,m}\\
(n\,00\,|z|\,m\,10) &= (m\,10\,|z|\,n\,00) = \frac{1}{\sqrt{3}}\,r_{n\,m}
\end{aligned}\right\} \quad (5.89)$$

In finding a matrix element of V we carry the angle functions as constants — it may be recalled that these angle functions depend on the heat motion of the molecules. Thus a matrix element of V is simply some combination of the matrix elements of $x_1, y_1, \ldots, z_2$. Now, for example, in the case of $V^{m\,10;\,n\,00}_{n\,00;\,m\,10}$ the matrix element of all the terms in V except $z_1 z_2 (1 - 3\cos^2\gamma)$ vanish. As a result:

$$V^{m\,10;\,n\,00}_{n\,00;\,m\,10} = \frac{e^2}{R^3}\,(n\;00\,|z|\,m\;10)^2\,(1 - 3\cos^2\gamma)$$

Finally, then Eq. (5.88) becomes:

$$\left.\begin{aligned}
b_1 &= \frac{a}{i\hbar}\,e^2 (n\,00\,|z|\,m\,10)^2 \int_0^\tau \frac{(1 - 3\cos^2\gamma)}{R^3}\,dt\\
b_2 = \bar{b}_3 &= -\frac{a}{i\hbar}\,\frac{3\sqrt{2}}{2}\,e^2 (n00\,|z|\,m10)^2 \int_0^\tau \frac{\cos\gamma\,(\cos\alpha - i\cos\beta)}{R^3}\,dt
\end{aligned}\right\} \quad (5.90)$$

Again, as in the classical case, we allow τ (the collision time) to approach infinity and integrate Eq. (5.90) to obtain:

$$\left.\begin{aligned}
b_1 &= \frac{a}{i\hbar}\,2\,e^2 (n\,00\,|z|\,m\,10)^2\,\frac{1}{v\,\varrho^2}\,(1 - \cos^2\gamma_1 - 2\cos^2\gamma_2)\\
b_2 = \bar{b}_3 &= -\frac{a}{i\hbar}\,\sqrt{2}\,e^2 (n\,00\,|z|\,m\,10)^2\,\frac{1}{v\,\varrho^2}\\
&\quad \times [\cos\gamma_1(\cos\alpha_1 - i\cos\beta_1) + 2\cos\gamma_2(\cos\alpha_2 - i\cos\beta_2)]
\end{aligned}\right\} \quad (5.91)$$

where Eq. (5.69) has been utilized for the $\alpha\beta\gamma$. From Eq. (5.91) it is apparent that the probability for excitation of the initially unexcited molecule is:

$$|b_1|^2 + |b_2|^2 + |b_3|^2 = a^2\,\frac{4\,e^4}{\hbar^2\,\varrho^4\,v^2}\,\sin^2\gamma_1\,(n\,00\,|z|\,m\,10)^4 \quad (5.92)$$

We may now carry out a second approximation to find a_1, still under the assumption $a_2 = a_3 = 0$:

$$|a_1|^2 + |b_1|^2 + |b_2|^2 + |b_3|^2 + |c_1|^2 = |a|^2 + |c|^2 \tag{5.93}$$

and, if we suppose $|c_1|^2 = |c|^2$, the change in $|a|^2$ may be determined from Eq. (5.93) as:

$$\left.\begin{aligned} \Delta|a|^2 &= |a_1| - |a|^2 = -(|b_1|^2 + |b_2|^2 + |b_3|^2) \\ &= -|a|^2 \frac{4e^4}{\hbar^2 \varrho^4 v^2} \sin^2\gamma_1 (n\,00\,|z|\,m\,10)^4 \\ &= -|a|^2 \frac{e^4}{m^2 \omega_{nm}{}^2} f_{nm}{}^2 \frac{1}{\varrho^4 v^2} \sin^2\gamma_1 \end{aligned}\right\} \tag{5.94}$$

where $f_{nm} = \dfrac{2m\,\omega_{nm}}{\hbar}(n00\;|z|\;m10)^2$, the oscillator strength of the mn transition.

The similarity of Eqs. (5.94) and (5.72) is quite apparent, especially when the $E - |a|^2$ analogy is considered.

We earlier set the condition $\varepsilon = E$ for determining ϱ_0.* Similarly, we now establish the criteria for ϱ_0 as $\Delta\,|a|^2 = |a|^2$, that is, the change in the probability of finding the system in the state $\psi_{n00}(1)\,\psi_{m10}(2)$ is equal to the probability of finding the system in this state. From Eq. (5.94) ϱ_0 may then be defined as:

$$\varrho_0 = \sqrt{\frac{e^1}{m\,\omega_{mn}} f_{nm} \frac{1}{v}} \tag{5.95}$$

where $\sin\gamma_1$ has again been given its maximum value.

To this point in the quantum mechanical development of their theory Furssow and Wlassow proceeded in a rather straightforward manner, but their subsequent machinations should at least be critically considered. We might recall that, when we viewed resonance broadening as a case of identical particle degeneracy, the two molecular problem was solvable, but the N-atomic case had no apparent solution. What Furssow and Wlassow did was extend the results of the two molecular case to that of many molecules, and if this procedure is not justifiable in the degeneracy picture, it should not be any more justifiable in the energy exchange picture. We must, therefore, agree and keep in mind the fact that this is simply an approximation, no matter how good a one.

If there are N molecules per unit of volume of our gas, then surely $2\pi\varrho\,\mathrm{d}\varrho\,v\,N$ of them appear per units of time lying a distance between ϱ and $\varrho + \mathrm{d}\varrho$ from the emitter. Furssow and Wlassow asserted this to mean that the total change of the probability per unit of time is:

$$\left.\begin{aligned} \frac{\mathrm{d}}{\mathrm{d}t}|a|^2 &= -|a|^2 \frac{e^4}{m^2\,\omega_{nm}} f_{mn}{}^2 \frac{1}{v^2} \sin^2\gamma_1 \int_{\varrho_0}^{\infty} \frac{2\pi\varrho\,\mathrm{d}\varrho}{\varrho^4} \\ \cdot vN &= -\pi|a|^2 \frac{e^2}{m\,\omega_{nm}} f_{nm} N \sin^2\gamma_1 \end{aligned}\right\} \tag{5.96}$$

* See *supra*, Eq. (5.73) and preceding.

If Eq. (5.96) is averaged over γ_1 and the resulting equation solved for $|a|^2$, the quantum mechanical analog of Eq. (5.74c) is obtained:

$$|a|^2 = |a_0|^2 \, e^{-\Upsilon_0 t} \tag{5.97}$$

where:

$$\Upsilon_0 = \frac{2\pi}{3} \frac{e^2}{m\,\omega_{nm}} f_{nm} N$$

Thus the state lifetime has been decreased by the energy exchange as predicted, and this decrease is dependent on the gas density.

Now let us write down the Hamiltonian for the system, excited molecule plus field:

$$H = H_a + H_f + S \tag{5.98a}$$

In Eq. (5.95a) H_a is the Hamiltonian for the molecule alone; H_f is the Hamiltonian of the field, and S is the Hamiltonian of the field-molecule interaction.

The Schrödinger equation is:

$$i\hbar \frac{\partial \psi}{\partial t} = (H_0 + S)\,\psi \tag{5.98b}$$

When we assume that $\psi = \sum_r a_r \psi_r$ where the ψ_r are the eigenfunctions of the unperturbed system, we may utilize the method by which we arrived at Eq. (5.85) to find the a_r. Let us recall that the symbol r is a quantum conglomerate of the form $nN_1N_2\ldots$ where, to begin with, n specifies the state of the molecule in the system and the N_i specify the states of the i-th radiation oscillator.

The analog of Eq. (5.85) is then:

$$i\hbar\,\dot{a}_{n'N_1'N_2'\ldots} = \sum_{nN_1N_2\ldots} H^{n'N_1'N_2'\ldots}_{nN_1N_2\ldots}\, a_{nN_1N_2\ldots}\, e^{i\frac{E_{n'N_1'N_2'\ldots} - E_{nN_1N_2\ldots}}{\hbar}t} \tag{5.99a}$$

where, of course, the $|a_i|^2$ still have a probability interpretation.

It is reasonable to assume here that at time $t = 0$ the molecule is in the excited state, and there are no photons in the field. This would mean that $a_{m00\ldots0_s0\ldots}(0) = 1$ while all other $a_{nN_1N_2\ldots}(0) = 0$. At some later time the molecule will be in the ground state and there will be a quantum of type s in the field where ν_s is the frequency of this quantum, and ν_{nm} is the frequency of the molecular transition mn. Under the assumption that all amplitudes do remain zero except $a_{m00\ldots0_s0\ldots}(t)$ during the time involved in this transition we obtain:

$$i\hbar\,\dot{a}_{n00\ldots0_s0\ldots}(t) = H^{n00\ldots1_s0\ldots}_{m00\ldots0_s0\ldots}\, a_{m00\ldots}(t)\, e^{i(\omega_{mn}-\omega_s)t} \tag{5.99b}$$

Since ω_s is the frequency of the photon which is present in the field after the transition of frequency ω_{mn} has taken place, ω_s is of the order of ω_{mn}.

It is clear that the probability amplitude $a_{m00\ldots}(t)$ will be the same as the probability amplitude of the initially excited molecule in the state designated by the quantum number m. In addition we may find $a(t)$ from Eq. (5.97):

$$a_{m00\ldots}(t) = a(t) = e^{-\frac{\Upsilon_0}{2}t} \tag{5.99c}$$

When this expression is substituted into Eq. (5.99b) and the resulting equation is integrated over time 0 to t, we will obtain:

$$a_{n00\ldots 1_s 0\ldots}(t) = H^{n00\ldots 1_s 0\ldots}_{m00\ldots 0_s 0\ldots} \frac{e^{i(\omega_s - \omega_{mn})t - \frac{\Upsilon_0}{2}t} - 1}{-\hbar\left[\omega_s - \omega_{nm} + i\left(\frac{\Upsilon_0}{2}\right)\right]} \tag{5.100}$$

After a time $t \to \infty$, the initially excited molecule will surely have undergone the transition to the ground state. Now the intensity of a frequency ω after this time will certainly be proportional to the probability of a photon of this frequency being present in the radiation field. These considerations lead to the following expression for the intensity distribution:

$$I(\omega) = \text{const}\,|a_{n00\ldots 1_s 0\ldots}(\infty)|^2 = \frac{\text{const}}{(\omega - \omega_{mn})^2 + \left(\frac{\Upsilon_0}{2}\right)^2} \tag{5.101}$$

From Eq. (5.101) it is apparent that our half-width is given by the following:

$$\delta_r = \Upsilon_0 = \frac{2\pi}{3} \frac{e^2}{m\,\omega_{mn}} f_{mn} N \tag{5.102}$$

It can be seen that Eq. (5.102) yields the same results for the half-width as Eq. (5.78), the latter of which equations was found for the case of distant collisions.

For the case of near collisions let us utilize the Heisenberg uncertainty principle in the form:

$$\Delta E\,\Delta t = \hbar$$

where we shall now assume ΔE to be the width of the state and Δt to be the mean lifetime of the state. Let us suppose that the life of a state is terminated by a collision so that $\Delta t = \tau$ where τ is the mean time between collisions. This means that the width of the state will now be:

$$\Delta E = \frac{\hbar}{\Delta t} = \frac{\hbar}{\tau} = 2\,\hbar\,\pi \frac{e^2}{m\,\omega_0} f_{nm} N$$

from Eq. (5.58).

Or:

$$\delta_s = \Delta\omega = \frac{\Delta E}{\hbar} = 2\pi \frac{e^2}{m\,\omega_0} f_{nm} N \tag{5.103}$$

Finally, we obtain for the half width:

$$\delta = \delta_r + \delta_s = \frac{8}{3}\pi \frac{e^2}{m\,\omega_0} f_{nm} N \tag{5.104}$$

In the main Furssow and Wlassow considered their results to hold only in the case of low gas densities. It may be recalled we have questioned the Furssow–Wlassow procedure in which they work out the broadening effect of one like molecule and then simply extend these results to the case of N similar molecules. In a later paper on the subject they investigated the self broadening of a spectral line in the case of high gas densities. They found it necessary to use a slightly different although comparable method of approach. We shall now discuss this in greater detail.

5.12. High pressure quantum resonance broadening

The theory was really predicated on a collision between two particles. Now if, as in the Schulz and Rompe[184,185] experiments, we take $\omega_0 = 1.02 \times 10^{16}$ sec^{-1}, $f = 1.3$, $T = 6000°$, we will obtain a ϱ of approximately 5.35×10^{-7}. Then with the pressure used by Schulz and Rompe $N \doteq 2.57 \times 10^{19}$ cm^{-3}. This means that in the "sphere of action" we will have $(4/3)\pi\varrho^3 N \doteq 17$ molecules. This in turn means that in the sphere of action as defined by the optical collision diameter at any time there will be approximately 17 particles. Thus, the approximation of a two particle collision is hardly reasonable.

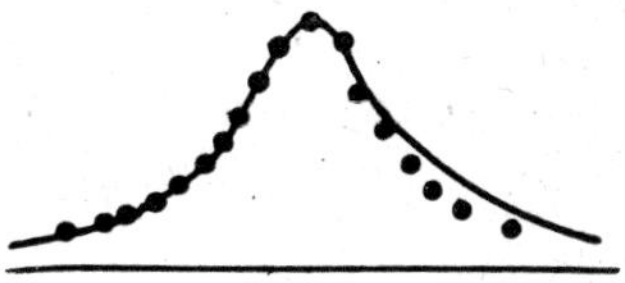

FIG. 5.2. The Schulz and Rompe treatment of an experimental trace (solid line) in order to obtain a half width. (After Schulz and Rompe[158])

A modification of the first theory is certainly called for, and, in this paper, Furssow and Wlassow[58] set out to calculate the broadening of the resonance level for this higher pressure case.

Let us begin by supposing the gas density sufficiently high that we can expect a large number n of molecules within the sphere of action. If we allow this system of $n + 1$ molecules to possess two levels, the resonance level and the ground state, the possible eigenfunctions to describe the system in the upper state will be:

$$\left.\begin{aligned} \psi_a &= u_1(0)\, u_0(1) \ldots u_0(n) \\ \psi_{b_1} &= u_0(0)\, u_1(1) \ldots u_0(n) \\ &\vdots \\ \psi_{b_n} &= u_0(0)\, u_0(1) \ldots u_1(n) \end{aligned}\right\} \qquad (5.105)$$

where we have assumed molecule 0 to be the initially excited atom.

The Hamiltonian of the dipole interaction will be:

$$U = \sum_{k \neq k'} \sum V(k, k') = \sum_{k=1}^{n} V(0, k) + \sum \sum V(k, k') \qquad (5.106)$$

The solution to the Schrödinger equations which results from this interaction Hamiltonian is surely:

$$\Psi = a\,\psi_a + \sum_{k=1}^{n} b_k\, \psi_{b_k} \qquad (5.107)$$

Using methods which have certnaily become familiar we find:

$$\left.\begin{aligned} i\hbar\,\dot{a} &= \sum_{k=1}^{n} (a\,|U|\,b_k)\, b_k \\ i\hbar\,\dot{b}_k &= (b_k\,|U|\,a)\, a + \sum_{k'} (b_k\,|U|\,b_{k'})\, b_k \end{aligned}\right\} \qquad (5.108)$$

Again under the initial conditions stipulating that molecule 0 is initially excited we obtain $a(t) = a(0) = 1$ and $b(0) = 0$. Subsequent to the time $t = 0$ we can expect the transfer of energy from the initially excited to the unexcited molecules to proceed as governed by the matrix element $(a\,|U|\,b_{k'})$. In addition Eq. (5.108) tells us that there will be a secondary transfer of energy among the initially unexcited molecules as governed by the matrix element $(b_k\,|U|\,b_k')$. We now introduce the approximation that the secondary energy transfer process as governed by $(b_k\,|U|\,b_{k'})$ can be disregarded. This allows us to rewrite Eq. (5.108) as:

$$i\hbar\dot{a} = \sum_{k=1}^{n} (a\,|U|\,b_k)\,b_k \qquad (5.109\text{a})$$

$$i\hbar\dot{b}_k = (b_k\,|U|\,a)\,a \qquad (5.109\text{b})$$

We next introduce an additional approximation to the effect that U is not time dependent. This is nothing more nor less than saying that the thermal motion of the molecules is so small as to be considered negligible during the energy transfer process*. As we shall see, at high pressure for the resonance level this is a reasonable approximation. Now let us take the time derivative of Eq. (5.109a) and substitute for b_k from Eq. (5.109b) in the resulting equation. This yields:

$$-\hbar\ddot{a} = \sum_{k=1}^{n} (a\,|U|\,b_k)^2 \qquad (5.110)$$

If we now ignore the angular dependence of the dipole interaction, there results:

$$\ddot{a} + p^2 a = 0 \qquad (5.111\text{a})$$

$$p^2 = a \sum_{k=1}^{n} \frac{\alpha}{R_k^{\,6}}; \qquad \alpha = \frac{e^4 f^2}{4\,m^2\,\omega_0^{\,2}} \qquad (5.111\text{b})$$

A solution of Eq. (5.111a) which satisfies the initial condition $a(0) = 0$ is:

$$a = \cos pt \qquad (5.112)$$

If we again assume that the condition $a = 0$ — that is, the energy has been transferred from the initially excited molecule — determines the collision time, τ, then from Eq. (5.112) there results:

$$\tau = \frac{\pi}{2\,p} \qquad (5.113)$$

We wish to develop a in a Fourier integral. We have only obtained an approximation for a, and another approximation which would be almost as good is the step function:

$$\left.\begin{aligned} a &- 1 \quad \text{for} \quad 0 \leq t < \tau \\ a &= 0 \quad \text{for} \quad 0 < t > \tau \end{aligned}\right\} \qquad (5.114)$$

* This is also equivalent to a statistical approximation for which see *supra*, Chap. 3.

In a quite straightforward manner we may write down the Fourier integral for a as:

$$a = \int_{-\infty}^{+\infty} g(\omega)\, e^{i\omega t}\, d\omega \tag{5.115a}$$

$$g(\omega) = \frac{1}{2\pi} \int_{-\infty}^{+\infty} a\, e^{-i\omega t}\, d\omega = \frac{1}{2\pi} \int_0^{\tau} 1 \cdot e^{-i\omega t}\, d\omega = \frac{1}{2\pi} \frac{1 - e^{-i\omega t}}{i\omega} \tag{5.115b}$$

The distribution of the energy in the level will surely be given by the absolute square of the Fourier amplitude $g(\omega)$:

$$|g(\omega)|^2 = \frac{1}{2\pi^2} \frac{1 - \cos \omega t}{\omega^2} \tag{5.116}$$

Eq. (5.116) gives us the desired result for a specific collision time τ. It would now appear reasonable to average this expression over all collision times in order to obtain the observed distribution in the level. Let us suppose that $w(\tau)\, d\tau$ is the probability that the collision time lies between τ and $\tau + d\tau$. Then for $J(\omega)$ we obtain:

$$J(\omega) = \int_0^{\infty} |g(\omega)|^2\, w(\tau)\, d\tau \tag{5.117}$$

Let us utilize the expression:

$$\nu = \frac{\pi^2}{4\tau^2} = p^2 \tag{5.118}$$

for a change in variable in Eq. (5.119) as follows:

$$w(\tau)\, d\tau = \frac{\pi\, w(\pi/2\sqrt{\nu})}{4\sqrt{\nu^3}}\, d\nu = -I'(\nu)\, d\nu \tag{5.119}$$

In order to find $I'(\nu)\, d\nu$ we may proceed in the following manner. Let us establish a configuration space of $3n$ dimensions whose volume is $\Gamma = V^n$ where V is the volume of our gas. Now if we temporarily disregard the intermolecular forces, we can expect equal probability for the occupation of any portion of this space. This means that the following relation will hold:

$$I'(\nu)\, d\nu = \frac{\Delta\Gamma}{\Gamma} \tag{5.120}$$

where $\Delta\Gamma$ represents that portion of configuration space for which our ν lies between ν and $\nu + d\nu$. This value of ν will, of course, depend on the distribution of the R_k in Eq. (5.111b). Thus:

$$\Delta\Gamma = (4\pi)^n \int \ldots \int R_1^2\, R_2^2 \ldots R_n^2\, dR_1\, dR_2 \ldots dR_n \tag{5.121}$$

The analogy to Margenau's Eq. (3.17a) is apparent. Our solution then is given by Eq. (3.32) as:

$$I'(\nu) = \frac{2}{3} \pi N \sqrt{\alpha}\, \frac{e^{-\frac{4\pi^3 \alpha N^2}{9\nu}}}{\sqrt{\nu^3}} \tag{5.122}$$

We may now transform back to our variable τ by utilizing Eq. (5.119). We obtain:

$$w(\tau) = \frac{2\,\Upsilon}{\sqrt{\pi}}\, e^{-2\Upsilon^2\tau^2} \tag{5.123}$$

$$\Upsilon = \frac{4}{3}\, N \sqrt{\pi\,\alpha} = \frac{2}{3}\sqrt{\pi}\,\frac{f e^2}{m\,\omega_0}\, N \tag{5.124}$$

Eqs. (5.113) and (5.123) may now be substituted into Eq. (5.117) and the results integrated to obtain:

$$I(\omega) = \frac{1}{2\,\pi^2}\,\frac{1 - e^{-(2/\omega\,\Upsilon)^2}}{\omega^2} \tag{5.125}$$

as the distribution of energies in the resonance level. For the breadth we obtain:

$$\delta = 2.54\,\Upsilon = 3\,\frac{f e^2}{m\,\omega_0}\, N \tag{5.126}$$

It is Eq. (5.126) which Furssow and Wlassow used to determine the resonance level width for comparison with experiment. In order to find the spectral line width, it would appear reasonable to assume that we must needs have a knowledge of not only the resonance level width but also the width of the level from which the radiating transition originates.* The phrase "from which" leads us to the next consideration.

Furssow and Wlassow felt in their second paper (they experienced no such qualms in their first) that their results should only be applied to transitions proceeding from upper levels to the resonance level, that is, these results should not be applied to the transition from the resonance level to the ground state. A short consideration renders this assertion quite plausible.

Let us suppose that the initially excited molecule undergoes a transition from the resonance level to the ground state with the accompanying emission of radiation. This emitted radiation may be absorbed by one of the unexcited molecules where absorption would not be possible were this radiation the result of a transition to some level above the ground state. This process, which should not be confused with the transfer of energy without accompanying radiation, has certainly not been considered in the theory, and, since it can be expected to have some effect on the line broadening, this theory should be somewhat effected in such cases.

The justification for the assumption of fixed atoms, that is, $d/dt\,(a\,|\,U\,|\,b_k) = 0$, which led to Eq. (5.110) appears worthy of note here. Let us consider the Rompe and Schulz case[184]. The pressure is 80 atmospheres and $T = 2800°$. The experimental width was found to be $\delta = 7 \times 10^{12}\ \text{sec}^{-1}$ so that the mean life of the resonance level is $\tau = 1/\delta = 1.4 \times 10^{-13}$ sec. The mean relative velocity

* The complexity, if not impossibility, of computing the width of a higher level may be inferred from the preceding work.

is $v = 1.3 \times 10^5$ cm/sec, which leads us to the conclusion that during the mean level life we may expect our atoms to move a distance $v\tau = 1.8 \times 10^{-8}$ cm, or the order of their own diameter. Thus, the approximation appears a reasonable one.

5.13. Review and summary

Although Resonance Broadening is of the same general class as the other broadening phenomena, the physical basis for the effect is sufficiently unique to warrant its treatment in a separate category. Classically, the broadening may be attributed to the exchange of energy between two like oscillators. Quantum mechanically the same sort of trade may be envisioned and directly treated or treated as a means for reducing level lifetime and hence width.

As in other specific types of broadening, such as van der Waals or Stark, the Interruption Theory is applicable to low densities or positions near line center while the Statistical Theory is applicable to high densities or positions in the line wing. Here, however, both theories yield a Lorentz-type symmetric line shape.

If the Statistical Theory is applicable, one may evaluate Eq. (3.113a) for g_3 and substitute this into Eq. (3.116c) to obtain:

$$\delta = 2 N g_3 = \frac{4}{3} \pi^2 C_3 N \qquad (5.127)$$

Here we have expressed the half-width in angular frequency units, and the resonance interaction constant is also to be expressed in angular frequency units.

If the Interruption Theory applies one may either obtain the simple Weisskopf result:

$$\delta = 4\pi C_3 N \qquad (5.128)$$

or the Lindholm result which includes distant collisions:

$$\delta = 2\pi^2 C_3 N \qquad (5.129)$$

A specific quantum treatment for binary interactions by Furssow and Wlassow led to the half-width:

$$\delta = \frac{16}{3} \pi C_3 N \qquad (5.130)$$

In all these equations for the half-width the constant may be specifically evaluated as:

$$C_3 = \frac{e^2 f_{nm}}{2 m \omega_0} \qquad (5.131)$$

CHAPTER 6

MOLECULAR BROADENING

WE SHALL now restrict ourselves to consideration of polyatomic molecules, and we trouble to do so since special phenomena are associated with this type molecule. These special phenomena, of course, arise from the unique characteristics of the polyatomic molecule, namely, their rotational and vibrational behavior. Here too we shall include broadening considerations which are particularly applicable to the microwave region of the spectrum. Although it is certainly true that this region is closely associated with polyatomic spectra, it is also true that such spectra do not provide the only source for such radiation.

In what is to follow we shall first discuss the interactions peculiar to molecules which have been found responsible for line broadenings. Next we consider Lindholm's thoughtful treatment of molecular broadenings for two particular molecules, a treatment which could well provide a model for similar treatments. Certain specific considerations of Margenau will be studied, after which we shall discuss the by now rather familiar Anderson theory. In this theory we remark the inclusion of diabaticity by the failure to exclude various possible induced transitions. We further remark its basically Interruption nature. Subsequently we shall detail the Van Vleck–Weisskopf Interruption treatment which supposes there to be a Maxwell–Boltzmann distribution of dipole moments after collision. This we have relegated to this chapter since the result is only appreciably different from the more familiar Interruption result when the low frequencies of the microwave region are under consideration. Quite naturally from this consideration follows the later work of Frölich, Gross, and Gora on the even more complex distributions which are predicated on the non-existence of thermal equilibrium leading to the invalidity of the Maxwell–Boltzmann distribution.

6.1. Early work on broadening of rotation-vibration lines

We might begin by remarking that the main types of foreign gas broadening as developed in Chapters 2 through 5 could be applied in general to the present molecular case of rotation and rotation-vibration lines. Indeed in an early work Kussmann[117] did just this, that is, he investigated the broadening of HCl rotation lines and checked the experimental results against the predictions of Holtzmark's Stark theory and Lorentz' Interruption theory. He concluded that the Lorentz theory agreed best and the subesquent work of Lasareff[119] and Grasse[63] on rotation-vibration lines appeared to bear this out. Thus, to this point, those interested in the broadening of rotation and rotation-vibration lines had found no reason for treating this type line any differently from electronic lines. The intimation that some different approach to the broadening of the former was in order appears to have arisen first in the work of Herzberg and Spinks[75]. These investigators found a decrease of the line width with higher values of the rotational quantum number J in the near infrared spectra of HCN. This discovery would surely indicate that rotation-vibration spectra should be treated differently than electronic spectra, but this clear indication was rather clouded when the work of Cornell and Watson[28] and Herzberg, Spinks, and Watson[76] failed to verify this variation with J. The hint had been given however, and it was certainly apparent that the possibility of J width dependence and other unique molecular phenomena should be theoretically investigated.

6.2. Interactions between rotating dipoles (directional effect)

We shall now consider as an introduction to our first broadening investigation, the interaction between two rotators, one or both of which may possess an electric dipole moment.*

First, let us carry out a rather obvious modification of Eq. (3.1) to obtain:

$$V = -\frac{1}{R^3} \sum_{ij} e_i e_j (2 z_i z_j - x_i x_j - y_i y_j) \tag{6.1}$$

In Eq. (6.1) we have denoted the system of charges and the coordinates of those charges belonging to one of the rotators by i, and the charges and coordinates belonging to the other by j. Here also we have allowed the interrotator axis to be the common axis of the two systems, and R now specifies the (large) separation of the two rotators.

Now let us specify the dipole moments of the rotators by μ_1 and μ_2 and introduce spherical polar coordinates for the two systems. In this case then, there results:

$$V = \frac{\mu_1 \mu_2}{R_3} [\sin \vartheta_1 \cos \varphi_1 \sin \vartheta_2 \cos \varphi_2 + \sin \vartheta_1 \sin \varphi_1 \sin \vartheta_2 \sin \varphi_2 - 2 \cos \vartheta_1 \cos \vartheta_2] \tag{6.2}$$

The conditions here must conform to those leading up to Eq. (3.4), that is, we assume the separation of the rotators to be large enough so that no charge overlap or exchange need be considered. Thus the perturbations resulting from the action of these two dipoles is given by Eq. (3.4). Before writing this result down, however, let us remark that the eigenvalues of the rotators in question are:

$$E_{JM} = \frac{\hbar^2}{2I} J(J+1) \equiv E_J$$

where J is the total angular momentum quantum number and M the magnetic quantum number.

* See references 82 and 193.

If we now designate initial states by a double prime, final states by a prime and indicate the two rotators by the subscripts one and two, we may write Eq. (3.4) for the present case as:

$$E^{(2)} = \sum_{\substack{J_1' J_2' \\ M_1' M_2'}} \frac{|(J_1'' M_1'' J_2'' M_2''| V | J_1' M_1' J_2' M_2'')|^2}{E_{J_1''} - E_{J_1'} + E_{J_2''} - E_{J_2'}} \tag{6.3}$$

where V is given by Eq. (6.2).

Since no magnetic field, in which the magnetic quantum numbers may interest themselves, is present, we simply average $E^{(2)}$ over these quantum numbers. Equal probability for them is assumed and there results:

$$\langle E^{(2)} \rangle = \frac{1}{(2J_1'' + 1)(2J_2'' + 1)} \sum_{M_1'' M_2''} E^{(2)} \tag{6.4}$$

since each J_i'' level is $(2J_i'' + 1)$-fold degenerate.

The eigenfunctions for our rotator system will be:

$$\psi(1, 2) = \psi_{J_1 M_1} \psi_{J_2 M_2} = N \, e^{\pm i M_1 \varphi_1} P_{J_1}^{M_1}(\cos\vartheta_1) \, e^{\pm i M_2 \varphi_2} P_{J_2}^{M_2}(\cos\vartheta_2) \tag{6.5}$$

where the $P_{J_i}^{M_i}$ are associated Legendre functions.

In Eq. (6.3) we shall need matrix elements of $\cos\vartheta$, $\sin\vartheta\cos\varphi$, and $\sin\vartheta\sin\varphi$. In order to determine these matrix elements let us introduce three recursion formulas relating the associated Legendre function as:

$$(1 - z^2)^{1/2} P_J^{|M|-1}(z) = \frac{1}{(2J+1)} P_{J+1}^{|M|}(z) - \frac{1}{(2J+1)} P_{J-1}^{|M|}(z) \tag{6.6a}$$

$$\left.\begin{aligned} (1 - z^2)^{1/2} P_J^{|M|+1}(z) = {} & \frac{(J + |M|)(J + |M| + 1)}{(2J+1)} P_{J-1}^{|M|}(z) \\ & - \frac{(J - |M|)(J - |M| + 1)}{(2J+1)} P_{J+1}^{|M|}(z) \end{aligned}\right\} \tag{6.6b}$$

$$z P_J^{|M|}(z) = \frac{(J + |M|)}{(2J+1)} P_{J-1}^{|M|}(z) + \frac{(J - |M| + 1)}{(2J+1)} P_{J+1}^{|M|}(z) \tag{6.6c}$$

where $z = \cos\vartheta$.

For future reference, let us note that:

$$\int_{-1}^{+1} P_{J''}^{|M''|}(z) P_{J'}^{|M''|}(z) \, dz = \left\{\begin{array}{ll} 0 & \text{for } J'' \neq J' \\ \dfrac{2}{(2J''+1)} \dfrac{(J'' + |M''|)!}{(J'' - |M''|)!} & \text{for } J'' = J' \end{array}\right. \tag{6.7a}$$

by means of which we may obtain the normalization factor for $\Theta_{JM}(\vartheta)$ as:

$$\Theta_{JM}(\vartheta) = \left[\frac{(2J+1)}{2} \frac{(J - |M|)!}{(J + |M|)!}\right]^{1/2} P_J^{|M|}(\cos\vartheta) = N P_J^{|M|}(\cos\vartheta) \tag{6.7b}$$

Let us evaluate a portion of the matrix element of $\sin\vartheta \sin\varphi$ as an example.

$$(J''M''\,|\sin\vartheta\sin\varphi|\,J'M')$$

$$= \frac{N''N'}{2\pi}\int_{-1}^{+1}\int_{0}^{2\pi} e^{iM''\varphi}\,P_{J''}^{|M''|}\sin\vartheta\sin\varphi\, e^{-iM'\varphi}\,P_{J'}^{|M'|}\,d\varphi\,dz \tag{6.8a}$$

$$= \frac{N''N'}{2\pi}\int_{-1}^{+1}\int_{0}^{2\pi} e^{iM''\varphi}\,e^{-iM'\varphi}\left(\frac{e^{i\varphi}-e^{-i\varphi}}{2i}\right)P_{J''}^{|M''|}\,P_{J'}^{|M'|}(1-z^2)^{1/2}\,d\varphi\,dz$$

The integration over φ in Eq. (6.8a) goes to zero due to the periodicity of φ unless $M' = M'' \pm 1$. We carry out the integration over φ for the case $M' = M'' + 1 (\Delta M = +1)$ to obtain:

$$\frac{N''N'}{2i}\int_{-1}^{+1} P_{J''}^{|M''|}\,P_{J'}^{|M''|+1}(1-z^2)^{1/2}\,dz \tag{6.8b}$$

Now for $M \geq 0$ it follows that $|M+1| = |M|+1$ and for $M < 0$, $|M+1| = |M|-1$. First the case $M \geq 0$ may be considered and Eq. (6.6b) utilized to obtain:

$$\left.\begin{aligned}
&\frac{N''N'}{2i}\int_{-1}^{+1} P_{J''}^{|M''|}\,P_{J'}^{|M''|+1}(1-z^2)^{1/2}\,dz\\
&= \frac{N'N''}{2i}\,\frac{(J'+|M''|)(J'+|M''|+1)}{(2J'+1)}\int_{-1}^{+1} P_{J''}^{|M''|}\,P_{J-1}^{|M''|}\,dz\\
&- \frac{N'N''}{2i}\,\frac{(J'-|M''|)(J'-|M''|+1)}{(2J'+1)}\int_{-1}^{+1} P_{J''}^{|M''|}\,P_{J'+1}^{|M''|}\,dz
\end{aligned}\right\} \tag{6.8c}$$

Where Eq. (6.8a) told us that M may only change by ± 1, Eq. (6.8c) tells us that J may only change by ± 1. The J change is here governed by the orthogonality of the associated Legendre functions. Let us further limit ourselves to the first integral in Eq. (6.8c) which only fails to disappear for $J' = J'' + 1$. We are then considering the case $\Delta M = +1$, $\Delta J = +1$, $\Delta M \geq 0$.

$$\left.\begin{aligned}
&\frac{N'N''}{2i}\,\frac{(J'+|M''|)(J'+|M''|+1)}{(2J'+1)}\int_{-1}^{+1} P_{J''}^{|M''|}\,P_{J'-1}^{|M''|}\,dz\\
&= \frac{1}{2i}\left[\frac{(2J''+1)(2J''+3)(J''-|M''|)!\,(J''-|M''|+2)!}{4(J''+|M''|)!\,(J''+|M''|)!}\right]^{1/2}\\
&\times\left[\frac{2}{(2J''+1)}\,\frac{(J''+|M''|)!}{(J''-|M''|)!}\right]\left[\frac{1}{(2J''+3)}\right]\\
&= \frac{1}{2i}\left[\frac{(J''+M''+1)(J''+M''+2)}{(2J''+1)(2J''+3)}\right]^{1/2}
\end{aligned}\right\} \tag{6.9a}$$

If we now go back and consider Eq. (6.8b) for $M < 0$, we again obtain Eq. (6.9a). Thus:

$$(J''M''\,|\sin\vartheta\sin\varphi|\,J'+1, M''+1) = \frac{1}{2i}\left[\frac{(J''+M''+1)(J''+M''+2)}{(2J''+1)(2J''+3)}\right]^{1/2} \tag{6.9b}$$

In an entirely analogous manner we may obtain all allowed matrix elements of $\cos\vartheta$, $\sin\vartheta\cos\varphi$, and $\sin\vartheta\sin\varphi$ using Eqs. (6.6). We finally obtain for the desired matrix elements:

$$(J''M''\,|\sin\vartheta\cos\varphi|\,J'M') = \frac{1}{2}A(+)\,\delta(M', M''+1) - \frac{1}{2}A(-)\,\delta(M', M''-1) \tag{6.10a}$$

$$(J''M''\,|\sin\vartheta\sin\varphi|\,J'M') = \frac{1}{2i}A(+)\,\delta(M', M''+1) + \frac{1}{2i}A(-)\,\delta(M', M''-1) \tag{6.10b}$$

$$(J''M'')\,|\cos\vartheta|\,J'M') = B\,\delta(M', M'') \tag{6.10c}$$

where:

$$\left.\begin{aligned} A(\pm) &= \left[\frac{(J''\pm M''+2)(J''\pm M''+1)}{(2J''+1)(2J''+3)}\right]^{1/2}\delta(J', J''+1) \\ &\quad - \left[\frac{(J''\mp M''-1)(J''\mp M'')}{(2J''+1)(2J''-1)}\right]^{1/2}\delta(J', J''-1) \end{aligned}\right\} \tag{6.10d}$$

$$\left.\begin{aligned} B &= \left[\frac{(J''+M''+1)(J''-M''+1)}{(2J''+1)(2J''+3)}\right]^{1/2}\delta(J', J''+1) \\ &\quad + \left[\frac{(J''+M'')(J''-M'')}{(2J''+1)(2J''-1)}\right]^{1/2}\delta(J', J''-1) \end{aligned}\right\} \tag{6.10e}$$

Now in order to find the matrix elements of V we merely add products of Eqs. (6.10) as indicated by Eq. (6.2). We need not write down this intermediate result. What we finally desire, although it may not still appear so, are the squares of the matrix elements of V. When we square the matrix elements of V, a consideration of Eq. (6.2) tells us that we will obtain cross product terms of the form:

$$(J_1''M_1''\,|\sin\vartheta_1\cos\varphi_1|\,J_1'M_1')\,(J_1''M_1''\,|\sin\vartheta_1\sin\varphi_1|\,J_1'M_1')$$

All terms of this form will disappear when the summation over M_1' and M_2' indicated in Eq. (6.3) and the averaging over M_1'' and M_2'' indicated in Eq. (6.4) are carried out.* Thus $|(J_1''M_1''J_2''M_2''|\,V\,|J_1'M_1'J_2'M_2')|$ contains three terms of the form:

$$(J_1''M_1''\,|\sin\vartheta_1\cos\varphi_1|\,J_1'M_1')^2\,(J_2''M_2''\,|\sin\vartheta_2\cos\varphi_2|\,J_2'M_2')^2$$

according to Eq. (6.2).

Next we carry out the summation over M_i' as indicated by Eq. (6.3):

$$\left.\begin{aligned} \sum_{M'}(J''M''\,|\sin\vartheta\cos\varphi|\,J'M')^2 &= \sum_{M'}(J''M''\,|\sin\vartheta\sin\varphi|\,J'M')^2 \\ &= \frac{1}{4}\,[A^2(+)+A^2(-)]^2 \end{aligned}\right\} \tag{6.11a}$$

$$\sum_{M'}(J''M''\,|\cos\vartheta|\,J'M')^2 = B^2 \tag{6.11b}$$

according to Eqs. (6.10).

* This may be verified by utilizing Eqs. (6.10).

We may quite easily show that:

$$\langle M^2 \rangle = \frac{1}{2J+1} \sum_{-J}^{+J} M^2 = \frac{1}{3} J(J+1)$$

by the use of which in conjunction with Eqs. (6.10) there results:

$$\langle A^2 \rangle = \frac{1}{(2J+1)} \sum_M A^2 = \frac{2}{3(2J+1)} [(J+1)\,\delta(J', J''+1) + J\,\delta(J', J''-1)] \tag{6.12a}$$

$$\langle B^2 \rangle = \frac{1}{2} \langle A^2 \rangle \tag{6.12b}$$

Very straight forward substitution indeed of Eqs. (6.12) into (6.11) and utilization of the results yields:

$$\left.\begin{aligned} &\sum_{\substack{M_1'' M_2'' \\ M_1' M_2'}} |(J_1''\, M_1''\, J_2''\, M_2''\,|\,V\,|\,J_1'\, M_1'\, J_2'\, M_2')|^2 \\ &\quad = \frac{2}{3}\,\frac{\mu_1 \mu_2}{R^6}\,\frac{1}{(2J_1''+1)\,(2J_2''+1)} [(J_1''+1)\,\delta(J_1', J_1''+1) \\ &\quad + J_1''\,\delta(J_1', J_1''-1)]\,[(J_2''+1)\,\delta(J_2', J_2''+1) \\ &\quad + J_2''\,\delta(J_2', J_2''-1)] \end{aligned}\right\} \tag{6.13a}$$

In order to obtain Eq. (6.4) we simply need to substitute the energy eigenvalues* and carry out the summation over J_i', the upper rotational states:

$$\langle E^{(2)} \rangle = -\frac{2}{3}\,\frac{\mu_1^2 \mu_2^2}{R^6}\,\frac{1}{(2J_1+1)\,(2J_2+1)} \left\{ \frac{(J_1+1)\,(J_2+1)}{(2(J_1+1)\,A_1 + 2(J_2+1)\,A_2} \right. \tag{6.13b}$$

$$\left. + \frac{(J_1+1)\,J_2}{2\,(J_1+1)\,A_1 - 2\,J_2\,A_2} + \frac{J_1(J_2+1)}{2\,J_1\,A_1 + 2(J_2+1)\,A_2} - \frac{J_1\,J_2}{2\,J_1\,A_1 + 2\,J_2\,A_2} \right\}$$

$$\left.\begin{aligned} &= -\frac{2}{3}\,\frac{\mu_1^2 \mu_2^2}{R^6}\,\frac{1}{(2J_1+1)\,(2J_2+1)} \left\{ \left[\frac{2A_1}{J_2+1} + \frac{2A_2}{J_1+1}\right]^{-1} \right. \\ &+ \left. \left[\frac{2A_1}{J_2} - \frac{2A_2}{J_1+1}\right]^{-1} - \left[\frac{2A_1}{J_2+1} - \frac{2A_2}{J_1}\right]^{-1} - \left[\frac{2A_1}{J_2} + \frac{2A_2}{J_1}\right]^{-1} \right\} \end{aligned}\right\} \tag{6.14}$$

where the primes have been dropped, and it is now understood that the J_i refer to the lower state.

If the two molecules are identical so that $A_1 = A_2$ and $\mu_1 = \mu_2$ Eq. (6.14) becomes:

$$\langle E^{(2)} \rangle = \frac{2}{3}\,\frac{\mu^4\,I}{\hbar^2\,R^6}\,\frac{J_1(J_1+1) + J_2(J_2+1)}{(J_1+J_2)\,(J_1+J_2+2)\,(J_1-J_2-1)\,(J_1-J_2+1)} \tag{6.15}$$

In order to consider the special case of the broadener in the ground state, let us refer to Section 3.3. To begin with a comparison of Eq. (6.5) (the rigid

* For a spherical rotator they are $(\hbar/2I)\,J(J+1)$.

rotator eigenfunction) and Eq. (3.7) (the hydrogen-like wave function) tells us that they differ only in the radial wave function $R_{nl}(r)$ of the latter. Thus, with the neglect of the matrix elements of r, the results of Section 3.3 should be equally applicable here. Let us then rewrite Eq. (3.6) with the help of Eqs. (4.6) as:

$$\left.\begin{aligned}\sum_{m\beta} |V_{10,\alpha\beta}|^2 &= \frac{e^4}{r^6}[y_{1\alpha}^{\;2}(1) + z_{1\alpha}^{\;2}(1) + 4\,x_{1\alpha}^{\;2}(1)]\,\frac{1}{3}\,r_{0\beta}^{\;2}(2)\,\delta_{l_\beta 1}\\ &= \frac{e^4}{r^6}[y_{1\alpha}(1) + z_{1\alpha}(1) - 2\,x_{1\alpha}(1)]^2\,\frac{1}{3}\,r_{0\beta}^{\;2}(2)\,\delta_{l_\beta 1}\\ &= |(1\,|r|\,\alpha)|^2\,|(0\,|V|\,\beta)|^2\end{aligned}\right\} \quad (6.16)$$

Now for our present case we may apply Eq. (6.16) by, say, taking a rigid rotator in the ground state as the perturber and another rigid rotator — not necessarily in the ground state — as the absorber. For this case, we would obtain from Eqs. (6.13) and (6.16):

$$\langle E^{(2)}\rangle = -\frac{2}{3}\frac{\mu_1^{\;2}}{R^6}\frac{1}{2\,J_2''+1} \quad (6.17)$$

$$\times\left\{\sum{}'\,|(00\,|e\,r|\,J_1'\,M_1')|^2\left(\frac{J_2''+1}{E_{J_1}-E_0-2(J_2''+1)\,B}+\frac{J_2''}{E_{J_1}-E_0+2\,J_2''B}\right)\right\}$$

In passing, let us note a corresponding expression for the dipole–dipole interaction of two symmetric top molecules possessed of dipoles as given by Carroll[2c]:

$$\left.\begin{aligned}\langle E^{(2)}\rangle = &-\frac{2}{3}\frac{\mu^4}{R^6}\frac{I_A}{\hbar^2}\left\{\frac{(J_1+1)^2-K_1^{\;2}}{(J_1+1)\,(2\,J_1+1)}\right.\\ &\times\left[\frac{(J_2+1)^2-K_2^{\;2}}{(J_2+1)\,(2J_2+1)\,(J_1+J_2+2)}+\frac{K_2^{\;2}}{J_2(J_2+1)\,(J_1+1)}\right.\\ &\left.+\frac{J_2^{\;2}-K_2^{\;2}}{J_2(2J_2+1)\,(J_1-J_2+1)}\right]+\frac{K_1^{\;2}\,K_2^{\;2}}{J_1(J_1+1)\,J_2^{\;2}\,(J_2+1)^2}\\ &+\frac{J_1^{\;2}-K_1^{\;2}}{J_1(2J_1+1)}\left[\frac{(J_2+1)^2-K_2^{\;2}}{(J_2+1)\,(2J_2+1)\,(J_2-J_1+1)}\right.\\ &\left.\left.-\frac{K_2^{\;2}}{J_1J_2(J_2+1)}-\frac{J_2^{\;2}-K_2^{\;2}}{J_2\,(2J_2+1)\,(J_1+J_2)}\right]\right\}\end{aligned}\right\} \quad (6.18)$$

6.3. Rotational resonance and the case $J_1 = J_1 = 0$

A consideration of Eq. (6.15) tells us that a rotational resonance condition sets in for identical rotators for $|J_1 - J_2| = 1$. Although the result is that the equation does not hold for this case, it certainly does indicate that strong interactions can be expected for, say, $|J_1 - J_2| < 2$ or 3. Further Eq. (6.14) indicates for dissimilar dipoles that an accidental rotational resonance condition comes about when $J_2A_2 = (J_1 + 1)\,A_1$ or $(J_2 + 1)\,A_2 = J_1A_1$. Actually, these cases should be treated as for two indistinguishable particles,† where we are unable to state which of the two is excited but only that the sytem of two contains one excited and one unexcited molecule. Margenau[138] considers this case as follows:

If we have the quantum numbers involved related by $J_1 = J - 1$ and $J_2 A_2 = (J_1 + 1) A_1$ the same energy results as for the case $J_1 = J$ and $(J_2 + 1) A_2 = J_1 A_1$, namely, $J^2 (A_1 + A_2) A_1/A_2$ — note that for identical rotators we simply let $A_1 = A_2$ which does not much affect our results. Now each state is twofold degerenate due to this effect, so we now average over this degeneracy as we have previonsly averaged over the spatial degeneracy (given by M). As a result of this averaging process the singular terms cancel each other out since they have opposite signs for the two cases and we obtain:

$$\langle E^{(2)} \rangle = -\frac{1}{6} \frac{\mu^4}{\hbar^2} \frac{I}{R^6} \frac{4 J^4 - J^2 + 1}{(4 J^2 - 1)^2} \tag{6.19}$$

in place of Eq. (6.15).

Let us bring out a rather obvious but nonetheless important facet of Eq. (6.15). A perusal of the numerator of this equation tells us that for $|J_1 - J_2| \geq 2$ the force averaged over all M between two similar dipoles is repulsive, while for other values of $|J_1 - J_2|$ this average force is attractive. In essence, we must bar $|J_1 - J_2| = 1$ so that the forces are attractive only for $J_1 = J_2$.

In actuality the resonance situation which exists here would lead one to ask whether under these conditions there may not be a first order contribution to the energy. The answer, of course, is that there is. We may recall that the manner of the earlier disappearance of the first order contribution was contingent on the summation over the spatial degeneracy parameter M. With a two rotator system we now have an additional degeneracy since the system has the same energy for $(J_1, J_2 + 1)$ and $(J_1 + 1, J_2)$ where $J_1 = J_2$.

As usual the problem in degenerate perturbation theory requires a solution of:

$$|V_{ij} - E^{(1)} \delta_{ij}| = 0 \tag{6.20}$$

and we may quite conveniently divide the V_j into four types:

$$(J_1'' M_1'', J_2'' + 1, M_2'' \,|V|\, J_1'' M_1', J_2'' + 1, M_2') \tag{6.21a}$$

$$(J_1'' + 1, M_1'' J_2'' M_2'' \,|V|\, J_1'' + 1, M_1' J_2'' M_2') \tag{6.21b}$$

$$(J_1'' M_1'', J_2'' + 1, M_2'' \,|V|\, J_1'' + 1, M_1' J_2'' M_2') \tag{6.21c}$$

$$(J_1'' + 1, M_1'' J_2'' M_2'' \,|V|\, J_1'' M_1', J_2'' + 1, M_2') \tag{6.21d}$$

where in all cases $J_1 = J_2$.

The matrix elements given by Eqs. (6.21a) and (6.21b) are the type with which we have previously dealt, and these may be expected to vanish. In addition, all V_{ii} vanish.

† This leads us to an interaction dependence as $1/R^3$. See *supra*, Chap. 5.

In the manner of Section 6.2 we may then obtain for the remaining elements:

$$\left.\begin{aligned}
&(J_1'' M_1'', J_2'' + 1, M_2'' \,|\, V \,|\, J_1'' + 1, M_1' J_2'' M_2') \\
&\quad = \frac{\mu^2}{R^3} \frac{1}{(2J+1)(2J+3)} \{-2\,[(J + M_1'' + 1)(J - M_1'' + 1) \\
&\quad \times (J + M_2'' + 1)(J - M_2'' + 1)]^{1/2}\, \delta(M_1'' M_1')\, \delta(M_2'' M_2') \\
&\quad + \frac{1}{2} [(J - M_1'' + 2)(J - M_1'' + 1)(J - M_2'')(J - M_2'' + 1)]^{1/2} \\
&\quad \times \delta(M_1', M_1'' - 1)\, \delta(M_2', M_2'' + 1) + \frac{1}{2} [(J + M_1'' + 2) \\
&\quad \times (J + M_1'' + 1)(J + M_2'')(J + M_2'' + 1)]^{1/2} \\
&\quad \times \delta(M_1', M_1'' + 1)\, \delta(M_2' M_2'' - 1)\}
\end{aligned}\right\} \quad (6.22)$$

After a fashion, which we shall detail,* the fact that $V_{ii} = 0$ requires that the mean value of E be zero, and the root mean square value of E may be found as:

$$[\langle (E^{(1)})^2 \rangle]^{1/2} = \left(\frac{2}{3}\right)^{1/2} \frac{J+1}{[(2J+1)(2J+3)]^{1/2}} \frac{\mu^2}{R^3} \quad (6.23)$$

Margenau[138] has considered the special resonance case $J_1 = 0$, $J_2 = 1$. Leaving out the fixed J_i from the matrix element symbol and noting that $M_1'' = M_2' = 0$ leads to:

$$\left.\begin{aligned}
(0\, M_2'' \,|\, V \,|\, M_1'\, 0) &= \frac{1}{3} \frac{\mu^2}{R^3} [-2\,\delta(M_2'' 0)\, \delta(M_1' 0) \\
&\quad + \delta(M_2'', -1)\, \delta(M_1', -1) + \delta(M_2'' 1)\, \delta(M_1'' 1)]
\end{aligned}\right\} \quad (6.24)$$

from Eq. (6.22).

Eq. (6.24) yields the secular determinant:

$$\begin{vmatrix}
-E^{(1)} & 1/3a & 0 & 0 & 0 & 0 \\
1/3a & -E^{(1)} & 0 & 0 & 0 & 0 \\
0 & 0 & -E^{(1)} & -2/3a & 0 & 0 \\
0 & 0 & -2/3a & -E^{(1)} & 0 & 0 \\
0 & 0 & 0 & 0 & -E^{(1)} & 1/3a \\
0 & 0 & 0 & 0 & 1/3a & -E^{(1)}
\end{vmatrix} = 0$$

of solution:

$$E^{(1)} = \pm \frac{1}{3} \frac{\mu^2}{R^3}, \quad \pm \frac{2}{3} \frac{\mu^2}{R^3} \quad (6.25)$$

In some instances, the perturbations may become of the order of magnitude of the unperturbed energies or of the differences of these unperturbed energies. In such cases, ordinary perturbation methods fail and some other approach such as the Variation Method is called for.

* See *infra*, Sec. 6.9.

Let us suppose we are, for some reason or other, possessed of the Hamiltonian H. We now choose a so-called variation function φ — it is only required that φ behave more or less as a wave function should — and we maintain that, if E_0 is the lowest energy eigenvalue of H, the following is true.*

$$E_0 \leq \int \varphi H \varphi \, d\tau = E \tag{6.26}$$

where φ has been supposed real, so that $\bar{\varphi} = \varphi$.

Now let us be a bit more specific and suppose H to be $H_0 + V$, where V is our familiar interaction. Next we expand the variation function in terms of the unperturbed eigenfunctions of our system:

$$\varphi = \sum_i a_i \psi_i \tag{6.27}$$

and we recall that $|a_j|^2$ will be the probability that the system is in the state ψ_j.

Since the right side of Eq. (6.26) is always greater than the desired value E_0, minimization of this integral is called for. This may be accomplished as follows:

$$E = \frac{\int \varphi H \varphi \, d\tau}{\int \varphi \varphi \, d\tau} = \frac{\sum_{ij} c_i c_j H_{ij}}{\sum_{ij} c_i c_j \delta_{ij}} \tag{6.28}$$

where:

$$H_{ij} = \int \psi_i H \psi_j \, d\tau \quad \text{and} \quad \delta_{ij} = \int \psi_i \psi_j \, d\tau$$

or:

$$E \sum_{ij} c_i c_j \delta_{ij} = \sum_{ij} c_i c_j H_{ij} \tag{6.29}$$

It should be quite apparent that the only entities in Eq. (6.29) which may be adjusted to minimize E are the c_k. We proceed to differentiate with respect to each c_k:

$$\frac{\partial E}{\partial c_k} \sum_{ij} c_i c_j \delta_{ij} + E \frac{\partial}{\partial c_k} \sum_{ij} c_i c_j \delta_{ij} = \frac{\partial}{\partial c_k} \left(\sum_{ij} c_i c_j H_{ij} \right) \tag{6.30}$$

and in order for a minimum to be attained, it appears necessary that $\dfrac{\partial E}{\partial c_k} = 0$ for each c_k. All of which leads to:

$$\sum_i c_i (H_{ik} - \delta_{ik} E) = 0 \quad \text{for} \quad k = 1, 2, \ldots \tag{6.31}$$

which set of equations in turn yields the secular determinant:

$$|H_{ik} - \delta_{ik} E| = 0 \tag{6.32}$$

It is also true that:

$$H_{ik} = E_{ik}{}^0 \delta_{ik} + V_i \leftrightarrow |V_{ik} + (E_i^0 - E) \delta_{ik}| = 0 \tag{6.33}$$

by definition.

Margenau[136] utilized this variational perturbation method to treat the resonance case $J_1 = J_2 = 0$. As our variation function we take:

$$\psi = c_0 \psi_1 + c_1 \psi_1 + c_2 \psi_2 + c_3 \psi_3 \tag{6.34}$$

* This follows directly from (1) the Schrödinger equation $H\psi_n = E_n \psi_n \leftrightarrow \int \psi_n H \psi_n d\tau = E_{nj}$; (2) $\psi_n = \sum a_{ni} \varphi_i$ and (3) $E_0 < E_1 < \cdots < E_n < \cdots$.

where $\psi_0 = \psi_{J_1 M_1}(1)\,\psi_{J_2 M_2}(2) = \psi_{00}(1)\,\psi_{00}(2)$ is the ground state function and the remaining ψ_i are those eigenfunctions with which this ground state function may combine under the influence of V. Of course:

$$\psi_1 = \psi_{10}(1)\,\psi_{10}(2); \quad \psi_2 = \psi_{1-1}(1)\,\psi_{11}(2) \qquad (6.35)$$
$$\psi_3 = \psi_{11}(1)\,\psi_{1-1}(2)$$

The E_1^0 required in Eq. (6.33) is:

$$E_0^0 = 0; \quad E_1^0 = E_2^0 = E_3^0 = 2\varepsilon = \frac{2\hbar^2}{I} \qquad (6.36)$$

From Eqs. (6.10) we may obtain:

$$V_{01} = -\frac{2}{3}a; \quad V_{02} = V_{03} = -\frac{1}{3}a; \quad V_{ik} = V_{ki} \qquad (6.37)$$

where $a = \mu^2/R^3$.

As a consequence Eq. (6.33) takes the form:

$$\begin{vmatrix} -E^{(1)} & -2/3a & -1/3a & -1/3a \\ -2/3a & 2\varepsilon - E^{(1)} & 0 & 0 \\ -1/3a & 0 & 2\varepsilon - E^{(1)} & 0 \\ -1/3a & 0 & 0 & 2\varepsilon - E^{(1)} \end{vmatrix} = 0 \qquad (6.38)$$

whose solution is:

$$E^{(1)} = \frac{\hbar^2}{I}\left[1 - \left(1 + \frac{2\mu^4 I^2}{3\hbar^4 R^6}\right)^{1/2}\right] \qquad (6.39)$$

Eq. (6.39) tells us the first order interaction energy for two like dipole linear rotators for the case $J_1 = J_2 = 0$. For large R, the radical in this equation may be expanded with the result:

$$E^{(1)} \doteq -\frac{\mu^4 I}{3\hbar^2 R^6} \qquad (6.40)$$

and for small R Eq. (6.39) becomes:

$$E^{(1)} \doteq -\left(\frac{2}{3}\right)^{1/2} \frac{\mu^2}{R^3} \qquad (6.41)$$

One can hardly help but note the comparison between Eq. (6.41) and the equation for the classical potential, $-2\mu^2/R^3$, of two dipoles aligned along R. The interaction law change from R^{-6} to R^{-3} is also rather striking.

Let us now sketch a resonance development of London[128] in just enough detail so that the results which we shall later utilize will be somewhat intelligible. This author used what we might as well call an order-of-magnitude technique to arrive at the secular determinant:

$$|V_{\Upsilon_i \Upsilon_j} + (E_{\Upsilon_i}{}^{(0)} - E_{\Upsilon_k}{}^{(0)} - E_k{}^{(1)})\,\delta_{\Upsilon_i \Upsilon_j}| = 0 \qquad (6.42)$$

We shall be able to define the new symbols $E_{\Upsilon_j}{}^{(0)}$ and $E_{\Upsilon_k}{}^{(0)}$ after another step in the development .

London considered only the eigenfunction $\psi_{J_1}^{M_1}\psi_{J_2}^{M_2}$ and the thirteen eigenfunctions with which this eigenfunction may combine, as always, under the influence of V. The possible combinations are governed by the selection rules $\Delta J_1 = \pm 1$, $\Delta J_2 = \pm 1$, $\Delta M_1 = -\Delta M_2 = \pm 1{,}0$. He did not consider the twelve functions with which each of these twelve functions could combine, etc., etc., and as a consequence, as Margenau noted, his results do not yield the actual roots of the secular equation. However, the root mean square of these roots will be the same as that of the actual solution.

Under these restrictions then, we shall utilize the eigenfunctions:

$$u_0 = \psi_{J_1}^{M_1}\psi_{J_2}^{M_2}$$

$$\left.\begin{array}{ll}
\left.\begin{array}{l}
u_1 = \psi_{J_1+1}^{M_1}\psi_{J_2+1}^{M_2}\\
u_2 = \psi_{J_1+1}^{M_1+1}\psi_{J_2+1}^{M_2-1}\\
u_3 = \psi_{J_1+1}^{M_1-1}\psi_{J_2+1}^{M_2+1}
\end{array}\right\}\Upsilon_1 &
\left.\begin{array}{l}
u_7 = \psi_{J_1+1}^{M_1}\psi_{J_2-1}^{M_2}\\
u_8 = \psi_{J_1+1}^{M_1+1}\psi_{J_2-1}^{M_2-1}\\
u_9 = \psi_{J_1+1}^{M_1-1}\psi_{J_2-1}^{M_2+1}
\end{array}\right\}\Upsilon_7 \\
\left.\begin{array}{l}
u_4 = \psi_{J_1-1}^{M_1}\psi_{J_2-1}^{M_2}\\
u_5 = \psi_{J_1-1}^{M_1+1}\psi_{J_2-1}^{M_2-1}\\
u_6 = \psi_{J_1-1}^{M_1-1}\psi_{J_2-1}^{M_2+1}
\end{array}\right\}\Upsilon_4 &
\left.\begin{array}{l}
u_{10} = \psi_{J_1-1}^{M_1}\psi_{J_2+1}^{M_2}\\
u_{11} = \psi_{J_1-1}^{M_1+1}\psi_{J_2+1}^{M_2-1}\\
u_{12} = \psi_{J_1-1}^{M_1-1}\psi_{J_2+1}^{M_2+1}
\end{array}\right\}\Upsilon_{10}
\end{array}\right\} \qquad (6.43)$$

from which we may define $E_{\Upsilon_i}^{(0)}$ and $E_{\Upsilon_k}^{(0)}$.

$E_{\Upsilon_k}^{(0)}$ is in all cases the zeroth-order energy going with u_0, that is, that of the spherical rotator. $E_{\Upsilon_i}^{(0)}$ are the zeroth-order energies going with $u_1, u_2, \ldots, u_{12}$. We may note that the energies associated with u_1, u_2, u_3, are the same as are those associated with u_4, u_5, u_6, and so forth. Thus:

$$\left.\begin{aligned}
E_1 &= E_{\Upsilon_1}^{(0)} - E_{\Upsilon_0}^{(0)} = (J_1 + J_2 + 2)\frac{\hbar^2}{I}\\
E_2 &= E_{\Upsilon_4}^{(0)} - E_{\Upsilon_0}^{(0)} = -(J_1 + J_2)\frac{\hbar^2}{I}\\
E_3 &= E_{\Upsilon_7}^{(0)} - E_{\Upsilon_0}^{(0)} = (J_1 - J_2 + 1)\frac{\hbar^2}{I}\\
E_4 &= E_{\Upsilon_{17}}^{(0)} - E_{\Upsilon_0}^{(0)} = (J_2 - J_1 + 1)\frac{\hbar^2}{I}
\end{aligned}\right\} \qquad (6.44)$$

A secular determinant of precisely the form Eq. (6.44) wherein the 2ε are now replaced by the E_i from Eq. (6.42) results. In addition, we let $V_{0i} = V_i$, and from this secular determinant, the equation:

$$R^6 E^{(1)} = \frac{a_1^2}{E^{(1)} - E_1} + \frac{a_2^2}{E^{(1)} - E_2} + \frac{a_3^2}{E^{(1)} - E_3} + \frac{a_4^2}{E^{(1)} - E_4} \qquad (6.45)$$

follows, where:

$$\frac{a_i^2}{R^6} = V_{3i}^2 + V_{3i-1}^2 + V_{3i-2}^2 \qquad (6.46)$$

For the resonance case ($J_1 > 0, J_2 > 0, J_1 = J_2 + 1$) Eq. (6.45) becomes:

$$R^6 E^{(1)} = \frac{a_1^2}{E_{(1)} - E_1} + \frac{a_2^2}{E_{(1)} - E_2} + \frac{a_3^2}{E_{(1)} - E_3} + \frac{a_4^2}{E^{(1)}} \tag{6.47}$$

For large R and $|E^{(0)}| \ll |E_i|$:

$$R^6 (E^{(1)})^2 + \left(\frac{a_1^2}{E_1} + \frac{a_2^2}{E_2} + \frac{a_3^2}{E_3}\right) E^{(1)} - a_4^2 = 0$$

whose solution is:

$$\left.\begin{aligned} E^{(1)} &= \frac{1}{2R^6}\left[-\left(\frac{a_1^2}{E_1} + \frac{a_2^2}{E_2} + \frac{a_3^2}{E_3}\right) \pm \sqrt{\left(\frac{a_1^2}{E_1} + \frac{a_2^2}{E_2} + \frac{a_3^2}{E_3}\right)^2 + 4a_4^2 R^6}\right] \\ &\doteq -\frac{1}{2}\left(\frac{a_1^2}{E_1} + \frac{a_2^2}{E_2} + \frac{a_3^2}{E_3}\right)\frac{1}{R^6} \pm \frac{a_4}{R^3} \end{aligned}\right\} \tag{6.48}$$

Eq. (6.48) gives the first-order perturbation energy for the rotational resonance case. We note the comparison with Eq. (6.25).

6.4. Interaction between a deformable and a rigid dipole (induction effect)

Let us next consider the interaction between two molecular models, one a rigid dipole rotator and the other a deformable dipole. Now by deformable, we mean that the latter dipole may undergo distorting vibrations resulting in its possession of vibrational energy levels and to it also we attribute electronic energy levels. In this case, we may carry through the treatment in exactly the manner of Sec. 6.3 as far as Eq. (6.13). To this point in the development, the new physical conditions introduce no changes. In Eq. (6.13) the subscript 1 now indicates the deformable dipole while the subscript 2 represents the rigid dipole. From this point, it is the summation over J_1' and J_2', as indicated in Eq. (6.3), which must be modified.

The summation over J_2' remains unchanged since the rigid dipole possesses only rotational energy, but now we must not only sum over the upper rotational states of dipole 2 (for the ground electronic vibration state) but also over all upper electronic-vibration-rotation states. The grand sum which results may be broken up into two sums, (1) the sum over all J_1' for the electronic-vibration ground state and (2) the sum over the remaining upper states for the deformable dipole. Both of these sums are to be combined, of course, with the sum over the rotational levels of the rigid dipole. Subsum (1) corresponds precisely with Eq. (6.14).† Subsum (2), on the other hand takes the form:

$$\left.\begin{aligned} \langle E^{(2)}\rangle &= \frac{2}{3}\frac{\mu_2^2}{R^6}\sum_{K_1 J_2'} \\ &\times \frac{|(0|e\,\mathbf{r}_1|k_1)|^2\,[(J_2''+1)\,\delta(J_2', J_2''+1) + J_2''\,\delta(J_2', J_2'-1)]}{[E_0 - E_{K_1} + E_{J_2''} - E_{J_2'}]\,(2J_2''+1)} \end{aligned}\right\} \tag{6.49}$$

† Except that now μ_1 will be a function of the electronic vibration level involved.

where now K_1 is a quantum number aggregate representing the upper state electronic and vibrational quantum numbers.

6.5. Interaction between a deformable rotator and an isotropic harmonic oscillator

We now take as our absorber an isotropic, three dimensional, harmonic oscillator and as our perturber a deformable rotator in the ground state. The eigenfunction describing such an oscillator is:

$$\psi(v_1 v_2 v_3) = \prod_{i=1}^{3} H_{v_i}(\beta^{1/2} x_i) \exp[-(\beta/2) x_i^2] \tag{6.50}$$

where $\beta = m\omega/\hbar$.

In order to simplify our considerations, let us suppose that $v_1'' = v_2'' = v_3''$. Eq. (6.15) may be utilized in order to write Eq. (6.3) as:*

$$E^{(2)} = \sum_{v' \Lambda} \frac{|(v_1''v_2''v_3''| V |v_1'v_2'v_3'|^2 |(0|er|\Lambda)|^2}{E_{v'} - E_{v''} + E_\Lambda - E_0} \tag{6.51}$$

where V is given by $V = (e/R^3)[2z - x - y]$.

Using the recursion formula for Hermite polynomials

$$\xi H_n(\xi) = \frac{1}{2} H_{n+1}(\xi) + 2n H_{n-1}(\xi)$$

we may show that, for the harmonic oscillator:

$$x_{v''v'} = y_{v''v'} = z_{v''v'} = \left\{ \begin{array}{ll} \sqrt{\dfrac{v''+1}{2\beta}} & \text{for } v' = v''+1 \\ \sqrt{\dfrac{v''}{2\beta}} & \text{for } v' = v''-1 \\ 0 & \text{for } v' \neq v'' \pm 1 \end{array} \right. \tag{6.52}$$

Quite obviously then, the three vibrational quantum numbers in the eigenfunction of Eq. (6.50) may only individually change by ± 1. The states with which $\psi(v_1''\, v_2''\, v_3'')$ may combine under the influence of V are listed in Table 6.1. The matrix elements $V_{v''v'}$ resulting from these combinations together with the corresponding values of $E_{v'} - E_{v''}$ are also given in the table.

TABLE 6.1

$\psi(v_1'v_2'v_3')$	$E_{v'} - E_{v''}$	$-\frac{R^3}{e} V_{v''v'}$
$\psi(v_1''+1, v_2'', v_3'')$	$h\nu$	$-(2\beta)^{-1/2}(v_1+1)^{1/2}$
$\psi(v_1'', v_2''+1, v_3'')$	$h\nu$	$-(2\beta)^{-1/2}(v_2+1)^{1/2}$
$\psi(v_1'', v_2'', v_3''+1)$	$h\nu$	$2(2\beta)^{-1/2}(v_3+1)^{1/2}$
$\psi(v_1''-1, v_2'', v_3'')$	$-h\nu$	$-(2\beta)^{-1/2}(v_1)^{1/2}$
$\psi(v_1'', v_2''-1, v_3'')$	$-h\nu$	$-(2\beta)^{-1/2}(v_2)^{1/2}$
$\psi(v_1'', v_2'', v_3''-1)$	$-h\nu$	$2(2\beta)^{-1/2}(v_3)^{1/2}$

* The special case (rigid rotator) considered in obtaining Eq. (6.16) could be extended to the rotator-vibrator, or to the rotator-vibrator possessing also a hydrogenlike electronic eigenfunction, etc.

Let us note the threefold degeneracy* that exists in the final states of the same energy in Table 6.1 — at least for our purposes here. This will require our dividing by three when taking the sum indicated in Eq. (6.51).

We may now utilize the results of Table 6.1 in Eq. (6.51) to obtain:

$$\left.\begin{aligned} \langle E^{(2)}\rangle &= \frac{1}{3}\frac{e^2}{R^6}\frac{\hbar}{2m\omega}\sum_A \\ &\times\left\{\frac{v_1+1+v_2+1+4(v_3+1)}{E_A-E_0+h\nu}+\frac{v_1+v_2+4v_3}{E_A-E_0-h\nu}\right\} \\ &\times|(0|V|A)|^2 = \frac{e^2}{R^6}\frac{\hbar\omega}{m\omega^2}\sum_A \\ &\times\left\{\frac{v+1}{E_A-E_0+h\nu}+\frac{v}{E_A-E_0+h\nu}\right\}|(0|er|A)|^2 \end{aligned}\right\} \tag{6.53a}$$

since $v_1'' = v_2'' = v_3''$.

If $E_A - E_0$ is large compared to $h\nu$ Eq. (6.53a) may be rewritten as:

$$\left.\begin{aligned} \langle E^{(2)}\rangle &\doteq \frac{e^2}{R^6}\frac{\hbar\omega(2v+1)}{m\omega^2}\sum_A\frac{|(0|er|A)|^2}{E_A-E_0} \\ &= 2\frac{e^2}{R^6}\frac{E_v}{m\omega^2}\sum_A\frac{|(0|er|A)|^2}{E_A-E_0} \end{aligned}\right\} \tag{6.53b}$$

where $E_v = \hbar\omega(v+1/2)$.

This is, of course, the dipole–dipole interaction only. Utilizing the same principles which have arisen in this section, Margenau[139] obtained for the interaction between two threefold harmonic oscillators in the ground state, the following:

$$E^{(2)} = -\frac{3}{4}\frac{\alpha^2 h\nu}{R^6} - \frac{15}{4}\frac{\alpha^3(h\nu)^2}{e^2R^8} - \frac{315}{32}\frac{\alpha^4(h\nu)^3}{e^4R^{10}} \tag{6.54}$$

where α is the polarizability of either oscillator. The first term represents the dipole–dipole interaction, the second the dipole–quadrupole, and the third the quadrupole–quadrupole.

6.6. Broadening by molecules with no permanent poles (dispersion effect)

Margenau[138] initiated the first attack on the Theory of Molecular Broadening in 1936. Of the many possible special cases of this phenomenon which could have been considered, he chose to investigate the case of broadening by foreign gases which possess no permanent poles since Watson and Hull[206] had previously obtained experimental data on this case indicating that, for certain cases, the rotation-vibration lines are broadened and shifted by the same amount as the electronic line.

Let us label the states of an absorber by K with the state in which we now find this absorber being designated k. The broadener states we label A with the lowest state 0. Now from Eq. (6.16) we may obtain the perturbation on the

* Actually each level is $1/2\,(v+1)\,(v+2)$-fold degenerate in general.

energy of the absorber averaged over all orientations of the absorber, as:

$$\langle E^{(2)}\rangle = -\frac{2}{3R^6}\sum_{A\varkappa}' \frac{|(e\,r)_{0A}|^2\,|(V)_{k\varkappa}|^2}{E_A - E_0 + E_\varkappa - E_k} \tag{6.55}$$

We have stipulated that E_0 is the lowest energy of the perturber. This means that $E_A - E_0$ will always be positive. In addition we have not allowed this perturber a permanent dipole moment. This means that no pure rotational transitions, which amount to the lowest energy transitions, are possible. Thus the 0.1 volt vibrational transitions are the lowest energy allowed.*

On the other hand, $E_\varkappa - E_k$ may be positive or negative. This means that $\langle E^{(2)}\rangle$ may be greater than or less than zero so that the spectral line may exhibit either a violet or a red asymmetry.

Now where we simply have dealt with a set of relatively widely spaced electronic levels in our monatomic considerations, we now deal with a modification of this situation. We still have the set of electronic levels, but on each of these is superposed a set of more closely spaced vibrational levels, and, in turn, on each of these vibrational levels is superposed a set of still more closely spaced rotational levels. When we are considering the visible and ultra-violet portion of the spectrum (but certainly not when we are considering the infrared) we may, as an approximation, consider only the electronic contributions to $E_\varkappa - E_k$ — they surely constitute the largest ones — and merely let the rotation-vibration levels superposed upon them amount to degeneracies accounted for by a summation in Eq. (6.55). It follows that the result will closely correspond to that for monatomic molecules. To be sure, this could only be expected to give us results for electronic band spectra, but it does tend to indicate that we should not expect too heavy a broadening or shift dependence on rotational or vibrational quantum numbers under the present physical conditions.

In order to consider the possible effect of molecular rotation let us approximate our absorber by a rigid rotator of dipole moment μ. Then Eq. (6.2) takes the form Eq. (6.17):

$$\left.\begin{aligned} \langle E_J^{(2)}\rangle &= -\frac{2}{3}\frac{\mu^2}{R^6}\frac{1}{2J+1} \\ &\times \left\{\sum_A |(e\,r_{0A})|^2 \left(\frac{J+1}{E_A - E_0 + 2(J+1)B} + \frac{J}{E_A - E_0 + 2JB}\right)\right\} \end{aligned}\right\} \tag{6.17}$$

where, as usual, $B = h^2/8\pi^2 I$.

Now I is usually of the order of 10^{-40}, and, with h of the order of 10^{-27} we see that B will be small compared to $E_A - E_0$ when the perturber has no permanent dipole. When B is neglected Eq. (6.5a) becomes:

$$\langle E^{(2)}\rangle = -\frac{2}{3}\frac{\mu^2}{R^6}\sum_A' \frac{|(e\,r)_{0A}|^2}{E_A - E_0} = -\frac{\alpha\,\mu^2}{R^6} \tag{6.56a}$$

* Here we suppose an electronic level separation of 1 volt, a vibrational separation of 0.1 volt, and a rotational separation of 0.01 volt, this as a helpful if inaccurate rule of thumb.

† We may recall, that in general, homonuclear diatomic molecules (H_2, etc.) are possessed of no permanent or vibration induced electric dipole moments.

If the dipole moment $\mu \doteq 10^{-18}$ the perturbation given by Eq. (6.5b) is about 1/20 of that for the electronic case. Now let us write Eq. (6.17) as:

$$\begin{aligned}\langle E_J^{(2)}\rangle &= -\frac{3}{2}\frac{\mu^2}{R^6}\frac{1}{(2J+1)}\sum_A{}' |(e\,r)_{0A}|^2 \\ &\times\left[\frac{J+1}{(E_A-E_0)\left[-1\dfrac{2(J+1)B}{E_A-E_0}\right]}+\frac{J}{(E_A-E_0)\left[1+\dfrac{2JB}{E_A-E_0}\right]}\right] \\ &= -\frac{\alpha\mu^2}{R^6}\frac{1}{(2J+1)}\left[(J+1)\left\{1+\frac{2(J+1)B}{E_A-E_0}+\frac{4(J+1)^2B^2}{(E_A-E_0)^2}+\cdots\right\}\right. \\ &\quad\left.+J\left\{1-\frac{2JB}{E_A-E_0}+\frac{4J^2B^2}{(E_A-E_0)^2}+\cdots\right\}\right] \\ &= -\frac{\alpha\mu^2}{R^6}\left[1+\frac{2B}{(E_A-E_0)}+\frac{4(J^2+J+1)B^2}{(E_A-E_0)^2}+\cdots\right]\end{aligned} \tag{6.56b}$$

It is apparent from Eq. (6.15b) that the portion of $\langle E_J^{(2)}\rangle$ which depends on J is very small due to the appearance of B^2 in the numerator and $(E_A - E_0)^2$ in the denominator involved.

Margenau carried out his considerations of the effect of molecular vibrations in an analogous manner. In this case we obtain from Eq. (6.53b):

$$\langle E_v^{(2)}\rangle = 3\frac{e^2 E_v}{m\,\omega^2 R^6}\alpha \tag{6.57}$$

From Eq. (6.57) we may note that a more marked dependence on the vibrational quantum number is exhibited than on the rotational quantum number as given by Eq. (6.17). On the other hand the perturbation given by Eq. (6.17) is about five times that given by Eq. (6.57).

On the basis of these investigations then, Margenau concluded that, within obtainable experimental accuracy, all rotation-vibration lines associated with the same electronic transition should show broadening and shift of about the same degree when the broadening agent is a non-polar molecule. In addition, these lines should be affected by foreign perturbers almost in the manner of the corresponding monatomic lines under similar circumstances.

As Watson[205] has noted, "dispersion forces" — the type considered — also appear to predominate in the self-broadening of non-polar molecular lines. Were the normal resonance forces between like monatomic molecules present, we would not normally expect this to be the case, but the closely spaced rotational levels of the polyatomic molecule render this predominance reasonable. As we have noted, the rotational levels specified by, say, just J are comparatively closely spaced energywise. As a consequence, a Maxwell–Boltzmann temperature distribution of molecules over these levels will not lead to the preferential population of the monatomic case, and they will be more equally divided among many levels. On the other hand, the allowed J change in transition is generally limited to ± 1 or 0 so that the probability of energy exchange with the resultant degeneracy is severely restricted. *In toto* then, this resonance effect would not appear to be primarily responsible for self-broadening, and we may safely look to these dispersion forces as broadening agents for non-polar molecules.

We have broadened molecular lines by non-polar molecules, and it would now appear a logical next step to broaden these lines by polar molecules. Let us consider first the work of Margenau and Watson on this phenomenon.

6.7. Broadening by linear dipole molecules

Before turning to the qualitative considerations of Margenau and Watson[196] in this regard, we consider briefly an earlier work of London[128] in which this author contributed much toward the systematizing of inter-molecular forces.

In what might be considered the natural course of events we have encountered the "Directional Effect" of Keesom[102], the "Static Induction Effect" of Debye, and the "Dispersion Effect" of London in Sections 6.3, 6.5, and 6.7 respectively. London, on the other hand, began his considerations by the introduction of these effects. He started from Eq. (3.4) and considered the contribution of this equation to consist of a sum of terms of the form:

$$E_{kl}^{(2)} = E_{kl}^{(rr)} + E_{kl}^{(rg)} + E_{kl}^{(gr)} + E_{kl}^{(gg)}$$

In this equation $E_{kl}^{(rr)}$ is associated with small "jumps" or quantum number changes on the part of both interacting molecules. $E_{kl}^{(rg)}$ and $E_{kl}^{(gr)}$ arise due to a small jump by one molecule and a large jump by the other, and, finally, $E_{kl}^{(gg)}$ results from large jumps by both.

$E_{kl}^{(gg)}$ corresponds to the Dispersion Effect and is present to a greater or lesser extent for all molecules. The contribution is due to the "kurzperiodischen Störungen" arising from the motion of the inner electrons. The potential arising from this is of the form $-1/R^6$, $-1/R^8$ and higher inverse powers of R.

$E^{(rr)}$ is the contribution of the Directional Effect of Section 6.3 and arises from the interactions of the dipoles (or quadrupoles) at various relative orientations. We have seen the manner in which this effect depends on $\perp 1/R^6$ for dipoles at large distances and $-1/R^3$ for the lowest state or, in the case of the suppression of degeneracy, in the metastable state at short distances (Section 6.4). For quadrupoles, a corresponding $\pm 1/R^{10}$ and $-1/R^5$ dependence results.

$E^{(gr)}$ and $E^{(rg)}$ result from the Static Induction Effect, and, as we have noted, depend on $-1/R^6$ or $-1/R^8$ for dipoles or quadrupoles respectively. London describes this effect as "due to the charge distribution of the entire molecules".

It is to one of these three effects than that Margenau and Watson[146] looked for an explanation of the broadening in linear dipole molecules such as HCN possessed of relatively large ($> 10^{-18}$) dipole moments. They believed the principal cause of broadening to be the directional effect, and we might enumerate the spectral prognostications toward which such an assumption leads.

To begin with we must hypothesize the unpolarizable (rigid) dipoles of Section 6.3, and we recall from Sections 6.3 and 6.4 that (1) the intermolecular forces are relatively weak except when the near resonance condition, $|J_1 - J_2| < 2$ or 3, sets in, and (2) the lowest rotational state is most strongly influenced by other molecules in the lowest or neighboring states.*

The lower state J_1'' we take as other than the lowest rotational state so that we can concern ourselves with (1). As has been mentioned, the rotational levels lie sufficiently close together so that a thermal distribution over them may be expected. Suppose J_1'' to designate a level near the maximum of this thermal distribution. Then a much greater number of perturbers can be expected to fulfill the condition $|J_1'' - J_2''| < 2$ or 3 than would be the case were the J_1'' level to be found toward the wings of this thermal distribution. Since the upper level will also enter into the line broadening one cannot say more as yet than that those most intense lines arising from J levels near the maximum of the thermal distribution should be broadened more than the remainder of the lines of the band.

In regard to (2) this indicates that a greater broadening of the line arising from the lowest rotational level is to be expected, but only under certain conditions. At normal temperatures, the thermal distribution should not populate levels lying near $J_1'' = 0$ very highly so that the effect may not be very pronounced. As the temperature is lowered, however, these low level populations will increase, and we should expect to see an increase in the broadening of the line with $J_1'' = 0$ over neighboring spectral lines. It might be noted also that, due to "greater flexibility downward", a violet asymmetry in this line may be expected.

We now devote ourselves to Margenau and Warren's[145] more quantitative considerations of the broadening effected by dipole interactions.

6.8. Interactions between symmetrical top dipole molecules

For our two interacting molecules we choose two identical symmetrical rotators each possessed of a permanent electric dipole moment μ oriented along the molecular figure axis. The interaction potential between the two rotators is again given by Eq. (6.2) where now $\mu_1 = \mu_2 = \mu$. The eigenfunctions for this system are given by:

$$\psi_{J_1 K_1 M_1 J_2 K_2 M_2} = \qquad (6.58)$$
$$N(J_1 K_1 M_1)\, N(J_2 K_2 M_2)\, \Theta_{J_1 K_1 M_1}(\vartheta_1)\, \Theta_{J_2 K_2 M_2}(\vartheta_2)\, e^{i(M_1\varphi_1 + M_2\varphi_2 + K_1\chi_1 + K_2\chi_2)}$$

for sufficient separation R. Equation (6.2) tells us that V is not dependent on the Eulerian angle χ so that $K_1'' = K_1'$ and $K_2'' = K_2'$ under the aegis of V, and the matrix of V is diagonal in K_1 and K_2 as a consequence.

* Cf., for example, *supra*, Eqs. (6.21).

Now in obtaining the matrix elements of V, the integrations over φ_i and χ_i may be carried out immediately to yield:

$$(J_1'' K_1'' M_1'' J_2'' K_2'' M_2'' \,|V|\, J_1'' K_1'' M_1' J_2'' K_2'' M_2')$$
$$= -\frac{\mu^2}{R^3} N_{M_1''} N_{M_2''} N_{M_1'} N_{M_2'}$$
$$\times \{32\pi^4 I_1(J_1'' K_1'' M_1'') I_1(J_2'' K_2'' M_2'') \delta(M_1' M_1'') \delta(M_2' M_2'')$$
$$- 8\pi^4 I_2(J_1'' K_1'' M_1'') I_3(J_2'' K_2'' M_2'') \delta(M_1' M_1'' - 1) \delta(M_2', M_2'' + 1)$$
$$- 8\pi^4 I_3(J_1'' K_1'' M_1'') I_2(J_2'' K_2'' M_2'') \delta(M_1', M_1'' + 1) \delta(M_2', M_2'' - 1)\} \tag{6.59a}$$

where:

$$I_1(J'' K'' M'') = \int_0^\pi \cos\vartheta\, \Theta_{J'' K'' M''}{}^2 \sin\vartheta\, d\vartheta = 2\int_0^1 (1 - 2x)\, \Theta_{J'' K'' M''}{}^2(x)\, dx \tag{6.59b}$$

$$\left.\begin{aligned} I_2(J'' K'' M'') &= \int_0^\pi \sin\vartheta\, \Theta_{J'' K'' M''}\, \Theta_{J'' K'', M''-1} \sin\vartheta\, d\vartheta \\ &= 4\int_0^1 x^{1/2}(1-x)^{1/2}\, \Theta_{J'' K'' M''}(x)\, \Theta_{J'' K'', M''-1}(x)\, dx \end{aligned}\right\} \tag{6.59c}$$

$$I_3(J'' K'' M'') = \int_0^\pi \sin\vartheta\, \Theta_{J'' K'' M''}\, \Theta_{J'' K'', M''+1} \sin\vartheta\, d\vartheta = I_2(J'' K'', M'' + 1) \tag{6.59d}$$

and I_1 has been evaluated by Reiche and Rademacher[174 175] — among others — and is given by:

$$N^2(M'')\, I_1(J'' K'' M'') = \frac{K'' M''}{4\pi^2 J''(J'' + 1)} \tag{6.60a}$$

Margenau and Warren calculated I_2 by the reduction method of Reiche and Rademacher with result:

$$N(M'')\, N(M'' - 1)\, I_2(J'' K'' M'') = \frac{\pm K''[(J'' - M'' + 1)(J'' + M'')]^{1/2}}{4\pi^2 J''(J'' + 1)} \tag{6.60b}$$

where the minus sign is to be taken if the numerically greater of K or M is positive while for the opposite case or if $K = -M$ the positive sign is to be taken.

Eqs. (6.50) may now be substituted into Eq. (6.59a) to obtain the result:

$$\left.\begin{aligned} (M_1'' M_2'' \,|V|\, M_1' M_2') &= \frac{\mu^2}{R^3} \frac{K_1 K_2}{J_1(J_1+1) J_2(J_2+1)} \\ &\times \Big\{-2 M_1'' M_2'' \delta(M_1' M_1'') \delta(M_2' M_2'') \pm \frac{1}{2} [(J_1 - M_1'' + 1) \\ &\times (J_1 + M_1'')(J_2 - M_2'')(J_2 + M_2'' + 1)]^{1/2} \delta(M_1', M_1'' - 1) \\ &\times \delta(M_2', M_2'' + 1) \pm \frac{1}{2} [(J_1 + M_1'' - 1)(J_1 - M_1'') \\ &\times (J_2 - M_2'' + 1)(J_2 + M_2'')]^{1/2} \delta(M_1', M_1'' + 1) \delta(M_2', M_2'' - 1)\Big\} \end{aligned}\right\} \tag{6.61}$$

TABLE 6.2

THE SECULAR DETERMINANT $|E\delta_{ij} - H_{ij}'| = 0$ FOR $V = H'$ AS GIVEN BY Eq. (6.61). THE BLOCKS INCREASE TO A MAXIMUM SIZE OF $(2J_2 + 1)$ BY $(2J_2 + 1)$ FOR THE $(2J_2 + 1)$-ST BLOCK FOR $J_1 > J_2$

$M_1' \cdot M_2'$ \ M_1'' M_2''	$(-J_1)(-J_2)$	$(-J_1)(-J_2+1)$	$(-J_1+1)(-J_2)$	$(-J_1)(-J_2+2)$	$(-J_1+1)(-J_2+1)$	$(-J_1+2)(-J_2)$	...	$(J_1-2)(J_2)$	$(J_1-1)(J_2-1)$	$(J_1)(J_2-2)$	$(J_1-1)(J_2)$	$(J_1)(J_2-1)$	$(J_1)(J_2)$
$(-J_1)(-J_2)$	$E-H$												
$(-J_1)(-J_2+1)$		$E-H$	$-H$										
$(-J_1+1)(-J_2)$		$-H$	$E-H$										
$(-J_1)(-J_2+2)$				$E-H$	$-H$	0							
$(-J_1+1)(-J_2+1)$				$-H$	$E-H$	$-H$							
$(-J_1+2)(-J_2)$				0	$-H$	$E-H$							
...							Increasing sized blocks symmetric about the secondary diagonal						
$(J_1-2)(J_2)$								$E-H$	$-H$	0			
$(J_1-1)(J_2-1)$								$-H$	$E-H$	$-H$			
$(J_1)(J_2-2)$								0	$-H$	$E-H$			
$(J_1-1)(J_2)$											$E-H$	$-H$	
$(J_1)(J_2-1)$											$-H$	$E-H$	
$(J_1)(J_2)$													$E-H$

$= 0$

From Eq. (6.61) it is apparent that we are dealing with a somewhat different situation than had previously arisen in our molecular model considerations in that the matrix elements of $V = H'$ over the *system* do not disappear. As a consequence, the first-order perturbation energy does not disappear. In addition, the problem is one of degenerate perturbation theory in which the degeneracy in M is partially removed in first order.* Our next step is then the solution of the secular determinant $|V_{ij} - E^{(1)}\delta_{ij}| = 0$ of the form Table 5.2. Since J and K are fixed and since M takes on all values from $-J$ to $+J$, this determinant will possess $(2J_1 + 1)(2J_2 + 1)$ rows and columns. This determinant is given in Table 6.2.

A consideration of Table 6.2 tells us that the determinant has broken into blocks, with identical blocks symmetrical about the secondary diagonal. This immediately tells us that every root of this determinant will be a double root.

A little consideration of the situation indicates that a general equation describing

* We note no dependence of the unperturbed energies as given by Eq. (2.37b) on M while the energy dependence in first order may be seen from Eqs. (4.141) and (6.61).

the level splitting due to this interaction is not to be found, but an idea as to the maximum splitting may be obtained, and the example of a special case may be presented.

We may write down from Eq. (6.61) the double root which lies at either end of the principal diagonal and where $J_1 = M_1$ and $J_2 = M_2$.

$$E^{(1)} = -\frac{2\,\mu^2}{R^3}\,\frac{K_1 K_2}{(J_1+1)\,(J_2+1)} \tag{6.62}$$

If we recall the linear Stark effect,

$$E^{(1)} = \mu\,E\,\frac{K\,M}{J\,(J+1)} \tag{6.63 a}$$

it is interesting to note the close resemblance which its maximum ($J = M$),

$$E^{(1)} = \mu\,E\,\frac{K}{J+1} \tag{6.63 b}$$

bears to our result Eq. (6.62).

Fig. (6.1) illustrates the level splitting with the rotator separation R and the magnetic quantum number M for the special case $J_1 = K_1 = 2$, $J_2' = K_2' = 1$ for the upper state and $J_1'' = K_1'' = 1$, $J_2'' = K_2'' = 1$ for the lower state.

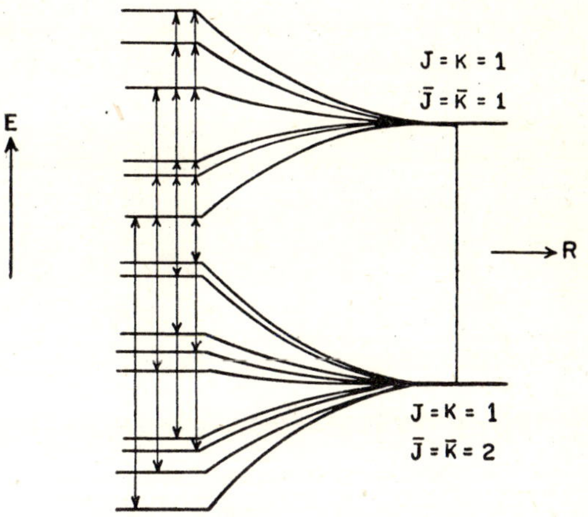

FIG. 6.1. Level splitting with molecular separation for the symmetric dipole interaction. (After Margenau and Warren[144])

Let us finally remark that since $K = 0$ for a spherical top rotator, the matrix elements of V in Eq. (6.61) would all be zero so that the effect considered would not take place.

Insofar as second order effects are concerned, we certainly are aware that they will depend on the inverse sixth power of the molecular separation. In addition, Margenau and Warren felt that, for finding those regions in which the first order perturbations predominate, London's second order results for diatomic molecules should be sufficient. Under these assumptions a limiting range of about 7A for the predominance of the forces which we have considered was imposed by these tauhors.

6.9. The broadening and shift dut to the symmetrical top dipole interaction

The secular determinant of Table 6.2 will have $(2J_1+1)\,(2J_2+1)$ roots which we may designate by $\varepsilon_1, \varepsilon_2, \ldots$, and, for convenience, let these roots be given in units of:

$$\frac{\mu^2}{R^3}\,\frac{K_1 K_2}{J_1(J_1+1)\,J_2(J_2+1)}$$

the "coefficient" in Eq. (6.61). The secular determinant leads to the secular equation possessed of these roots so we may write,

$$|V_{ij} - \varepsilon\,\delta_{ij}| = \prod_i (\varepsilon - \varepsilon_i) = \varepsilon^n - a\,\varepsilon^{n-1} + b\,\varepsilon^{n-2} - \cdots + x = 0 \tag{6.64}$$

and let us recall the relations which exist among the roots and coefficients of a polynomial and which were discussed in connection with Eq. (5.9). These relations first lead us to:

$$a = \sum_i \varepsilon_i = \sum_i V_{ii} \tag{6.66a}$$

Eq. (6.65a) may be evaluated specifically using Eq. (6.61) as:

$$a = \sum_{M_1=-J_1}^{J_1} \sum_{M_2=-J_2}^{J_2} 2\,M_1\,M_2 = 0 \tag{6.65b}$$

Thus, the weigthed mean of the energy perturbation is zero which in turn means that we would expect no shift of the spectral line, a situation which also arises in connection with the R^{-3} interaction in atomic resonance broadening (Statistical and Interruption).

As a measure of the broadening of the line due to this interaction, let us now find the root mean square spread of the perturbed energy levels. The coefficient b is given by:

$$b = \sum_{i>j} (V_{ii}\,V_{jj} - V_{ij}^2) = \frac{1}{2}\sum_i \sum_j V_{ii}\,V_{jj} - \frac{1}{2}\sum_i V_{ii}^2 - \sum_{i>j} V_{ij}^2 \tag{6.66a}$$

or by:

$$b = \sum\sum_{i>j} \varepsilon_i\,\varepsilon_j = \frac{1}{2}\sum_i \varepsilon_i \sum_j \varepsilon_j - \frac{1}{2}\sum_i \varepsilon_i^2 = -\frac{1}{2}\sum_i \varepsilon_i^2 \tag{6.66b}$$

by virtue of Eq. (6.65). The substitution for the V_{ij} in Eq. (6.66a) from Eq. (6.61) results in:

$$\left.\begin{aligned} b &= 2 \sum_{M_1'' M_1'} \sum_{M_2'' M_2'} M_1''\,M_2''\,M_1'\,M_2' - 2 \sum_{M_1'' M_2''} M_1''^2\,M_2''^2 \\ &\quad - \frac{1}{4} \sum_{M_1'' M_2''} (J_1^2 - M_1''^2 + J_1 + M_1'')\,(J_2''^2 - M_2''^2 + J_2 - M_2'') \\ &= 0 - \frac{2}{9} J_1(J_1+1)\,(2J_1+1)\,J_2(J_2+1)\,(2J_2+1) \\ &\quad - \frac{1}{4}\left[(J_1^2 + J_1)\,(2J_1+1) - \frac{1}{3} J_1(J_1+1)\,(2J_1+1)\right] \\ &\quad \times \left[(J_2^2 + J_2)\,(2J_2+1) - \frac{1}{3} J_2(J_2+1)\,(2J_2+1)\right] \\ &= -\frac{1}{3} J_1(J_1+1)\,(2\,J_1+1)\,J_2(J_2+1)\,(2\,J_2+1) \end{aligned}\right\} \tag{6.67}$$

Eqs. (6.66b) and (6.67) lead to the result:

$$\langle \varepsilon_i^2 \rangle^{1/2} = \left[\frac{2}{3} J_1 (J_1 + 1) J_2 (J_2 + 1) \right]^{1/2} \qquad (6.68a)$$

which in more normal energy units is:

$$\langle (E^{(1)})^2 \rangle^{1/2} = \left(\frac{2}{3} \right)^{1/2} \frac{\mu^2}{R^3} \frac{K_1 K_2}{[J_1 (J_1 + 1) J_2 (J_2 + 1)]^{1/2}} \qquad (6.68b)$$

In Chapter 5 it was shown that when a resonance interaction, resulting from a potential of the form

$$E^{(1)} = \pm B/R^3$$

occurs, the Statistical Theory decrees an approximate half-width:

$$\delta \doteq \frac{4}{3} \pi^2 \frac{B}{h} N$$

A reasonable approximation to the half width arising from these considerations may surely be expected if we equate B to

$$\left(\frac{2}{3} \right)^{1/2} \mu^2 \frac{|K_1 K_2|}{[J_1 (J_1 + 1) J_2 (J_2 + 1)]^{1/2}}$$

There then results:

$$\delta \doteq \frac{\pi^2 \mu^2 |K_1 K_2| N}{h [J_1 (J_1 + 1) J_2 (J_2 + 1)]^{1/2}} \qquad (6.69)$$

It is true, of course, that we have here found the level width while we are really desirous of the spectral line width. The level here $(J_1 K_1 J_2 K_2)$ would have to be considered in conjunction with another possible level $(J_1 \pm 1, 0;\ K_1 \pm 1, 0;\ J_2 K_2)$. One may perhaps consider the K_1 and J_1 appearing in Eq. (6.69) as the mean of the quantum numbers for the two states.

We might conclude by noting that the effect as given by Eq. (6.69) is around 20 times as weak as the self-broadening effect in monatomic resonance lines.

6.10. Broadening by the linear dipole molecule HCN

The first attempt at a rigorous theoretical interpretation of the self broadening of the absorption lines of a linear dipole molecule was carried out by Lindholm[123] for the *HCN* molecule. This investigator found that, as he put it, the Directional Effect ($\sim R^{-6}$) alone,* was not sufficient to account for the strong, J-dependent broadening observed, so that he considered also the Resonance Effect ($\sim R^{-3}$) in conjunction with the former. As we have remarked† what Lindholm called the rotational Resonance Effect is in actuality a special case of the Directional Effect. Let us keep this fact mind, but for convenience of consideration we shall use the Lindholm nomenclature for the two interaction relations (R^{-3} and R^{-6}) in this section.

* See *supra*, Sec. 8.8.
† See *supra*, Sec. 8.8.

In essence, Lindholm uses the energy perturbations due to the Directional and Resonance Effects within the framework of the Weisskopf theory of Interruption Broadening to determine the broadening of the *HCN* lines in a manner which we now consider.

According to Eq. (5.53) the half width of the lines will be given by:

$$\delta = \varrho^2 \langle v \rangle N$$

for a homogeneous gas. For a heterogeneous gas containing N_1 molecules with optical collision diameters ϱ_1, N_2 with diameters ϱ_2, etc.

$$\delta = \langle v \rangle \sum_i N_i \varrho_i^2 \tag{6.70}$$

if all molecules are of the same mass, thus allowing a common $\langle v \rangle$*. Let us next consider the necessity for introducing Eq. (8.70).

From Lindholm's experimental results, it is quite apparent that a marked dependence on J is present in the line width. In the Weisskopf theory, this can only come about through some variation in N and ϱ with J_1 (the absorber J value) and J_2 (the perturber J value). Thus, we hypothesize a dependence of ϱ on J_1 and J_2 which a little consideration immediately bears out. To begin with a frequency perturbation $\Delta\nu = b/R^6$ (which is brought about by the Directional Effect $|J_1 - J_2| \neq 1$) leads to an optical collision diameter:

$$\varrho = \left(\frac{3\pi^2}{4} \frac{b}{\langle v \rangle}\right)^{1/5} \tag{4.9'}$$

On the other hand, a frequency perturbation $\Delta\nu = \pm B/R^3$ (which may arise from the Resonance Effect $|J_1 - J_2| = 1$) occasions an optical collision diameter:

$$\varrho = \left(\frac{4\pi B}{\langle v \rangle}\right)^{1/2} \tag{5.57b'}$$

This then is at least a part of the general manner in which ϱ may depend on J_1 and J_2, and let us now specificize this into usefulness. Lindholm dealt only with the *P*-Branches ($\Delta J = -1$) of the two *HCN* bands which he considered. Insofar as the vibrational quantum numbers involved are concerned $v_1'' = v_2'' = 0$. We now let:

$$\Delta = J_2'' - J_1'' \tag{6.71a}$$

so that:

$$J_2'' - J_2' = J_2'' - (J_1'' - 1) = \Delta + 1 \tag{6.71b}$$

for the upper state since we consider only the *P*-Branch. Δ is now a convenient parameter for the determination of the type effect to be considered. (1) For $\Delta = \pm 1$ we obtain the Resonance Effect in the ground state and the Directional

* Lindholm defines $\langle v \rangle$ as the mean relative velocity. Here again the choice at $\langle v \rangle$ is (1) a matter of the particular author's taste and (2) a matter of the author's ability to justify his taste.

Effect in the upper State. (2) For $\Delta = 0, -2$ there results the Directional Effect in the ground state and the Resonance Effect in the upper state. (3) Finally, for all other Δ the Directional Effect occurs in both states. The case of small J† was not considered by Lindholm. The optical collision diameters for these three cases may now be determined from Eqs. (2.9′) and (5.57b′)

Case (1):

Frequency perturbation: $$\Delta\nu = b/R^6 \pm B/R^3 \tag{6.72a}$$

$$\int 2\pi\,\Delta\nu\,dt = \left|\frac{3\pi^2 b}{4\langle v\rangle}\frac{1}{\varrho^5} \pm \frac{4\pi B}{\langle v\rangle}\frac{1}{\varrho^2}\right| = 1 \tag{6.72b}$$

Case (2):

Same as Case (1).

Case (3):

Frequency perturbation: $$\Delta\nu = b/R_6 \tag{6.73a}$$

$$\varrho = \sqrt[5]{\frac{3\pi^2}{4}\frac{b}{\langle v\rangle}} \tag{6.73b}$$

In Cases 1 and 2, ϱ may be found as roots of Eq. (6.72b), in particular there will be roots ϱ_+ for the repulsive case (positive sign) and ϱ_- for the attractive case (negative sign). In all cases of multiple roots ϱ the highest valued root will be taken as significant. Further, we shall suppose there to be equal amounts of resonance repulsion and attraction so that for ϱ^2 from Eq. (6.72b) we shall write $\frac{\varrho_+^2 + \varrho_-^2}{2}$. Now the ϱ values furnished by the three possible sets of physical conditions may be substituted into Eq. (6.70) with the result:

$$\delta = 2.2\,(\langle v\rangle)^{3/5} \sum_i b_i^{3/5} N_i + \langle v\rangle \sum_j N_j \frac{\varrho_{+j}^2 + \varrho_{-j}^2}{2} \tag{6.74}$$

Let us first consider the Directional Effect contributions. The energy perturbation of a level for this case has been given by Eq. (6.15). Lindholm approximated this equation by the expression:

$$E_{J_1 J_2}^{(2)} = \frac{1}{R^6}\frac{\mu^4 I}{3\hbar^2(\Delta^2 - 1)} \tag{6.75}$$

which he noted is asymptotic for large J, exact for $J_1 = J_2$, 86% of the correct value for $J_1 = 1$, $J_2 = 3$, and 92% of the correct value for $J_1 = 2$, $J_2 = 4$.

We are desirous of obtaining b for Eq. (6.74) from Eq. (6.75). b will be given by:

$$b = \frac{R^6}{h}\left[E_{J_1' J_2'}^{(2)} - E_{J_1'' J_2''}^{(2)}\right] = \frac{4\pi^2\mu^4 I}{3h^3}\left[\frac{1}{(\Delta+1)^2 - 1} - \frac{1}{\Delta^2 - 1}\right] \tag{6.76}$$

according to Eqs. (6.71).

† See *supra*, Sec. 6.4.

In order to find N_i a Maxwell–Boltzmann distribution of the molecules over the rotational levels is assumed. The rotational energy of these linear molecules is:

$$E_J = h\,c\,B\,J(J+1)$$

where:

$$B = \frac{h}{8\pi^2 I c}$$

so that the level population will contain the term $e^{-\frac{hcBJ(J+1)}{kT}} = e^{-E/kT}$. In addition each rotational level is $(2J+1)$-fold spatially degenerate as determined by M. Thus, if there are N molecules per cubic centimeter, the number in the rotational state described by J_i will be:

$$N\begin{pmatrix}\%\text{ molecules}\\ \text{in } J_i\end{pmatrix} = N\left(\frac{(2J_i+1)\,e^{-\frac{hcBJ_i(J_i+1)}{kT}}}{\sum\limits_{J=0}^{\infty}(2J+1)\,e^{-\frac{hcBJ(J+1)}{kT}}}\right) = N\,\frac{w_i}{\sum\limits_{J=0}^{\infty} w_J} \tag{6.77}$$

Eqs. (6.76) and (6.77) may now be substituted into the first term of Eq. (6.74) with the result:

$$\left.\begin{aligned}\delta_D &= 2{\cdot}2\,(\langle v\rangle)^{3/5}\left(\frac{4\mu^4\pi^2 I}{3h^3}\right)^{2/5}\frac{N}{\sum\limits_{J_2=0}^{\infty} w_{J_2}}\sum_{J_2=0}^{\infty}{}' w_{J_2}\\ &\times\left[\frac{1}{(\Delta+1)^2-1}-\frac{1}{\Delta^2-1}\right]^{2/5}\end{aligned}\right\} \tag{6.78}$$

where the prime on the summation sign decrees the dropping of those J_2 values for which $\Delta = \pm 1, 0, -2$ (Resonance Effect).

Lindholm utilized the values $\mu = 2.65\times10^{-18}$, $\langle v\rangle = 67560$, $N = 1.937\times10^{19}$ and $I = 18.70\times10^{-40}$. For these values of the constants, the first twenty or thirty terms in the sum must be considered. For J_2 values greater than this, the exponential factor in Eq. (6.77) leads to sum contributions small enough to be neglected.

Now our attention may be devoted to the resonance contributions to Eq. (6.74). Lindholm utilized a mildly modified version of the London development leading to Eq. (6.48). London's molecular model, if you will, consisted of a rigid rotator. Lindholm's modifications amounted to a model change to a rotating-vibrating molecule. As we have seen in Chapter 2, this means that the eigenfunctions will now be of the form:

$$\psi_{vJM} = \psi_v\,\psi_{JM} \tag{6.79}$$

Interactions will arise between rotation and vibration, but the only one with which we shall be required to concern ourselves is the change in the moment of inertia with vibrational state. The rotational energies will now be:

$$E_J = B_v\,h\,c\,J(J+1) \tag{6.80}$$

We shall return to these considerations at a later point, but let us first consider the case of both molecules in the ground state with $\Delta = \pm 1$. For the vibrational quantum numbers, the same Eq. (6.48) is valid as it stands. London had calculated the matrix elements requisite for the evaluation of a_1, a_2, and a_3 and had obtained the

$$-\frac{1}{2}\left(\frac{a_1^2}{E_1}+\frac{a_2^2}{E_2}+\frac{a_3^2}{E_3}\right) \doteq -\frac{1}{R^6}\,\frac{\pi^2 \mu^4 A}{6\,h^2} \tag{6.81a}$$

Lindholm proceeded to evaluate a_4 as follows: From Eq. (6.46):

$$\frac{a_4^2}{R^6} = V_{12}^2 + V_{11}^2 + V_{10}^2$$

He obtained the V_i after the usual fashion from which there resulted:

$$\begin{aligned} a_4 = {} & \frac{\mu^2}{2(2J+1)(2J+3)} \\ & \times \sqrt{\begin{aligned}&(J+M_1)(J+M_1+1)(J+M_2+2)+(J-M_1)(J-M_1+1)\\ &\times(J-M_2+1)(J-M_2+2)+16\,[(J+1)^2-M_1^2]\,[(J+1)^2-M_2^2]\end{aligned}} \end{aligned} \tag{6.82}$$

Next a_4 must needs be averaged over all orientations of both molecules as determined by M_1 and M_2 where $(J+1) \geq M \geq -(J+1)$ and $J \geq M_2 \geq -J$ as decreed by the resonance condition. For the mean square value of a_4 he obtained, as was to be expected, Eq. (6.23):

$$\langle a_4^2\rangle = \frac{2}{3}\,\mu_4\,\frac{(J+1)^2}{(2J+1)(2J+3)} = \frac{\mu^4(2J+2)^2}{6(2J+1)(2J+3)} \tag{6.23}$$

Eq. (6.23) would appear almost completely independent of J for large J. (1) Lindholm supposed that this would also be the case for $\langle a_4\rangle$. (2) For sufficiently large J it is a reasonable approximation to replace $\sum\limits_{M_1=-(J+1)}^{J+1}$ and $\sum\limits_{M_2=-J}^{J}$ by $\int\limits_{-J}^{J} dM_1$, and $\int\limits_{-J}^{J} dM_2$ respectively. (3) Only the highest order of magnitude terms under the radical in Eq. (6.82) are to be retained. Utilizing these three approximations Eq. (6.82) may be employed and we may let $M_1 = xJ$, $M_2 = yJ$ to obtain:

$$\langle a_4\rangle = \frac{\mu^2}{32}\int\limits_{-1}^{+1}\int\limits_{-1}^{+1} dx\,dy\,\sqrt{18+(18\,x^2y^2+18\,x\,y \quad 14\,x^2-14\,y^2)} \doteq \frac{\mu^2}{2.52} \tag{6.81b}$$

after series expansion of the integrand and integration of the first five terms.

For small J direct calculation using Eq. (6.82)

$$J = 0 \rightarrow \langle a_4\rangle = \frac{\mu^2}{2.25}; \qquad J = 1 \rightarrow \langle a_4\rangle = \frac{\mu^2}{2.48}$$

17*

indicates a rapid convergence toward Eq. (6.81b). Let us then substitute Eqs. (6.81) into Eq. (6.48) with the result:

$$E^{(1)} = -\frac{\pi^2 \mu^4 I}{6 h^2} \frac{1}{R^6} \pm \frac{\mu^2}{2.52} \frac{1}{R^3} \tag{6.83}$$

Eq. (6.83) then gives the *first-order* interaction energy for this resonance condition, in the lower state (Case 1). Eqs. (6.72) must now be solved for ϱ. In Eq. (6.72b) B is simply given by the coefficient of R^{-3}. In Eq. (6.83) b is a slightly different case, however. Since in the upper state we have the Directional Effect taking place, b_{upper} will be given by the coefficient of R^{-6} in Eq. (6.75). On the other hand, the Resonance Effect of the lower state furnishes b_{lower} which is the coefficient of the R^{-6} term in Eq. (6.83). Finally:

$$b = b_{\text{upper}} - b_{\text{lower}} \tag{6.84}$$

The values of B and b thus obtained may now be used in Eq. (6.72b) to yield the ϱ values listed in Table 6.3.

TABLE 6.3

ϱ_+ AND ϱ_- FOR CASE (1), RESONANCE IN LOWER STATE AND DIRECTIONAL EFFECT IN UPPER STATEV. ALSO ϱ FOR RESONANCE ALONE IN THE LOWER STATE.

(After Lindholm[123])

Δ	ϱ from Eq. (6.58b′)	ϱ_+	ϱ_-
+1	2.82×10^{-7}	2.96×10^{-7}	2.58×10^{-7}
−1	2.82×10^{-7}	3.06×10^{-7}	1.59×10^{-7}

From Table 6.3 the more powerful influence of the Directional Effect for $\Delta = -1$ is apparent. Substitution from this table into Eq. (6.74) yields:

$$\delta_R = \langle v \rangle \frac{N}{\sum_{J'=0}^{\infty} w_{J'}} [w_{J+1}\, 7.71 \times 10^{-14} + w_{J-1}\, 5.95 \times 10^{-14}] \tag{6.85}$$

where again the w_τ are defined by Eq. (6.77).

Finally, Case 2 remains to be investigated, and we now utilize the molecular model of Eqs. (6.79) and (6.80). In correspondence with the u_0 of Eq. (6.43) let us take

$$u_0' = \psi_{J+1}{}^{M_1} \psi_v(1)\, \psi_J{}^{M_2} \psi_{v'}(2),$$

and now there will be twenty-four additional eigenfunctions of this form which may combine with this one — we assume that v is the ground vibrational state and v' some resonating upper vibrational state.

In considering the matrix elements of V we first note that, in Eq. (6.2), the μ_i will be functions of the vibrational coordinates. Corresponding to each of London's matrix elements $V_i = V_{0i}$ there are now two matrix elements. As an example, to V_1 there correspond:

$$V_{01}' = V_1' = \int \psi_{J+1}{}^{M_1} \psi_v(1)\, \psi_J{}^{M_2} \psi_{v'}\, V\, \psi_{J+2}{}^{M_1} \psi_J(1)\, \psi_{J+1}{}^{M_2} \psi_{v'}(2)\, d\tau \tag{6.86a}$$

$$V_1'' = \int \psi_{J+1}{}^{M_1} \psi_v(1)\, \psi_J{}^{M_2} \psi_{v'}(2)\, V\, \psi_{J+2}{}^{M_1} \psi_{v'}(1)\, \psi_{J+1}{}^{M_2} \psi_{v'}(2)\, d\tau \tag{6.86b}$$

Let us look at Eqs. (6.86) rather carefully, since some important physical phenomena are inferred by them. The rotational resonance condition, $|J_1 - J_2| = 1$, is fulfilled by all four two-molecule eigenfunctions. In Eq. (6.86a) the individual vibrational quantum numbers remains the same for both system state functions appearing in the matrix element. Eq. (6.86b) presents a different case, however. In this matrix element, the vibrational quantum number for molecule one changes from v to v' under the aegis of V while the vibrational quantum number for molecule two changes from v' to v. Thus, one or more quanta of vibrational energy are *exchanged* in this process, and resonance in the sense of Chapter 5 sets in. It seems important to clearly differentiate between this exchange type resonance and the rotational type of Eq. (6.15). Perhaps this type differentiation is not too satisfactory, for $|J_1 - J_2| = 1$ also implies an exchange in that the two molecules may exchange one quantum of rotational energy between themselves. The semantics of the situation should hardly trouble us, however, if we have a clear picture of the physical phenomena involved. In order to determine one of the reasons for neglecting exchange resonance, the vibrational portions of Eqs. (6.86) may be written as:

$$V_{v1}' = \int \psi_v(1)\, \mu_1\, \psi_v(1)\, d\tau \int \psi_{v'}(2)\, \mu_2\, \psi_{v'}(2)\, d\tau = \mu^2 \qquad (6.87a)$$

$$V_{v1}'' = \int \psi_v(1)\, \mu_1\, \psi_{v'}(1)\, d\tau \int \psi_v(2)\, \mu_2\, \psi_{v'}(2)\, d\tau \qquad (6.87b)$$

Eq. (6.87a) tells us that $V_i' = V_i$, London's matrix elements. Eq. (6.87b) is nothing more nor less than the square (the molecules are identical) of the matrix element of the electric dipole moment for the vibrational transition $v \to v'$. In other words, it is proportional to the intensity of the rotation-vibration band involved. For the bands which Lindholm considered, the intensity is very low which means that Eq. (6.87b) will be small. This in turn will cause Eq. (6.86b) to be small which allows us to neglect the exchange resonance. Another consideration leads to the weakness of exchange resonance. The dispersion f-value, oscillator strength, or what have you, enters this resonance broadening as we have seen in Chapter 5. In this case, we apply the "Principle of Spectroscopic Stability" in reverse and portion out pieces of the dispersion f-value* for a monatomic transition to myriads of rotation vibration levels. The net result of all this is a further decrease in the exchange resonance effect which we now proceed to neglect. This means that we are again only concerned with twelve matrix elements of the form Eq. (6.86a).

The perturber — we may distinguish it since we neglect exchange resonance — will hereinafter be in the ground vibrational state and the absorber in an upper vibrational state for our considerations of Case (2): $\Delta = 0, -2$. If molecule one is the absorber and two the perturber, then:

$$\left.\begin{aligned} E(1) &= B'\, h\, c\, J'(J'+1) \\ E(2) &= B''\, h\, c\, J''(J''+1) \end{aligned}\right\} \qquad (6.88)$$

* "Do you really eat babies, Mr. Swift?" "No, nor do I portion out pieces of dispersion f-values."

according to Eq. (6.80) where B' is the value of the rotational "constant" in the upper state and B'' its value in the lower state.

The eigenfunction which we select to correspond to u_0 of Eq. (6.43) is:

$$u_0' = \psi_{Jv'}^{M_1} \psi_{J+1,v}^{M_2} \tag{6.89}$$

The twelve eigenfunction with which u_0' may combine are then analogous to the twelve functions of Eq. (6.43), for example:

$$u_1' = \psi_{J+1,v}^{M_1} \psi_{J+2,v}^{M_2}, \quad \text{etc.}$$

Having obtained the requisite eigenfunctions, we may write down the equivalent of Eq. (6.44) using Eq. (6.88):

$$\left.\begin{aligned} E_1' &= 2hc(B' + B'')(J+1) + 2B''hc \\ E_2' &= -2hc(B' + B'')J - 3B''hc \\ E_3' &= -2hc(B' + B'')J + B''hc \\ E_4' &= 2hc(B' - B'')(J+1) \end{aligned}\right\} \tag{6.90}$$

Now $|B' - B''| \ll B', B''$, since the variation of rotational constant with vibrational quantum number should not be too great. Thus, although $|E_1'|$, $|E_2'|$, $|E_3'| \gg |E^{(1)}|$, this inequality does not hold between $E^{(1)}$ and E_4'. As a consequence, $E^{(1)}$ may be dropped from the first three denominators on the right of Eq. (6.45) but not from the fourth so that for Case 2 we obtain:

$$R^6 E^{(1)} = -\left(\frac{a_1^2}{E_1'} + \frac{a_2^2}{E_2'} + \frac{a_3^2}{E_3'}\right) + \frac{a_4^2}{E^{(1)} - E_4'} \tag{6.91}$$

where a_i is again given by Eq. (6.46).

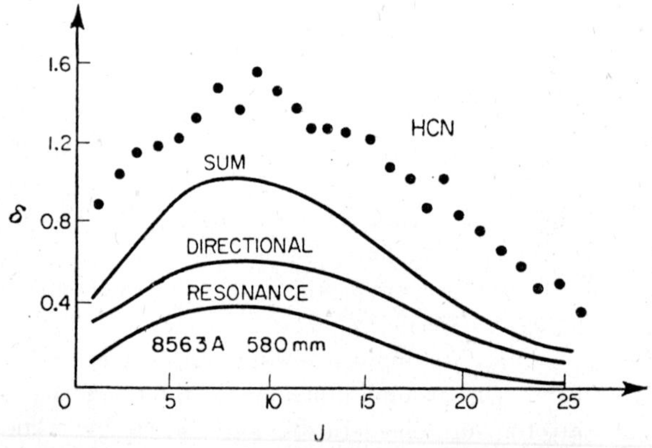

Fig. 6.2. Half width as a function of J. The solid lines represent theoretical contributions as indicated. (After Lindholm[125])

The value of the first term on the right is available from Eq. (6.48), $\langle a_4^2 \rangle$ may be taken as $\mu^4/6$ from Eq. (6.23) with the result:

$$6R^6(E^{(1)})^2 - E^{(1)}\left(6R^6 E_4 - \frac{6\pi^2\mu^4 I}{3h^2}\right) - \left(\frac{6\pi^2\mu^4 I E_4}{3h^2} + \mu^4\right) = 0$$

of solution:

$$E^{(1)} = \frac{1}{2}\left\{E_4 - \frac{\pi^2 \mu^4 I}{3\,h^2 R^6} \pm \sqrt{E_4^{\,2} + \frac{2\,E_4 \pi^2 \mu^4 I}{3\,h^2 R^6} + \frac{4\,\mu^4}{6\,R^6}}\right\} \tag{6.92}$$

in which the term in R^{-12} under the radical has been dropped.

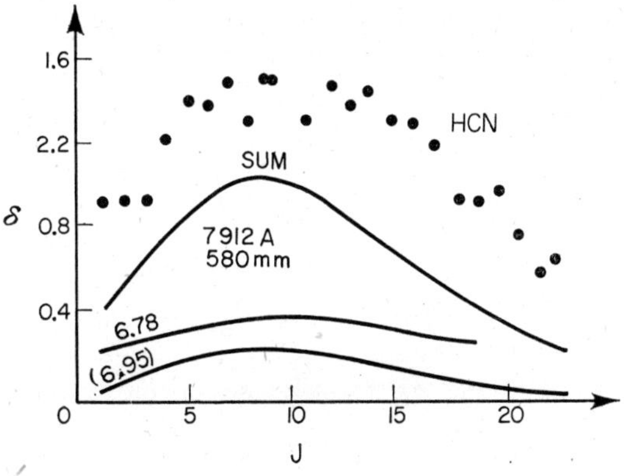

FIG. 6.3. The individual Directional Effect contributions are indicated. (After Lindholm[123])

The second term within the curly braces in Eq. (6.92) gives us b_{upper}. b_{lower} is again obtained from Eq. (6.75).

From which:

$$\frac{3\pi^2 b}{4} = \frac{3\,\pi^2}{4}\,(b_{\text{upper}} - b_{\text{lower}}) = \frac{7}{8}\,\frac{\pi^4 \mu^4 I}{h^3} \tag{6.93}$$

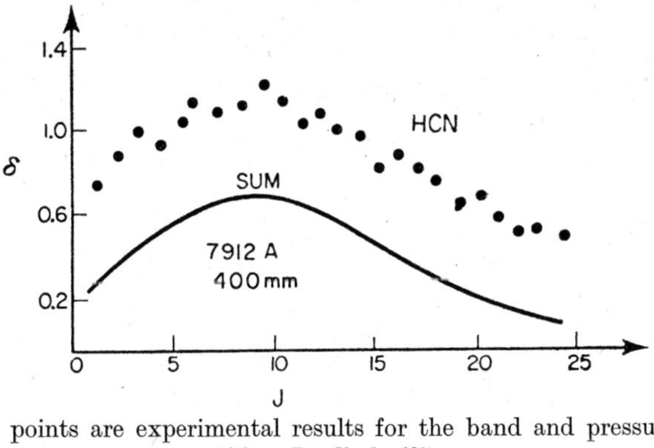

FIG. 6.4. The points are experimental results for the band and pressure indicated. (After Lindholm[123])

It is again necessary to solve Eq. (6.72b) for ϱ. If we let $E_1^{(1)}$ be $E^{(1)}$ with the b_{upper} contribution deleted, $4\pi B/\langle v\rangle\, 1/\varrho^2$ may be replaced in Eq. (6.72b) by $(2\pi/h)\int E_1^{(1)}\,dt$ so that this equation becomes: (6.94)

$$\left|\frac{7\,\pi^4 \mu^4 I}{8\,h^3 \langle v\rangle}\,\frac{1}{\varrho^5} + 2\,\pi \int_{-\infty}^{+\infty} \left[\sqrt{\frac{E_4^2}{4\,h^2} + \frac{E_4 \pi^2 \mu^4 I}{6\,h^4 \varrho^6} + \frac{\mu_4}{6\,h^2 \varrho^6}} + \frac{E_4}{2\,h}\right] dt\right| = 1$$

In his first paper on the subject[123] Lindholm concluded that B contributed little to the optical collision diameter* so that he neglected the resonance contribution in this case. This small contribution would indicate that a sharp resonance is required for effect. This sharpness is reduced by the variation of the rotational constant for the two states, since this variation will in turn cause a variation in the rotational level separations for the two vibrational states.

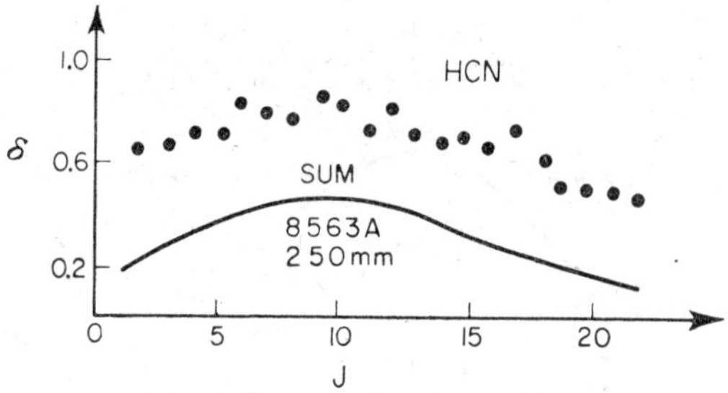

FIG. 6.5. (After Lindholm[123])

Lindholm then obtained the Case 2 contribution as the additional Directional Effect contribution:

$$\delta_D = \langle v \rangle \frac{N}{\sum_{J'=0}^{\infty} w_{J'}} [w_J \, 4.78 \times 10^{-14} + w_{J-2} \, 3.43 \times 10^{-14}] \qquad (6.95)$$

with the w_τ defined as before.

The total half-width is now the sum of Eqs. (6.78), (6.92), and (6.85). Fig. 6.2 gives the Directional and Resonance Effect contributions as well as the sum of

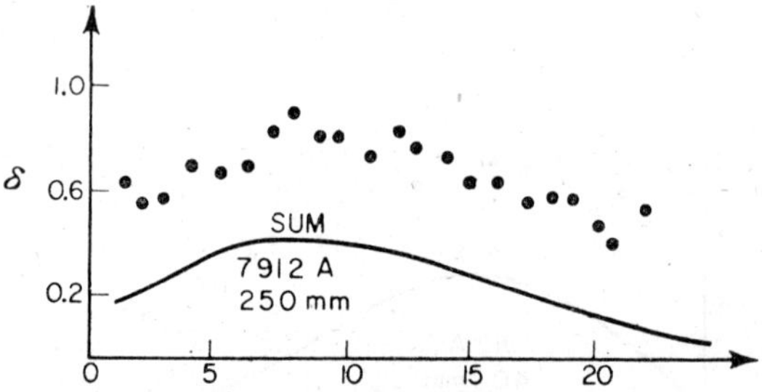

FIG. 6.6. (After Lindholm[123])

the Directional Effect plus the Resonance Effect. Fig. (6.3) gives the individual Directional Effect contributions as well as the sum of the total Directional Effect and Resonance Effect. Figs. 6.4, 6.5, and 6.8 simply give the total predicted widths in comparison with Lindholm's observed widths.

In a later paper[124] Lindholm concluded that his approximation Eq. (6.95), was not sufficient and that Eq. (6.94) should be utilized for a numerical calculation of ϱ for each J value. This has the effect of raising the plot of Eq. (6.95) in Fig. 6.2 0.06 cm^{-1} for $J = 2$, 0.12 cm^{-1} for $J = 5$, 0.11 cm^{-1} for $J = 10$, 0.05 cm^{-1} for $J = 16$, and 0.02 cm^{-1} for $J = 20$.

* See *supra*, Eq. (6.72b).

6.11. Broadening in the diatomic dipole molecule HCl

Lindholm also carried out an investigation similar to that of HCN for the HCl molecule[124]. This molecule together with HCN provide two of the rare examples of simple linear molecules possessed of large dipole moments. This fact led to Lindholm's choice of it as an object of his considerations.

The vibrational transition involved was $0 \rightarrow 4$ resulting in the band at 9152 A. In this case he considered both the P and the R branches in contradistinction to the HCN investigation.

We begin by considering case (3) of Section 10, the Directional Effect in both upper and lower states. In arriving at Eq. (6.76) for this case for HCN the constancy of he moment of intertia in the two vibrational states was inferred. Since this would be a much worse approximation for a diatomic molecule, the use of Eq. (6.75) for the HCl energy perturbation with the perturber in the lower vibrational state and the absorber in the upper vibrational state is ruled out. Eq. (6.75) may still be used, however, when both perturber and absorber are in the lower state. Margenau's Eq. (6.14) may be used for the interaction energy with the absorber in the upper state. We again let the subscripts one and two denote the absorber and the perturber respectively and I_0 and I_4 the moments of inertia in the ground vibrational state and in the vibrational state having $v = 4$ respectively. Eqs. (6.14) and (6.75) are used to obtain b from an equation of the form Eq. (6.76); ϱ is subsequently found from Eq. (6.73b) and the result for the P-Branch follows* from Eq. (6.74):

$$\delta_D = 2{\cdot}2\,(\langle v\rangle)^{3/5}\left(\frac{8\pi^2\mu^4 I_0}{3h^3}\right)^{2/5}\frac{N}{\sum\limits_{J_2=0}^{\infty} w_{J_2}}\, w_{J_2}\sum_{J_2=0}^{\infty}{}' \left\{ \left[\frac{1}{(2J_1-1)(2J_1+1)}\left\{\left(\frac{1}{J_1}+\frac{I_0/I_4}{J_2+1}\right)^{-1}+\left(\frac{1}{J_1-1}-\frac{I_0/I_4}{J_2+1}\right)^{-1} - \left(\frac{1}{J_1}-\frac{I_0/I_4}{J_2}\right)^{-1}-\left(\frac{1}{J_1-1}+\frac{I_0/I_4}{J_2}\right)^{-1}\right\}+\frac{1}{2(\Delta^2-1)}\right]^{2/5}\right\} \quad (6.96)$$

in which all symbols are defined as in Section 10.

One may obtain a quite similar expression for the R-Branch ($\Delta J = +1$). In addition to Eq. (6.96) Lindholm imposed the restriction on δ_D that $\varrho \geq 0.30 \times 10^{-7}$ cm, the Landolt–Börnstein value of the gas kinetic diameter. The resulting line widths with J-value for the Directional Effect alone (Case 3) are given in Fig. 6.7.

Case 1 of Section 10 may now be considered. One again obtains Eq. (6.84) for the Resonance Effect in the lower state. The treatment of the Directional Effect in the upper state of the absorber differs now from Section 10 in that we again use Eq. (6.14) to determine b_{upper} of Eq. (6.84). Having found b_{upper}, Eqs. (6.83) and (6.84) may be used as in Section 10 to arrive at two equations

* Cf. *supra*, Eq. (6.78).

(one for the P-Branch and one for the R-Branch) similar to Eq. (6.85). The resulting δ_R vs. J curves for Case 1 appear in Fig. 6.7.

Finally Eq. (6.94) (and a similar equation for the R-Branch) was utilized by Lindholm in the treatment of Case 2. By numerical solution of these two equations for ϱ for each value of J in the P and R-Branches and subsequent utilization of them in an equation of the form Eq. (6.70) (with the N_i given by Eq. (6.77)) this author obtained the Case 2 contributions as given by Fig. 6.7.

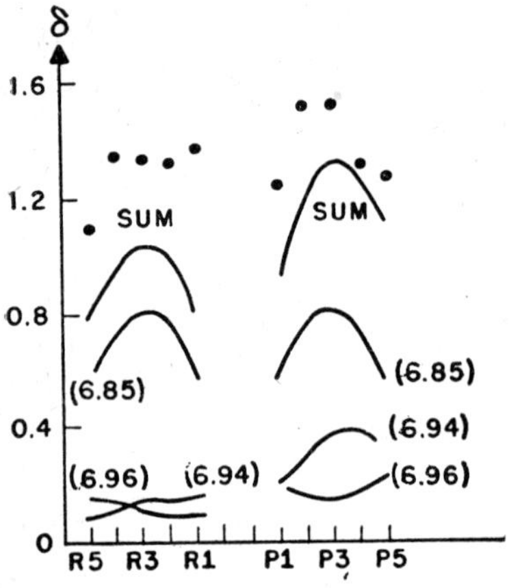

FIG. 6.7. The broadening of HCl as a function of J. Theoretical contributions arise fromequations as indicated. (After Lindholm[124])

Fig. 6.7 gives also the sum of the three contributions to the line width as well as Lindholm's experimental results for the HCl widths considered. Although the agreement between theory and experiment appears confirmatory, Lindholm investigated the possible contributions of the Dispersion and Induction Effects as being responsible, rather than experimental error, for the disagreement observed. In general he concluded that the effect of these two phenomena was negligible except possibly for $J = 0$, thus explaining the $R(1)$ discrepancy apparent in the figure.

The pure rotation line widths may also be calculated for HCl. All absorbing transitions in this case arise from $\Delta J = +1$, and no change in the moment of inertia with rotational state occurs. For the Directional Effect then, we obtain Eq. (6.78) with the bracketed expression replaced by:

$$\left[\frac{1}{(\Delta - 1)^2 - 1} - \frac{1}{\Delta^2 - 1}\right] \tag{6.97}$$

Since there is no change in the moment of inertia with rotational state, we may here obtain sharp resonance both for the upper and lower of a pair of states. Thus one would obtain in place of Eqs. (6.85) and (6.95):

$$\delta_R = \langle v \rangle \frac{N}{\sum\limits_{J_2=0}^{\infty} w_{J_2}} [1.364 \times 10^{-14}(w_{J_1-} + w_{J_1+1}) + 1.357 \times 10^{-14}(w_{J_1} + w_{J_1+1})] \tag{6.98}$$

6.12. Broadening of linear dipole molecules according to Foley

Foley[46] considered the broadening of the rotation-vibration lines. For an inverse third power dependence of the interaction energy on the molecular separation, for example, $p = 3$ in Eq. (2.114) and this equation becomes:

$$\delta = 8\pi N \sum_k \frac{\gamma_k}{2} \int_0^\infty \frac{d\alpha \sin^2 \alpha}{\alpha^2} = 4\pi N \sum_k \gamma_k \frac{\pi}{2} = 4\pi^3 N \langle \beta \rangle \tag{6.99}$$

where we may recall that γ arises from $\Delta\nu = \hbar\gamma/r^3$ while β arises from $\Delta\nu = \hbar\beta/r^3$.

TABLE 6.4

(After Foley[46])

l	HCl δ cm^{-1}	l	δ sec. order	Resonance	HCN Total	Shift
0	0.39	0	0.32	0.54	0.86	+ 0.12
1	0.60	1	0.39	0.92	1.31	+ 0.14
2	0.72	2	0.44	1.33	1.87	+ 0.14
3	0.70	3	0.48	1.69	2.17	+ 0.14
4	0.58	4	0.50	1.98	2.48	+ 0.13
5	0.42	5	0.53	2.20	2.73	+ 0.12
6	0.27	6	0.54	2.35	2.89	+ 0.10
7	0.158	7	0.54	2.41	2.95	+ 0.065
8	0.80	8	0.55	2.41	2.96	+ 0.040
9	0.038	9	0.56	2.34	2.90	+ 0.010
10	0.016	10	0.55	2.24	2.79	− 0.018
		11	0.54	2.06	2.60	− 0.043
		12	0.53	1.87	2.40	− 0.067
		13	0.51	1.67	2.17	− 0.082
		14	0.50	1.46	1.96	− 0.10
		15	0.48	1.25	1.73	− 0.11
		16	0.44	1.05	1.49	− 0.11
		17	0.42	0.87	1.29	− 0.12
		18	0.38	0.71	1.09	− 0.11
		19	0.36	0.56	0.92	− 0.11
		20	0.32	0.44	0.76	− 0.10
		21	0.30	0.34	0.64	− 0.10
		22	0.26	0.26	0.52	− 0.088

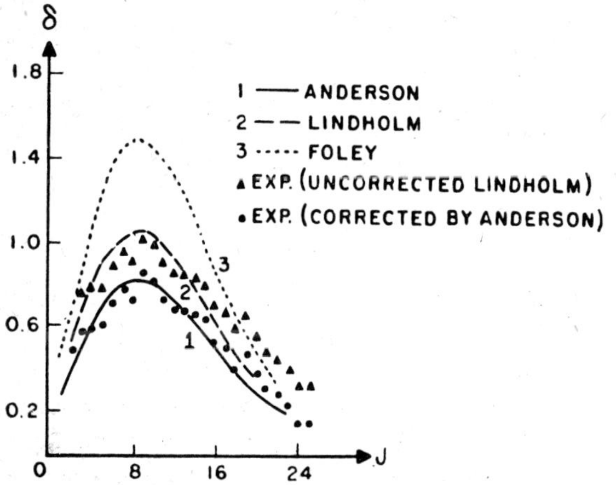

FIG. 6.8. A comparison of three theories with corrected and uncorrected experimental data. (After Anderson[3])

Now Foley used Eq. (6.23) — the root mean square value of the Rotational Resonance interaction — as $\langle\beta\rangle$ in Eq. (6.99) to obtain the half-width due to this effect. One might question this procedure, however, since Foley's definition of γ^* is the difference between two β's defining interactions for two states. At any rate, the resonance half-widths obtained in this manner are listed in Table 6.4.

* Cf. *supra*, Eq. (2.139).

Foley utilized Eq. (2.114) with $p = 6$ for the case of the Directional Effect, and for this case the interaction constants were obtained by him from Eq. (6.15). The results of his calculations for HCN and HCl are given in Table 6.4.

Fig. 6.8 gives a comparison between the δ vs. J curve for Foley as well as Lindholm. It would appear that Lindholm's curve more closely approximates the experimental results than Foley's which Anderson[3] attributed to possible errors in Foley's calculations. Actually there is no reason for much variation in these two sets of results. If Foley was desirous of working within the framework of his theory rather than Weisskopf's there appears to be no particular reason for his not using Lindholm's careful method of doing so. Had he so done it seems reasonable to suppose that the result would have been to raise Lindholm's curves of Fig. 6.8 (for p equals three or six in Eq. (4.47) Foley's half-widths exceed Weisskopf's by factors of 1.57 and 1.21 respectively) perhaps to better agreement with experimental data rather than to distort these curves.

6.13. An application of symmetrical top dipole broadening: Ammonia

The development of microwave spectroscopic techniques in recent years has naturally led to a quickened interest on the part of many investigators in pure rotational spectra and the line broadening particular to this spectra. In this and the next several sections we shall consider some of the theories which have been advanced recently aimed directly at microwave broadening effects or at microwave and infrared effects.

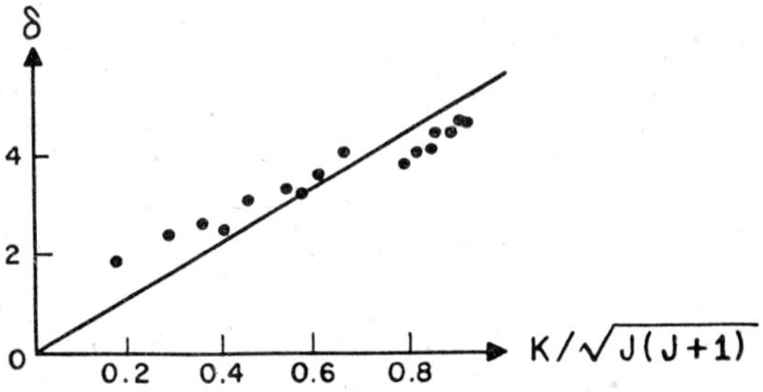

FIG. 6.9. Half widths of the ammonia inversion line. The experimental results are indicated by points. (After Margenau[140])

Absorbing transitions may take place between the two members of an inversion doubled vibrational level such as possessed by the symmetric top ammonia. Taking, for example, the ground vibrational state and the possible inversion transition within this state, the variation of the inversion doublet separation with J and K will be responsible for the occurrence of many spectral lines as a result of this transition.

Bleaney and Penrose[13] have furnished the microwave field with some excellent results on the widths of these ammonia inversion lines. Some of the results of their experiments are contained in Fig. 6.9. Bleaney[10] used an interruption type approach to this broadening under the assumption that the disturbing ammonia molecule causes a certain perturbation of the absorbing ammonia molecule due to the field of its dipole. A collision is declared as having occurred at a certain separation or interaction energy, and, under an application of the normal Interruption Theory, Bleaney determines the half-widths as proportional to $3\sqrt{K^2/J^2 + J}$. Although a cursorily satisfactory agreement between theory and experiment was thus obtained, Margenau[140] has made a telling point which casts more than a little doubt on the meaningfulness of this agreement.

In order that this interruption approach succeed in explaining the broadening, it was found that an interaction energy (corresponding to a minimum collision-defining separation) of around twice the inversion doublet level separation be used. "If this were to be taken literally it would imply that the molecule could go on absorbing even when the perturbation is larger than its natural frequency or indeed when the natural frequency is negative. In suggesting this the impact theory rather defeats itself."[140]

Margenau thus decided to use the interaction between two symmetric dipole rotators* in conjunction with his Statistical theory to account for this broadening. His results were in good, if not spectacular, agreement with those of experiment, but, equally important, they were based on tenable theoretical considerations.

To begin with, the interaction energy has already been found as:

$$V(J_1 K_1 J_2 K_2 \Lambda) = \frac{\mu^2}{R^3} \frac{K_1 K_2}{J_1(J_1+1)\, J_2(J_2+1)} \varepsilon_\Lambda \qquad (6.100)$$

where ε_Λ is the expression (in M_1 and M_2) in curly braces in Eq. (6.61). Here Λ, and hence ε_Λ, may take on $(2J_1+1)(2J_2+1)$ values corresponding to the number of values available to M_1 and M_2.

Eq. (6.65b) has already told us that an interaction of the form Eq. (6.100) will result in no spectral line shift so we turn our attention solely to the widths.

If the doublet transition "normally" results in the absorption of radiation of frequency ν_0, then at separation R between two ammonia molecules, the absorber would absorb radiation of frequency:

$$\nu = \nu_0 + \frac{1}{h}(V' - V'') = \nu_0 + \frac{B_{\Lambda\mu}}{R^3} \qquad (6.101a)$$

where, from Eq. (8.69):

$$B_{\Lambda\mu} = \frac{\mu^2}{h} \frac{K_1 K_2}{J_1(J_1+1)\, J_2(J_2+1)} (\varepsilon_\Lambda - \varepsilon_\mu) = b(\varepsilon_\Lambda - \varepsilon_\mu) \qquad (6.101b)$$

Perhaps it should be emphasized again that in the transition from one member of the inversion doubled vibrational level to the other $J_1 K_1 J_2 K_2$ do *not* change their values although the dipole changes its orientation causing M_i to change. If this were not the case Eq. (6.70b) would obviously not be justified. Margenau ignored the selection rules on the basis of the large number of possibilities for J_1 and J_2 for all but small J_1 and J_2. For this reason Λ_μ is replaced by σ where σ takes on all values from one to $n = [(2J_1+1)(2J_2+1)]^{1/2}$.

We are now in a position, after assuming binary encounters and the resulting low pressure,† to carry out the probability calculation leading to a statistical line shape.

* See See *supra*, Sec. 6.

† Bleaney and Penrose used quite low pressures so that this should be a reasonable approximation.

We now divide space into private cells of volume $1/N$, and to each of these we assign a perturber. Next, n of these cells are taken, one for each value of $(1 \leq \sigma \leq n)$. In the σ-th cell:

$$\nu = \nu_0 + \frac{B_\sigma}{R^3} = \nu_0 + f \tag{6.102}$$

according to Eq. (6.101a) at separation R. We presume the cells to be spherical, and the probability that the frequency lies near f (the portion of the σ-th space contributing this frequency being a spherical shell of radius R) is:

$$I_0(f)\,\mathrm{d}f \propto \frac{4\pi}{3}\left(\frac{\mathrm{d}}{\mathrm{d}f} R^3\right)_\sigma \mathrm{d}f \tag{6.103}$$

Let us take $f > 0$ which means that one half the n cells will contribute to the probability since Eq. (6.101b) is symmetrical as indicated by the zero value for its mean. In order to obtain the intensity at a frequency separation f from the line center it is now merely necessary to sum Eq. (6.103) over all $n/2$ values of σ (for $f > 0$) with the result:

$$I(f) = \left|\frac{N}{n}\sum_\sigma^{n/2}\frac{4}{3}\pi\left(\frac{\mathrm{d}R^3}{\mathrm{d}f}\right)_\sigma\right| = \frac{4\pi}{3}\frac{N}{n}\left|\sum_\sigma^{n/2}\frac{\mathrm{d}}{\mathrm{d}f}\left(\frac{B_\sigma}{f}\right)\right| \tag{6.104}$$

where N/n, the normalizing factor, is simply the reciprocal of the volume contained in the n cells.

Again utilizing the symmetry of the B_σ arising from the disappearance of the mean value

$$\frac{\mathrm{d}}{\mathrm{d}f}\sum_\sigma^{n/2}\frac{|B_\sigma|}{f} = \frac{1}{2}\frac{\mathrm{d}}{\mathrm{d}f}\sum_\sigma^{n}\frac{|B_\sigma|}{f}$$

so that Eq. (6.104) becomes:

$$I(f) = \frac{2\pi}{3}\frac{N}{f^2}\frac{\sum_\sigma |B_\sigma|}{n} \tag{6.105}$$

In normalizing I_f, the limits of integration are $B_\sigma/R_{\max}^3$ ($R_{\max}^3 = 3\pi N/4$) and infinity, where the lower limit is different in each term of the sum in Eq. (6.105) gives the positive wing ($f > 0$) of the distribution which, however, is mirrored in the negative wing ($f < 0$). Consequently the normalization integral $I(f)$ must be equal to one half.

On the basis of the large value of n, Margenau assumed a Gaussian distribution* of the B_σ. By virtue of this Gaussian assumption we may equate the mean of the absolute values to the standard deviation or root mean square value of the deviation from average times $\sqrt{2/\pi}$.

$$\frac{1}{n}\sum_\sigma |B| = \sqrt{\frac{2}{\pi}}\left[\frac{1}{n}\sum_\sigma (B_\sigma - B)^2\right]^{1/2} = \sqrt{\frac{2}{\pi}}\left[\frac{1}{n}\sum_\sigma B_\sigma^2\right]^{1/2} \tag{6.106a}$$

and from Eq. (6.101):

$$\sum_\sigma B_\sigma^2 = b^2\sum_{A\mu}(\varepsilon_A - \varepsilon_\mu)^2 = 2b^2 n^{1/2}\sum_A \varepsilon_A^2 = 2b^2 n\,\langle\varepsilon^2\rangle \tag{6.106b}$$

where $\langle\varepsilon^2\rangle$ is the mean square energy and has been given by Eq. (6.68a).

* This should not be supposed to imply a Gaussian distribution of intensities in the spectral line.

Eq. (6.106b) may now be substituted into Eq. (6.106a) and the result in turn substituted into Eq. (6.105) to yield:

$$I(f) = \frac{4}{3}\left(\frac{2\pi}{3}\right)^{1/2} \frac{\mu^2}{h} \frac{K_1 K_2}{[J_1(J_1+1)\,J_2(J_2+1)]^{1/2}} \frac{N}{f^2} \tag{6.107}$$

which quite obviously holds only in the line wing.

If we assume a dispersion distribution, one obtains for large $f = \nu - \nu_0$:

$$I(f) \rightarrow \frac{1}{\pi} \frac{\delta}{(\nu - \nu_0)^2}$$

so that:

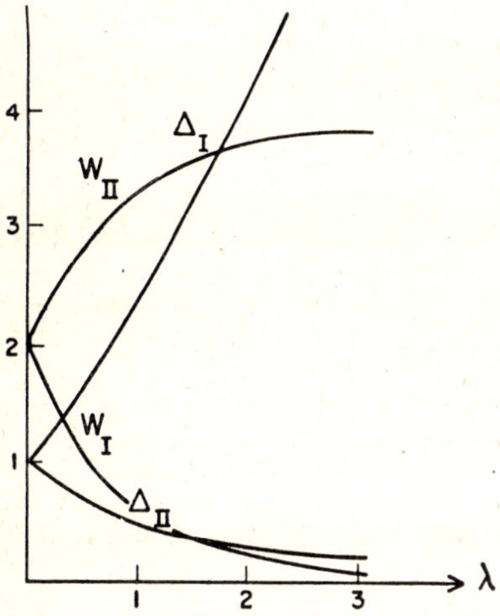

FIG. 6.10. The intensities and shifts of the NH_3 line split and shifted by the interaction of two linear vibrators with mirror potentials. (After Margenau[141])

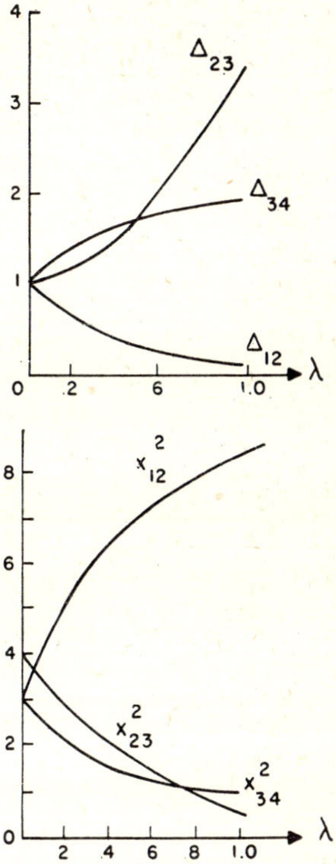

FIG. 6.11. The effect of the interaction between three linear vibrators with mirror potentials. (After Margenau[141])

$$\delta = 2\left(\frac{2\pi}{3}\right)^{3/2} \frac{\mu^2}{h} N \frac{K_1 K_2}{[J_1(J_1+1)\,J_2(J_2+1)]^{1/2}} \tag{6.108}$$

Finally the average value of the involved combination of perturber quantum numbers over a Maxwell–Boltzmann distribution† is taken:

$$\langle K_2 J_2(J_2+1)\rangle = \frac{\sum_{J_2 K_2} [K_2(J_2^2 + J_2)]\, g(J_2 K_2)\, e^{-W(J_2 K_2)/kT}}{\sum_{J_2 K_2} g(J_2 K_2)\, e^{-W(J_2 K_2)/kT}} \tag{6.109}$$

In Eq. (6.109) the $W(J_2 K_2)$ are the symmetric rotator energies. The $g(J_2 K_2)$ are the statistical weights of the levels. Smith[188] has carried out the calculation

† See *supra*, Eq. (6.77).

indicated by Eq. (6.109) to obtain 0.54 for $T = 20°$C. For ammonia $\mu = 1.44 \times 10^{-18}$ e. s. u. On this basis Margenau obtained the result:

$$\frac{\delta}{p} = 1.13 \times 10^{-3} \frac{K_1}{[J_1(J_1+1)]^{1/2}} \text{ cm}^{-1} \text{ (mm Hg)}^{-1} \text{ at } 20\,°\text{C} \tag{6.110}$$

and the straight line in Fig. 6.9 yields the results of applying this equation.

This then is Margenau's low pressure (and the theory has not here developed for any but low pressures) result for ammonia inversion lines, and its success for this case is manifestly apparent from Fig. 6.11. Some of the modifications which higher pressures invoke will become apparent in Margenau's more refined treatment of interactions between potential hill molecules which we consider next.

6.14. Interactions between linear vibrators with mirror potentials

The shift to the red with increase of pressure of the NH_3 inversion line, which had recently been observed[11, 188] provoked the Margenau investigation[141] of the phenomenon which we shall consider in this and the two succeeding sections. First let us note that in applying the term "linear" to NH_3 (or other mirror potential molecules) we simply wish to imply that the mirror potential of the molecule is a function only of the separation of the N atom from the H plane.

In considering this problem we again take the eigenfunctions for the two doublet members of a vibrational level as:

$$\psi_1 = [2(1-\Upsilon)]^{-1/2}(\psi_- - \psi_+) \tag{6.111a}$$

$$\psi_2 = [2(1+\Upsilon)]^{-1/2}(\psi_- + \psi_+) \tag{6.111b}$$

where again $E_1 > E_2$.

Now two interacting molecules of this type are to be considered.

Margenau chooses to treat the case in which the dipoles of the two molecules are aligned along the same axis. If we let x_i be the separation of the mean positive charge from the center of mass of a molecule, the resulting interaction between the two molecules under these circumstances is:

$$V = -2e^2 \frac{x_1 x_2}{R^3} \tag{6.112}$$

If $H(1)$ and $H(2)$ are the unperturbed Hamiltonians for the two molecules and $V(1, 2)$ the Hamiltonian of the interaction, then the present problem may be approached by minimizing

$$H(1, 2) = H(1) + H(2) + V(1, 2) \tag{6.113}$$

after the fashion of Section 6.3. Eq. (6.27) for the present case is:

$$\left.\begin{aligned} \varphi &= A\psi_1(1)\psi_1(2) + B\psi_2(1)\psi_2(2) + C\psi_1(1)\psi_2(2) + D\psi_2(1)\psi_1(2) \\ &= A\psi_A + B\psi_B + C\psi_C + D\psi_D \end{aligned}\right\} \tag{6.114}$$

All diagonal elements of V vanish, which can be made easily apparent. As we have noted an integrand must be symmetric. Now a linear combination of position coordinates is antisymmetric. Thus, ψ_2 is symmetric so $x\psi_2$ is antisymmetric and $\psi_2 x\psi_2$ is antisymmetric. ψ_1 is antisymmetric so $x\psi_1$ is symmetric and $\psi_1 x\psi_1$ is antisymmetric. Thus, the integrals involved in the diagonal elements of V disappear. As a consequence the only allowed elements of V are:

$$V_{AB} = V_{CD} = -\frac{2e^2}{R^3} x_{12}^2 \quad (6.115a)$$

where:

$$x_{12} = \int \psi_1 x \psi_2 \, dx = -\frac{x}{(1-\Upsilon^2)^{1/2}} \quad (6.115b)$$

and

$$\bar{x} = \int x \psi_+^2 \, dx = -\int x \psi_-^2 \, dx \quad (6.115c)$$

Thus, the nonvanishing matrix elements of $H(1, 2)$ are:

$$\left.\begin{aligned} H_{AA} &= 2E_1; \quad H_{BB} = 2E_2; \quad H_{CC} = H_{DD} = E_1 + E_2 \\ H_{AB} &= H_{CD} = V_{AB} \end{aligned}\right\} \quad (6.116)$$

Eq. (6.116) leads immediately to the secular determinant:

$$\begin{vmatrix} 2E_1 - E & y & 0 & 0 \\ y & 2E_2 - E & 0 & 0 \\ 0 & 0 & E_1 + E_2 - E & y \\ 0 & 0 & y & E_1 + E_2 - E \end{vmatrix} = 0 \quad (6.117)$$

in which we have replaced the V_{ij} by y.

As is apparent from Eq. (6.117) the determinant has been split into two, two by two blocks, each of which has two solutions. One simply solves for the $E^{(i)}$ and subsequently obtains the constants A through D in Eq. (6.114) from equations of the form Eq. (6.31). After this fashion one obtains as first order energies and eigenfunctions:

$$\left.\begin{aligned} E^{(1)} &= 2E_1 + R(E_1 - E_2) \\ \Phi^{(1)} &= (1+S^2)^{-1/2} [\psi_1(1)\psi_1(2) + S\psi_2(1)\psi_2(2)] \end{aligned}\right\} \quad (6.118a)$$

$$\left.\begin{aligned} E^{(3)} &= 2E_2 - R(E_1 - E_2) \\ \Phi^{(3)} &= (1+S^{-2})^{-1/2} [\psi_1(1)\psi_1(2) - S^{-1}\psi_2(1)\psi_2(2)] \end{aligned}\right\} \quad (6.118b)$$

$$\left.\begin{aligned} E^{(2)} &= E_1 + E_2 + y \\ \Phi^{(2)} &= \frac{1}{\sqrt{2}} [\psi_1(1)\psi_2(2) + \psi_2(1)\psi_1(2)] \end{aligned}\right\} \quad (6.118c)$$

$$\left.\begin{aligned} E^{(4)} &= E_1 + E_2 - y \\ \Phi^{(4)} &= \frac{1}{\sqrt{2}} [\psi_1(1)\psi_2(2) - \psi_2(1)\psi_1(2)] \end{aligned}\right\} \quad (6.118d)$$

$$\left.\begin{aligned} R &= \left[1 + \left(\frac{y}{E_1 - E_2}\right)^2\right]^{1/2} - 1 \\ S &= \frac{E_1 - E_2}{y} R \end{aligned}\right\} \quad (6.118e)$$

A cursory consideration of Eq. (6.118d) is sufficient to show that the eigenfunction in this equation is the only one which is antisymmetric (changes sign) on exchange of the coordinates one and two. As a consequence $\Phi^{(4)}$ cannot combine with any of the other eigenfunctions and may be neglected.

Now $\Phi^{(1)}$ and $\Phi^{(3)}$ are symmetric with respect to a reflection of their coordinates through the origin (even parity). On the other hand, $\Phi^{(2)}$ is antisymmetric (odd parity). The matrix elements of $X = x_1 + x_2 \Lambda$, the electric dipole moment (odd parity), determine the intensities of possible transitions among the states whose energies and eigenfunctions we have ascertained. Since X is antisymmetric, it may only combine with a symmetric and an antisymmetric pair of eigenfunctions. This means that under the influence of X, $\Phi^{(1)}$ and $\Phi^{(2)}$ as well as $\Phi^{(3)}$ and $\Phi^{(2)}$ may combine, but $\Phi^{(1)}$ and $\Phi^{(3)}$ may not. If we let Δ_i and w_i respectively be the energy difference and probability of such a combination, there results, from Eqs. (6.118):

$$\Delta_I = E^{(1)} - E^{(2)} = (E_1 - E_2)\,[(1 + \Lambda^2)^{1/2} + \Lambda] \tag{6.119a}$$

$$\Delta_{II} = E^{(2)} - E^{(3)} = (E_1 - E_2)\,[(1 + \Lambda^2)^{1/2} - \Lambda] \tag{6.119b}$$

$$\Lambda = \frac{|y|}{E_1 - E_2} = \frac{2\,\mu_2}{(1 - \Upsilon^2)\,R^3}\,\frac{1}{(E_1 - E_2)} \tag{6.119c}$$

$$w_1 \doteq |\textstyle\int \Phi^{(1)}\,X\,\Phi^{(2)}\,\mathrm{d}x|^2 = 2\,\frac{(S+1)^2}{S^2+1}\,{x_{12}}^2 \tag{6.119d}$$

$$w_2 \doteq |\textstyle\int \Phi^{(2)}\,X\,\Phi^{(3)}\,\mathrm{d}x|^2 = 2\,\frac{(S-1)^2}{S^2+1}\,{x_{12}}^2 \tag{6.119e}$$

wherein all symbols have previously been explained.

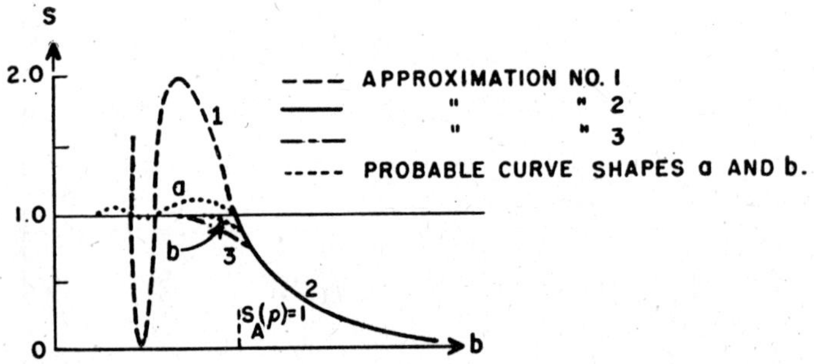

FIG. 6.12. Approximations to $S(\varrho)$ in Anderson's Theory. (After Anderson[3])

The results of Eqs. (6.119) are rather self-evident, and appear in Fig. 6.10. The frequencies appearing in this figure are given in units of the inversion doublet separation. From this figure also the manner in which the line splitting increases with decreasing particle separation is apparent. The decrease in the intensity with decreasing separation for the higher frequency is also notable. From these considerations, however, one would expect that with increasing pressure the inversion line frequency should decrease as indeed it does. One further point might be made to the effect that the line width could hardly be expected to vary linearly with the pressure, as Margenau has noted, under the conditions prevalent here — two frequencies varying differently with pressure.

6.15. Rotating linear dipoles with mirror potentials

We now allow our dipoles to rotate as symmetric tops in addition to the behavior which we prescribed in the last section as a next approximation to the actual molecular phenomenon. It appears apparent that our Hamiltonian will now take the form:

$$H = H_0 + V = H_r(1) + H_r(2) + H_v(1) + H_v(2) + Uf(\vartheta, \varphi) \quad (6.120)$$

where H_r is the Hamiltonian of a symmetric top rotator; H_v is the $H(i)$ of Eq. (6.113) and $2f(\vartheta, \varphi)$ is the bracketed expression of Eq. (6.2).

As our unperturbed eigenfunctions let us choose:

$$\Phi_N(x_1\, x_2)\, \psi_{J_1 K_1 M_1}(\vartheta_1\, \varphi_1\, \chi_1)\, \psi_{J_2 K_2 M_2}(\vartheta_2\, \varphi_2\, \chi_2) \quad (6.121\text{a})$$

Section 14 indicates that a good choice for Φ_N will be:

$$\left.\begin{aligned} \Phi_1 &= \psi_1(1)\, \psi_1(2) \\ \Phi_2 &= \sqrt{1/2}\, [\psi_1(1)\, \psi_2(2) + \psi_2(1)\, \psi_1(2)] \\ \Phi_3 &= \psi_2(1)\, \psi_2(2) \end{aligned}\right\} \quad (6.121\text{b})$$

There is no question but that H_0 is diagonal with respect to Eqs. (6.121) since the eigenfunctions of this latter equation are the eigenfunctions of the Hamiltonian operator H_0. The eigenvalues of this operator we know to be:

$$(NJ_1\, K_1\, M_1\, J_2\, K_2\, M_2\, |H_0|\, NJ_1\, K_1\, M_1\, J_2\, K_2\, M_2) = E_N + E_{J_1 K_1} + E_{J_2 K_2} \quad (6.122)$$

in which the E_N are given by the diagonal elements in Eq. (6.116), and the E_{JK} are the rotational energies.

V, on the other hand, is not diagonal and its matrix elements are given by:

$$\left.\begin{aligned} &(N''J_1\, K_1\, KM_1''\, J_2\, K_2\, M_2''\, |V|\, N'J_1\, K_1\, M_1'J_2\, K_2\, M_2') \\ &= (N''\, |U|\, N')\, (J_1\, K_1\, M_1''J_2\, K_2\, M_2''\, |f|\, J_1\, K_1\, M_1'J_2\, K_2\, M_2') \\ &= (N''\, |U|\, N')\, (M_1''M_2''\, |f|\, M_1'\, M_2') \end{aligned}\right\} \quad (6.123)$$

In Eq. (6.123) the matrix elements of f which are involved with changes in the quantum numbers J and K are neglected. This ignoration entails a good approximation since the levels having different J and K are so greatly separated in comparison to the separation of the two members of an inversion doublet that they may be neglected. Neglecting M, the spatial degeneracy parameter, and its possible change would be something else again.

Eqs. (6.112), (6.115), and (6.121b) may now be used directly to obtain:

$$||U| = ||(N''\,|U|\,N')|| = -\frac{2\,e^2}{R^3}\,x_{12}^{\;2}\begin{pmatrix} 0 & 0 & 1 \\ 0 & 1 & 0 \\ 1 & 0 & 0 \end{pmatrix} \qquad (6.124)$$

In addition Eq. (6.122) may be written as:

$$||T|| = \begin{pmatrix} E_r + 2\,E_1 & 0 & 0 \\ 0 & E_r + E_1 + E_2 & 0 \\ 0 & 0 & E_r + 2\,E_2 \end{pmatrix} \qquad (6.125)$$

Then from Eqs. (6.120) and (6.122) through (6.125):

$$H = ||T||\;||(M_1''M_2''\,|1|\,M_1'M_2')|| + ||U||\;||(M_1''M_2''\,|f|\,M_1'M_1')|| \qquad (6.126)$$

A quite standard and well known procedure exists for diagonalizing $||(M_1''M_2''\,|f|\,M_1'M_2')||$ while leaving $||(N_1''M_2''\,|1|\,M_1'M_2')||$ a unit matrix. This procedure is carried out to obtain a matrix for H which consists of three by three blocks along the principal diagonal of the form:

$$||T + f_i U|| = \begin{pmatrix} E_r + 2\,E_1 & 0 & f_i\,y \\ 0 & E_r + E_1 + E_2 + f_i\,y & 0 \\ f_i\,y & 0 & E_r + 2\,E_2 \end{pmatrix} \qquad (6.127)$$

whose eigenvalues must be determined.

In order to accomplish this Margenau let $\varepsilon = E - (E_1 + E_2) - E_r$, that is, he let ε be the difference between the perturbed energy E and the intermediate of the three unperturbed levels (see, for example, Eq. (6.125)). Solutions are then required of:

$$\begin{pmatrix} E_1 - E_2 - \varepsilon & 0 & f_i\,y \\ 0 & f_i\,y - \varepsilon & 0 \\ f_i\,y & 0 & -(E_1 - E_2) - \varepsilon \end{pmatrix} = 0 \qquad (6.128)$$

which turn out to be

$$\varepsilon_i^{(1)} = [(E_1 - E_2)^2 + f_i^2\,y^2]^{1/2} = (E_1 - E_2)\,(R_i + 1) \qquad (6.129\text{a})$$

$$\varepsilon_i^{(2)} = f_i\,y \qquad (6.129\text{b})$$

$$\varepsilon_i^{(3)} = -\,[(E_1 - E_2)^2 + f_i\,y^2]^{1/2} = -\,(E_1 - E_2)\,(R_i + 1) \qquad (6.129\text{c})$$

where we now define R_i as:

$$R_i = \left[1 + \left(\frac{f_i y}{E_1 - E_2}\right)^2\right]^{1/2} - 1$$

We may note the agreement between Eqs. (6.118) and (6.129) when E_r is neglected. Let us consider just how the problem stands at this point.

The matrix of f in Eq. (6.126) contains $(2J_1 + 1)(2J_2 + 1)$ rows and columns since this is the possible number of values of M_1 and M_2. This in turn means that there will be this number of diagonal elements. Note that in Eq. (6.126) each element of the matrix of f is multiplied by the three by three matrix $||U||$ and similarly for the unit matrix. In the diagonalized version of Eq. (6.126) there will be $(2J_1 + 1)(2J_2 + 1)$ matrices of the form Eq. (6.128) along the principal diagonal leading to $3(2J_1 + 1)(2J_2 + 1)$ roots of the form Eq. (6.129). This means quite simply that due to the interaction considered in this section each inversion doublet will be split into $3(2J_1 + 1)(2J_2 + 1)$ energy levels.

The symmetries of the eigenfunctions associated with Eqs. (6.129) are the same as those associated with the corresponding eigenvalues of Eqs. (6.28). Thus, the corresponding states may combine here. In place of the two spectral lines resulting from the considerations of Section 15 we now have $(2J_1 + 1)(2J_2 + 1)$ lines of increased frequency and a like number of lines of diminished frequency. If A is replaced by $f_i A$ in Eqs. (6.119), thus redefining also S, the transition probabilities for the present case result. As may be noted from Eq. (6.2), f_i is never greater than unity so that in general we may expect the level shift indicated by Eqs. (6.129) to be somewhat less, on the whole, than the shift of the previous section.

We shall return to a consideration of the results of this section after a brief study of multiple encounters, but now let us merely mention an earlier point which this section bears out before turning our attention to a three molecule interaction.

In Eqs. (6.129a) and (6.129c) ε has a dependence on quadratic $f_i y$ while in Eq. (6.129b) a linear relationship exists. Then for the case $f_i y \ll 1$ we may consider the upper and lower levels, as given by the first and third of Eqs. (6.129), as fixed while the intermediate level shifts its position in linear relation to $f_i y$. This in turn means that the two inversion frequencies change in value as $\pm f_i y$, and we may recall that this result was forecast in Section 14 for the case of low pressure. The more accurate low pressure treatment of this section tends then to bear out this earlier result.

6.16. Three molecule interactions and the NH_3 inversion line shift

In order to determine whether a consideration of many molecule interactions tends to change the theoretical prognostications in the preceding sections, Margenau investigated the case of the interaction between three very symmetrically arranged linear vibrators possessed of mirror potentials. The simplest configuration of three such models is a placement of them at the vertices of an equilateral triangle with the axes of their dipoles aligned parallel to the opposing faces of this triangle. This was done; no rotation was allowed, and the resulting interaction effect on the energy levels was determined after the fashion of Section 15. Let us sketch this solution very briefly.

Eq. (6.112) becomes:

$$U = -\frac{5}{4}\frac{e^2}{R^3}(x_1x_2 + x_1x_3 + x_2x_3) \tag{6.130}$$

and the symmetric eigenfunctions corresponding to those in Eq. (6.114)

$$\left.\begin{array}{llll} \psi_A = \psi_1(1)\,\psi_1(2)\,\psi_1(3) \equiv 111 & & E_A = 3E_1 \\ \psi_B = 112 + 121 + 211 & & E_B = 2E_1 + E_2 \\ \psi_C = 122 + 212 + 221 & & E_C = E_1 + 2E_2 \\ \psi_D = 222 & & E_D = 3E_2 \end{array}\right\} \tag{6.131}$$

which lead to:

$$||U|| = \begin{pmatrix} 0 & 0 & 32 & 0 \\ 0 & 62 & 0 & 32 \\ 32 & 0 & 62 & 0 \\ 0 & 32 & 0 & 0 \end{pmatrix} \tag{6.132}$$

where:

$$z = -\frac{5}{4(1-\Upsilon^2)}\frac{\mu^2}{R^3}$$

The determinantal analogy to Eq. (6.119)

$$\begin{vmatrix} 3E_1 - E & 0 & 32 & 0 \\ 0 & 2E_1 + E_2 + 2z - E & 0 & z \\ z & 0 & E_1 + 2E_2 + 2z - E & 0 \\ 0 & 32 & 0 & 3E_2 - E \end{vmatrix} = 0 \tag{6.133}$$

of solution:

$$E^{(1)} = 3E_1 + \eta_1(E_1 - E_2); \quad \Phi^{(1)} = \left(1 + \frac{\eta_1{}^2}{3\Lambda^2}\right)^{-1/2}\left[\psi_A + \frac{\eta_1}{3\Lambda}\psi_C\right] \tag{6.134a}$$

$$E^{(2)} = 3E_2 + \eta_2(E_1 - E_2); \quad \Phi^{(2)} = \left(1 + \frac{\eta_2{}^2}{3\Lambda^2}\right)^{-1/2}\left[\frac{\eta_2}{3\Lambda}\psi_B + \psi_D\right] \tag{6.134b}$$

$$E^{(3)} = 3E_1 + \eta_3(E_1 - E_2); \quad \Phi^{(3)} = \left(1 + \frac{\eta_3{}^2}{3\Lambda^2}\right)^{-1/2}\left[\psi_A + \frac{\eta_3}{3\Lambda}\psi_C\right] \tag{6.134c}$$

$$E^{(4)} = 3E_2 + \eta_4(E_1 - E_2); \quad \Phi^{(4)} = \left(1 + \frac{\eta_4{}^2}{3\Lambda^2}\right)^{-1/2}\left[\frac{\eta_4}{3\Lambda}\psi_B + \psi_D\right] \tag{6.134d}$$

where:

$$\Lambda = \frac{z}{E_1 - E_2}; \quad \eta_1 = [(1-\Lambda)^2 + 3\Lambda^2]^{1/2} - (1-\Lambda);$$

$$\eta_2 = [(1+\Lambda)^2 + 3\Lambda^2]^{1/2} + (1+\Lambda)$$

$$\eta_3 = -[(1-\Lambda)^2 + 3\Lambda^2]^{1/2} - (1-\Lambda); \quad \eta_4 = -[(1+\Lambda)^2 + 3\Lambda^2]^{1/2} + (1+\Lambda)$$

The allowed matrix elements of $X \equiv x_1 + x_2 + x_3$ together with the corresponding energy differences are:

$$X_{12}^2 = \frac{\left(\Lambda\eta_1 + \Lambda\eta_2 + \frac{2}{3}\eta_1\eta_2\right)^2}{\left(\Lambda^2 + \frac{1}{3}\eta_1^2\right)\left(\Lambda^2 + \frac{1}{3}\eta_2^2\right)}\overline{x^2}; \quad \Delta_{12} = (3 + \eta_1 - \eta_2)(E_1 - E_2) \tag{6.135a}$$

$$X_{14}^2 = \frac{\left(\Lambda\eta_1 + \Lambda\eta_4 + \frac{2}{3}\eta_1\eta_4^2\right)}{\left(\Lambda^2 + \frac{1}{3}\eta_1^2\right)\left(\Lambda^2 + \frac{1}{3}\eta_4\right)^2}\overline{x^2}; \quad \Delta_{14} = (3 + \eta_1 - \eta_4)(E_1 - E_2) \tag{6.135b}$$

$$X_{23}^2 = \frac{\left(\Lambda\eta_2 + \Lambda\eta_3 + \frac{2}{3}\eta_2\eta_3\right)^2}{\left(\Lambda^2 + \frac{1}{3}\eta_2^2\right)\left(\Lambda^2 + \frac{1}{3}\eta_3^2\right)}\overline{x^2}; \quad \Delta_{23} = (-3 + \eta_2 - \eta_3)(E_1 - E_2) \tag{6.135c}$$

$$X_{34}^2 = \frac{\left(\Lambda\eta_3 + \Lambda\eta_4 + \frac{2}{3}\eta_3\eta_4\right)^2}{\left(\Lambda^2 + \frac{1}{3}\eta_3^2\right)\left(\Lambda^2 + \frac{1}{3}\eta_4^2\right)}\overline{x^2}; \quad \Delta_{34} = (3 + \eta_3 - \eta_4)(E_1 - E_2) \tag{6.135d}$$

The results of Eqs. (6.135) are plotted in Fig. 6.11. Since the transition probability X_{14}^2 is very small it has not been indicated on these plots, and, correspondingly, Δ_{14} has been neglected. As was to be expected, the inversion frequency has now been split into four components. Of the three we are here considering, it is to be noted that, as in the two particle collision case, the component of decreasing energy is also the component of increasing intensity while the reverse now holds for both the components of increasing energy. A comparison of Figs. 6.10 and 6.11 serves to illustrate a notable difference between the two and three particle cases. The dominant (intensitywise) frequency decreases much more rapidly with Λ in the three molecule case than it does in the two molecule case. This "fact" leads Margenau to an interesting conclusion.

As we have seen, the larger number of interacting particles has led to a greater frequency perturbation. Through Eq. (6.129) the effect of rotation is to decrease this spread. Compensating effects appear to be in the wind, and the results of Section 15 are at least suggested as valid under these conditions.

Under the assumption that the resonance frequency could be approximated by the mean frequency, as given by $h\bar{\nu} = w_I\Delta_I + w_{II}\Delta_{II}$, and that $\Lambda \ll 1$, Margenau obtained for the shifted frequency:

$$\bar{\nu} \doteq \nu_0(1 - \Lambda^2) \tag{6.136a}$$

Since:

$$\Lambda = \frac{2\mu^2}{(E_1 - E_2)R^3} \doteq \frac{2\mu^2 N}{E_1 - E_2} \doteq 0.6\,p$$

where p is the ammonia pressure in atmospheres, $E_1 - E_2 \doteq 0.8\ \mathrm{cm}^{-1}$, $\mu = 1.44 \times 10^{-18}$, and the temperature is taken as 0 °C. Eq. (6.136a) becomes:

$$\bar{\nu} \doteq \nu_0(1 - 0.36\,p^2) \qquad (6.136\,\mathrm{b})$$

This can only be expected to hold at low pressures since (1) it has been obtained under the assumption of a binary collision or a two particle interaction, and (2) the factor Λ has been assumed as small. The "dominant" frequency may also be written approximately for low pressures from the binary theory as:

$$\nu_{\mathrm{II}} = \nu_0(1 - 0.6\,p) \qquad (6.136\,\mathrm{c})$$

and this may be compared with the experimental results.

From Fig. 6.19 the result is $0.44\ \mathrm{cm}^{-1}$ and from the experimental data[73] $0.5\ \mathrm{cm}^{-1}$ for the dominant frequency at one atmosphere. It would appear from this agreement then that this approximation and the theory from which it arose is good up to pressures of this order. As Margenau has demonstrated, it definitely fails for ammonia above this pressure, however.

Thus, in this and the preceding two sections we have considered Margenau's method of approach in successfully accounting for the shift of the inversion line at low pressures through the assumption of binary interactions. We have also noted the much more rapid decrease of the energy of the "dominant" frequency* with Λ for the three particle case than for the two particle case.† The actual failure of the two particle theory for higher pressures stems from an inability on the part of this theory to cause a rapid enough decrease of this frequency with Λ. Thus, it would appear that the many particle interactions are at least pointing in the direction of the required correction for higher pressures.

6.17. Anderson's line broadening theory

What Mizushima has described as an elaborate approach was utilized by Anderson[3] in what we may, with some justification, term an Interruption Broadening consideration. Our consideration of his theory will perhaps be somewhat sketchy in comparison to certain of our other studies.**

Anderson begins by making the assumption of a classical path and taking as his intensity expression:

$$I(\omega) = \mathrm{const}\ \omega^4\ Tr \left\| \varrho_0 \int_{-\infty}^{+\infty} \mathrm{d}t\, \mathrm{e}^{i\omega t} \mu_z(t) \int \mathrm{d}t'\, \mathrm{e}^{i\omega t'} \mu_z(t') \right\| \qquad (6.137)\dagger\dagger$$

Let us suppose we have a matrix relation as follows:

$$\mu(t) = \mathrm{U}^{-1} \mu_0\, \mathrm{U} \qquad (6.138)$$

Then we have carried out a similarity transformation, and a factor concerning this transformation for which we shall have some little use is the invariance

* See Fig. 6.11.

† See Fig. 6.10.

** A coverage and extension of the Anderson theory has been carried out by Tsao and Curnutte[198].

†† See *supra*, Sec. 4.15.

of the trace under such a transformation. In Eq. (6.138) $\mathbf{U}^{-1}$ is the inverse matrix* such that $\mathbf{U}^{-1}\mathbf{U} = 1$. We shall be primarily interested in unitary matrices which are defined by the relation $\mathbf{U}^{-1} = \mathbf{U}\dagger$ where $\mathbf{U}\dagger$ is the adjoint matrix.** If the matrix $\mathbf{U}$ in Eq. (6.134) is unitary then the transformation of Eq. (6.138) is unitary.

It so happens that Eq. (6.138) not only served to illustrate some facts about matrices but also is a relation among the time development matrix (*TDM*), the moment matrix $\boldsymbol{\mu}(t)$, and the matrix, $\boldsymbol{\mu}_0$, of $\boldsymbol{\mu}(t)$ at $t = 0$. Although we shall not dwell on this relation one may surely intuitively grasp the manner in which the *TDM* shapes the behavior with time of the moment matrix from the value of that matrix at time zero. §

A modified *TDM* may now be defined such that:

$$\mathbf{T} = \mathbf{U}_0^{-1}\,\mathbf{U} \leftrightarrow \mathbf{U} = \mathbf{U}_0\,\mathbf{T} \qquad (6.139\text{a})$$

where:

$$\mathbf{U}_0 = \exp\left[-\frac{i}{\hbar}\,\mathbf{H}_0\,\mathbf{T}\right] \qquad (6.139\text{b})$$

When Eqs. (6.139) are substituted into $\left(i\hbar\dfrac{\partial}{\partial t} - H\right) = 0$ the result is:

$$i\hbar(\dot{\mathbf{U}}_0\mathbf{T} + \mathbf{U}_0\dot{\mathbf{T}}) = (\mathbf{H}_0 + \mathbf{H}_1)\,\mathbf{U}_0\,\mathbf{T} \leftrightarrow i\hbar\,\mathbf{U}_0\,\dot{\mathbf{T}} = \mathbf{H}_1\,\mathbf{U}_0\mathbf{T} \leftrightarrow i\hbar\,\dot{\mathbf{T}} = (\mathbf{U}_0^{-1}\mathbf{H}_1\,\mathbf{U}_0)\,\mathbf{T} \qquad (6.140\text{a})$$

and when they are substituted into Eq. (6.138) one finds:

$$\boldsymbol{\mu}(t) = \mathbf{T}^{-1}\exp\left(\frac{i}{\hbar}\,H_0\,t\right)\boldsymbol{\mu}_0\exp\left(-\frac{i}{\hbar}\,H_0\,t\right)\mathbf{T} \qquad (6.140\text{b})$$

a typical matrix element of which is:

$$\left.\begin{aligned}(m\,|\mu(t)|\,n) &= \sum_{ijkl}(m\,|T^{-1}|\,i)\left(i\left|\exp\left(\frac{i}{\hbar}\,H_0\,t\right)\right|k\right)(k\,|\mu_0|\,j)\\ &\quad\cdot\left(j\left|\exp\left(-\frac{i}{\hbar}\,H_0\,t\right)\right|l\right)(l\,|T|\,n)\\ &= \sum_{kl}(m\,|T^{-1}|\,k)\exp(-i\,\omega_{kl}t)\,(k\,|\mu_0|\,l)\,(l\,|T|\,n)\end{aligned}\right\} \qquad (6.140\text{c})\ \S\S$$

where all but the diagonal elements of U_0 have disappeared due to the diagonal quality of this matrix and where $\hbar\,\omega_{kl} = (E_l - E_k)$.

* We shall cease to use the symbol || || as designating a matrix and simply use boldfaced type from this point on.

** We here refer to the complex conjugate transpose ($a_{ij} = \bar{a}_{ji}$) which is sometimes termed the "associate matrix".

§ See *supra*, Sec. 6.15.

§§ If a_{ij} is an element of $\mathbf{A}$, etc., we may recall the labelling rule for the multiplication of matrices: $(\mathbf{ABCD})_{nm} = \sum_{jkl} a_{nj}\,b_{jk}\,c_{kl}\,d_{lm}$, and so on for the element in the product matrix for larger products.

We directly substitute Eq. (6.140c) into Eq. (6.137) as a typical matrix element of $\mu(t)$. What we desire is the element of the product matrix of $\mu(t)$ and the density matrix. We take the proper element and then sum over the diagonal elements of this product matrix in order to obtain the desired trace. Finally then:

$$\left.\begin{aligned} I(\omega) = \text{const}\cdot\omega^4 \sum_{abcdefg} (a|\varrho_0|b) \int_{-\infty}^{+\infty}\int_{-\infty}^{+\infty} dt'\, e^{iw(t-t')} \\ \times (b|T^{-1}(t)|c)(c|\mu_0|d)\exp(i\,\omega_{cd}\,t) \\ \times (d|T(t)|e)(e|T^{-1}(t')|f)(f|\mu_0|g)\exp(-i\,\omega_{fg}t')(g|T(t')|a) \end{aligned}\right\} \quad (6.141)$$

We might stop to remember at this point that all the operations which have been carried out so far and all those which will follow are simply for the purpose of transforming Eq. (6.137) into some form which will tell us explicitly the shift, the shape and the width of the spectrum line which we are to expect.

We now modify Eq. (6.141) by substituting a relation of the form Eq. (6.138) for the density matrix at $t = 0$, $\boldsymbol{\varrho}_0 = \mathbf{T}^{-1}\boldsymbol{\varrho}\,\mathbf{T}$ and then, substituting the new time development matrix,

$$\mathbf{T}(t \to t') = \mathbf{T}(t')\,\mathbf{T}^{-1}(t)$$

we obtain a function of the form $Tr(\mathbf{T}^{-1}\,\mathbf{Z}\,\mathbf{T})$ which since we know the trace to be invariant under a similarity transformation, is equivalent to $Tr\,\mathbf{Z}$. Thus, there finally results:

$$\left.\begin{aligned} I(\omega) = \text{const}\cdot\omega^4 \int_{-\infty}^{+\infty} dt \int_{-\infty}^{+\infty} dt'\, e^{iw(t-t')} \sum_{abcde} \\ \times (a|\varrho(t)|b)(b|\mu_0|c)\exp(-i\,\omega_{bc}t)\cdot(e|T^{-1}(t\to t')|d) \\ \times (d|\mu_0|e)\exp(-i\,\omega_{de}t')(e|T(t\to t')|a) \end{aligned}\right\} \quad (6.142)$$

By using the substitution $t' = t + \tau$ Eq. (6.142) may be re-expressed in "Correlation Function Form"† as:

$$I(\omega) = \omega^4\,\text{const} \sum_{de} \int_{-\infty}^{+\infty} d\tau \exp[i(\omega - \omega_{de})\,\tau]\,\varphi_{de}(\tau) \quad (6.143\text{a})$$

$$\left.\begin{aligned} \varphi_{de}(\tau) = \int_{-\infty}^{+\infty} dt \sum_{abc} \exp[-i(\omega_{bc} + \omega_{de})\,t]\,(a|\varrho(t)|b)(b|\mu_0|c) \\ \times (c|T^{-1}(t)|d)(d|\mu_0|e)(e|T(t)|a) \end{aligned}\right\} \quad (6.143\text{b})$$

Up to this point no restrictions on the theory in the form of modifying approximations have been made, but at this point Anderson found it necessary to introduce the condition which changes the subsequent treatment into an Interruption type treatment. The assumption to be made is that the time between collisions is much greater than the time of collision. This assumption when applied to a consideration of the matrix elements of $(\mathbf{U}_0^{-1}\,\mathbf{H}_1\,\mathbf{U}_0)$ leads one to

† See *supra*, Chap. 4.

the conclusion that these matrix elements and hence Eq. (6.143b) disappear unless:

$$\omega_{bc} + \omega_{de} = 0 \qquad (6.144)$$

and the two frequencies appear to be resonating.

Eq. (6.144), of course, has the advantage of greatly simplifying Eq. (6.143b) by removing the exponentials in question from it. Anderson noted that Eq. (6.143b) had the look of a time average. Let us now investigate the consequences of this conclusion.

To begin with, we specify the density matrix by a typical matrix element as:

$$(n\,|\varrho_B|\,m)\,\delta_{nm}\,g_n \exp(-E_n/kT) / \sum_n g_n \exp(-E_n/kT) \qquad (6.145)$$

This would appear to be the reasonable time to introduce the degenerate indices as well as the explicit fact that there are two molecules present both of whose states must be considered. Although the electric dipole moment could refer also to molecule 2 (the perturber), we suppose that it does not so that this operator is diagonal in the states of the perturber. We then (1) replace the integral in Eq. (6.145b) by an average, (2) replace $\varrho(t)$ by its specific value, (3) introduce degenerate indices, and (4) recognize the existence of perturber and emitter to obtain from Eqs. (6.145):

$$I = \sum_{if} \int_{-\infty}^{+\infty} d\tau \exp[i(\omega_{if} - \omega)\tau]\,\frac{1}{g_i}\,(J_i\,|\varrho_B|\,J_i)\,\varphi_{if}(\tau) \qquad (6.146\text{a})$$

$$\left.\begin{aligned} \varphi_{if}(\tau) = \sum_{J_2} \frac{1}{g_2} (J_2\,|\varrho_B|\,J_2) \sum_{J_2'} \sum_{M_i M_i' M_f M_f' M_2 M_2'} (J_i M_i\,|\mu_z|\,J_f M_f) \\ < (J_f M_f J_2 M_2\,|T^{-1}(\tau)|\,J_f M_f' J_2' M_2')\,(J_f M_f'\,|\mu_z|\,J_i M_i') \\ \times (J_i M_i' J_2' M_2'\,|T(\tau)|\,J_i M_i J_2 M_2) >_{\text{over}\,t} \end{aligned}\right\} \qquad (6.146\text{b})$$

It is certainly reasonable to suppose that the collisions which the emitter undergoes occur isotropically insofar as collision direction is concerned. As a consequence, the average of $\mathbf{T}^{-1}\mu_z\mathbf{T}$ will be independent of the actual placement of the z-axis. We are thus free to conclude that for a rotation of the reference frame specified by:

$$\mu_z = \alpha_1 \mu_x' + \alpha_2 \mu_y' + \alpha_3 \mu_z' \qquad (6.147\text{a})$$

the average transforms as:

$$\langle \mathbf{T}^{-1}\mu_z\mathbf{T}\rangle = \alpha_1 \langle \mathbf{T}^{-1}\mu_x'\mathbf{T}\rangle + \alpha_2 \langle \mathbf{T}^{-1}\mu_y'\mathbf{T}\rangle + \alpha_3 \langle \mathbf{T}^{-1}\mu_z\mathbf{T}\rangle \qquad (6.147\text{b})$$

so that this average transforms as a vector. We shall not demonstrate it, but it is known[224] that the matrix elements of this averaged matrix, due to its vector component behavior, are determined by the transformation property so that we may write:

$$(J_f M_f J_2 M_2\,|\langle T^{-1}\mu_z T\rangle|\,J_i M_i J_2' M_2') = (J_f M_f\,|\mu_z|\,J_i M_i)\,\delta_{J_2 J_2'}\,\delta_{M_2 M_2'}\,F(\tau) \qquad (6.148)$$

The substitution of Eq. (6.148) into Eq. (6.146b) results in:

$$\varphi(\tau) = \sum_{M_i M_f} |(J_i M_i | \mu_z | J_f M_f)|^2 F(\tau) = Tr[\mu_z^{if} \mu_z^{fi}] F(\tau) \qquad (6.149)$$

since the square arises by matrix multiplication definition; the summation over M_2 cancels the g_2 in the numerator of Eq. (6.146b), and:

$$\sum_{J_2} (J_2 | \varrho | J_2) = 1$$

The next step in the solution amounts to determining $F(\tau)$ from a differential equation for this quantity which may be obtained. Then an expression for the shift and shape may be obtained.

We now let:

$$\mathbf{T}(\tau \to \tau + \mathrm{d}\tau) = \mathbf{T}(\tau + \mathrm{d}\tau)\, \mathbf{T}^{-1}(\tau) \qquad (6.150)$$

and note that, since the collisions are random, the events occurring in the time intervals τ and $\mathrm{d}\tau$ are independent and consequently the TDM's for these two periods commute. We make the substitution as indicated by Eq. (6.150), and then utilize Eq. (6.149) to carry through a series of steps quite similar to those which we have already carried out to finally obtain:

$$\varphi(\tau + \mathrm{d}\tau) = \sum_{J_2} \frac{(J_2 | \varrho | J_2)}{g_{J_2}} \sum_{J_2'} Tr \langle \mu_z^{if} [T^{-1}]^{f2, ff'2'} \mu_z^{fi} [T]^{ii'2', i2} \rangle F(\tau) \qquad (6.151)$$

where:

$$[T]^{f2, ff'2'} = (J_f M_f J_2 M_2 | T | J_f M_f' J_2' M_2') \qquad (6.152)$$

We now suppose that different types (this means collisions of different optical collision diameter and direction) of collisions are designated by different values of σ. Now the probability that a collision will occur during the time interval $\mathrm{d}\tau$ and lie in the type range $\mathrm{d}\sigma$ is given:

$$p(\mathrm{d}\sigma \text{ in } \mathrm{d}\tau) = N v \,\mathrm{d}\sigma\, \mathrm{d}\tau \qquad (6.153)$$

Next $\mathrm{d}\varphi(\tau)$ is average dover these collisions as:

$$\mathrm{d}\varphi(\tau) = \mathrm{d}\tau N v \int \mathrm{d}\sigma [\varphi(\tau + \mathrm{d}\tau) - \varphi(\tau)] \qquad (6.154)$$

A substitution may be made for $\mathrm{d}\varphi(\tau)$ from Eq. (6.149) for $\varphi(\tau)$, and from Eq. (6.151a) for $\varphi(\tau + \mathrm{d}\tau)$. These substitutions result in the differential equation for $F(\tau)$:

$$\frac{\mathrm{d}}{\mathrm{d}\tau} F(\tau) = -N v \sigma F(\tau) \qquad (6.155\mathrm{a})$$

where:

$$\sigma = \int \mathrm{d}\sigma \left[1 - \frac{\sum_{J_2} \frac{(J_2 | \varrho | J_2)}{g_{J_2}} \sum_{J_2'} Tr[\mu_z^{if} \{T^{-1}(\mathrm{d}\sigma)\}^{f2, ff'2'} \mu_z^{fi} \{T(\mathrm{d}\sigma)\}^{ii'2, i2}]}{Tr[\mu_z^{if} \mu_z^{fi}]} \right] \qquad (6.155\mathrm{b})$$

or, with:

$$1 = \sum_{J_2} (J_2 |\varrho| J_2): \quad \sigma = \sum_{J_2} (J_2 |\varrho| J_2) \, \sigma_{J_2} \tag{6.155c}$$

where:

$$\sigma_{J_2} = \int \mathrm{d}\sigma \left[1 - \frac{\sum_{J_2'} Tr\,[\mu_z{}^{if} \{T^{-1}(\mathrm{d}\sigma)\}^{f2,ff'2'} \mu_z{}^{fi} \, T(\mathrm{d}\sigma)^{ii'2,i2}}{g_{J_2} \, Tr\,[\mu_z{}^{if} \mu_z{}^{fi}]} \right] \tag{6.155d}$$

The solution of Eq. (6.155a) is:

$$F(\tau) = \mathrm{e}^{-Nv\sigma\tau} \tag{6.156a}$$

where:

$$\sigma = \sigma_r + i\sigma_i \tag{6.156b}$$

In order to obtain the intensity distribution in the broadened spectral line, we substitute Eqs. (6.156a) and (6.149) into Eq. (6.146a). Integration subsequently yields:

$$I(\omega) = \sum_{J_i J_f} \frac{(J_i |\varrho| J_i)}{g_{J_i}} \sum_{M_i M_f} |(J_i M_i |\mu_z| J_f M_f)|^2 \frac{Nv\sigma_r}{(\omega - \omega_{fi} + Nv\sigma_i)^2 + (Nv\sigma_r)^2} \tag{6.157}$$

From Eq. (6.157) we may immediately obtain the line half-width and shift as:

$$\delta = \frac{Nv}{\pi c} \sigma_r \ \mathrm{cm}^{-1} \tag{6.158a}$$

$$D = -\frac{Nv}{2\pi c} \sigma_i \ \mathrm{cm}^{-1} \tag{6.158b}$$

It would now appear that were σ in a form which was amenable to calculation the problem would be essentially solved, and, indeed, this remains the major step to solution.

By a series of manipulations involving vector addition coefficients and certain relations among them* the traces in Eq. (6.155d) may be evaluated. Choosing the axis of quantization (simply the axis along which the spatial degeneracy parameter M has meaning) so that it coincides with the z-axis allows the expression of the trace in the denominator of Eq. (6.155d) in terms of these vector addition coefficients and the subsequent evaluation of this trace.

In evaluating the trace in the numerator a few preliminary considerations are requisite. This trace contains the matrix T — and its inverse, of course — so that for a given optical collision diameter ϱ this trace must needs be averaged over all directions.† Now the trace is independent of the axis of quantization, however, so that the axis of quantization may be taken along the optical collision diameter and the trace subsequently averaged over all directions of this axis

* These coefficients are described in detail in Chapter III of Reference 224 and in Chap. XVII of Reference 241. They become of specific interest only in particular cases.

† Jablonski would undoubtedly object to this inferred (but unstated) assumption of rectilinear motion.

which we have specified as along the z-axis. The averaged trace in question may be calculated with the result that Eq. (6.155d) takes the form:

$$\sigma_{J_2} = \int_0^\infty 2\pi\varrho\, d\varrho\, S(\varrho) \tag{6.159a}$$

$$\left.\begin{aligned} S(\varrho) = 1 & \\ - \sum_{\substack{M_i M_i' M_f \\ M_f' M_2 M_2' \\ M}} \sum_{J_2'} & \frac{(J_f 1 M_f M \,|J_f 1 J_i M_i)\,(J_f 1 M_f' M\,|J_f 1 J_i M_i')}{(2J_i + 1)\,(2J_2 + 1)} \\ & \times (J_f M_f J_2 M_2\,|T^{-1}(\varrho)|\,J_f M_f' J_2' M_2') \\ & \times (J_i M_i' J_2' M_2')\,|T(\varrho)|\,J_i M_i J_2 M_2) \end{aligned}\right\} \tag{6.159b)**}$$

where the first two symbols on the right of Eq. (6.159b) are vector addition coefficients. The differential $d\sigma$, which indicates a certain range of optical collision diameter, and optical collision diameter direction has been replaced by $2\pi\varrho\,d\varrho$ since $S(\varrho)$ has already been averaged over various collisions directions.

The next step in carrying out a calculation of σ is the computation of **T**, and a successive approximation procedure has been used where, to begin with:

$$\mathbf{T} = \mathbf{T}_0 + \mathbf{T}_1 + \mathbf{T}_2 \ldots; \qquad \mathbf{T}_0 = 1 \tag{6.160a}$$

so that, from Eq. (6.138a):

$$\mathbf{T}_1(t) = -\frac{i}{\hbar} \int_{-\infty}^{t} (\mathbf{U}_0^{-1}\, \mathbf{H}_c(t)\, \mathbf{U}_0)\, dt \tag{6.160b}$$

$$\mathbf{T}_2(t) = -\frac{1}{\hbar^2} \int_{-\infty}^{t} (\mathbf{U}_0^{-1}\, \mathbf{H}_c(t')\, \mathbf{U}_0)\, dt' \int_{-\infty}^{t'} (\mathbf{U}_0^{-1}\, \mathbf{H}_c(t'')\, \mathbf{U}_0)\, dt'' \tag{6.160c}$$

If H_c commutes with itself at different times then

$$\mathbf{T}(t) = \exp\left[-\frac{i}{\hbar} \int_{-\infty}^{t} (\mathbf{U}_0^{-1}\, \mathbf{H}_c(t)\, \mathbf{U}_0)\, dt\right] \tag{6.161}$$

and Eqs. (6.160) would simply be the power series expansion for the exponential in Eq. (6.161). It so happens that $\mathbf{H}_c$ does not so commute, but Anderson has carried out calculations which tend to indicate that the assumption of this commutation proves a good approximation. As a consequence then, if we take only the first three terms in the expansion of Eq. (6.161), there results for **T**:

$$\mathbf{T}(\varrho) \doteq \mathbf{T}_0 + \mathbf{T}_1 + \mathbf{T}_2$$

$$\mathbf{T}_0 = 1; \quad \mathbf{T}_1 = -i\mathbf{P}; \quad \mathbf{T}_2 = \frac{1}{2}\mathbf{P}^2 \tag{6.162a}$$

where:

$$\mathbf{P} = \frac{1}{\hbar} \int_{-\infty}^{+\infty} (\mathbf{U}_0^{-1}\, \mathbf{H}_c\, \mathbf{U}_0)\, dt \tag{6.162b}$$

** For example: $(J_i M_2\,|\mu_z|\,J_f M_f) = (J_f 1 M_f O|J_f 1 J_i M_i)\,F'$, where F is independent of M, so that $Tr[\mu_z^{if}\,\mu_z^{if}] = F'^2 \sum_{M_i M_f} (J_f 1 M_f O\,|J_f 1 J_i M_i)(J_i 1 M_i O|\,J_i 1 J_f M_f)$.

An approximation type solution for $S(\varrho)$ may now be carried out:

$$S = S_0 + S_1 + S_2 + \cdots \tag{6.163a}$$

where the successive approximations to S arise from successive powers of the **P** matrix (Eq. (4.162)) in Eq. (6.159b). $\mathbf{T}_0 = \mathbf{P}^0$ yields:

$$S_0(\varrho) = 0 \tag{6.163b}$$

In the first-order approximation to $S(\varrho)$ $\mathbf{T}_0^{-1}$, $\mathbf{T}_1$ and $\mathbf{T}_0$, $\mathbf{T}_1^{-1}$ both contribute to the result:

$$\left.\begin{aligned} S_1(\varrho) = i \Bigg[\sum_{M_i M_2} & \frac{(J_i M_i J_2 M_2 \,|P|\, J_i M_i J_2 M_2)}{(2J_i + 1)(2J_2 + 1)} \\ - \sum_{M_f M_2} & \frac{(J_f M_f J_2 M_2 \,|P|\, J_f M_f J_2 M_2)}{(2J_f + 1)(2J_2 + 1)} \Bigg] \end{aligned}\right\} \tag{6.163c}$$

For the second order approximation the possibilities $\mathbf{T}_0^{-1}$, $\mathbf{T}_2$ and $\mathbf{T}_0$, $\mathbf{T}_2^{-1}$ yield:

$$\left.\begin{aligned} S_2(\varrho)_{\text{outer}} = \frac{1}{2} \Bigg[\sum_{M_i M_2} & \frac{(J_i M_i J_2 M_2 \,|P^2|\, J_i M_i J_2 M_2)}{(2J_i + 1)(2J_2 + 1)} \\ + \sum_{M_f M_2} & \frac{(J_f M_f J_2 M_2 \,|P^2|\, J_f M_f J_2 M_2)}{[2J_f + 1)(2J_2 + 1)} \Bigg] \end{aligned}\right\} \tag{6.163d}$$

while $\mathbf{T}_1^{-1}$, $\mathbf{T}_1$ result in:

$$\begin{aligned} S_2(\varrho)_{\text{middle}} = \sum_{\substack{M_i M_i' M_f \\ M_f' M_2 M_2' \\ M}} \sum_{J_2'} & \frac{(J_f 1 M_f M \,|J_f 1 J_2 M_2)(J_f 1 M_f' M \,|\, J_f 1 J_i M_i')}{(2J_i + 1)(2J_2 + 1)} \\ & \times (J_f M_f J_2 M_2 \,|P|\, J_f M_f' J_2' M_2')(J_i M_i' J_2' M_2' \,|P|\, J_i M_i J_2 M_2) \end{aligned} \tag{6.163e}$$

so that:

$$S_2(\varrho) = S_2(\varrho)_{\text{outer}} + S_2(\varrho)_{\text{middle}} \tag{6.163f}$$

In Eqs. (6.163) we note that a summation over J_2' not occurring in Anderson's Eqs. (54) is present. This is due to the fact that Anderson had assumed T as diagonal in J_2. This is certainly true if (a) T causes no transitions among the nondegenerate states of the perturber, or (b) changes in the quantum numbers of the perturber may only occur simultaneously with changes in that of the emitter. This latter corresponds to the resonance effect, of course. Under these assumptions we may drop this summation.

Let us also notice that $S_1(\varrho)$ contributes only toward line shift while the $S_2(\varrho)$ contributes only to line breadth, not that these facts should prove startling.

The theory in general form is now complete and stands ready to be applied, but before we go into Anderson's applications of it to actual cases, let us consider some necessary restrictions to it, a summarization of its development, and the approximations contained within it.

As has been the case with the varied and sundry theories which we have considered, the theory breaks down for small values of ϱ. This transpires in the following manner. H_c in Eq. (6.162b) depends — for the dipole-dipole interaction which is the only one which Anderson considered* — on r^{-3} in first order and r^{-6} in second order. This means that for sufficiently small r, P will become as large as you like. The expansion of S in terms of P, Eq. (6.163a), is only valid, however, for small P. Thus does the theory, through S, break down for small ϱ. Anderson treats this in a manner which we may now detail.

To begin with Anderson assumed that for very small values of ϱ the collisions are so strong that they terminate the radiation by causing the molecule to proceed to some different nondegenerate state or when the molecule remains in the same state they result in an arbitrary phase shift that averages to zero. In either case then Eq. (6.159b) contains only the first term unity, so that while $S_2(\varrho)$ is given by Eq. (6.163f) for values of ϱ greater than ϱ_1 where $S_2(\varrho_1) = 1$, it is simply given by one for lesser values of ϱ.

Anderson also tried two other approximations for $S(\varrho)$, namely:

$$S_{\#1}(\varrho) = 1 - \cos(2 S_2(\varrho))^{1/2}$$

and

$$S_{\#3}(\varrho) = 1 - \exp(-2 S_2(\varrho))$$

These three possibilities for S_2 are illustrated in Fig. 6.12, (p. 274) and we might note here that Anderson found the best agreement with experiment to arise from $S_{\#2}$.

Let us obtain Anderson's "region of resonance parameter" k. Now if $H_c = Kr^{-3}$ or Kr^{-6} and (the rectilinear assumption) $r^2 = \varrho^2 + v^2t^2$

$$\left.\begin{aligned}(a\,|P|\,b) &= \int_{-\infty}^{+\infty} \mathrm{d}t \exp(i\,\omega_{ab}\,t)\,(a\,|H_1(t)|\,b) \\ &= K \int \mathrm{d}t \exp(i\,\omega_{ab}\,t)\,(\varrho^2 + v^2 t^2)^{-3/2\,\mathrm{or}\,-3} \\ &= \frac{k}{\varrho^2\,v \text{ or } \varrho^5\,v} \int \mathrm{d}x\, \mathrm{e}^{ikx} (H\,x^2)^{-3 \text{ or } -3/2}\end{aligned}\right\} \qquad (6.164\text{a})$$

where:

$$x = \frac{v\,t}{\varrho}; \qquad k = \frac{\varrho\,\omega_{ab}}{v} \qquad (6.164\text{b})$$

The parameter k tells us whether the molecules have undergone fast collisions and/or whether the ω_{ab} perhaps refers to degenerate levels. Either of the two integrals appearing in Eq. (6.164b) has about the same value for values of k up to about one as it has for k zero. Within this region then $(0 < k < 1)$ transitions will occur between the states a and b as though these two states were resonant (or degenerate). Anderson explains this by an appeal to the uncertainty principle. For $k < 1$ the velocity may be supposed sufficiently high that, with Δt the time of collision, there will be an uncertainty in energy (from $\Delta E \Delta t > h$) such that ω_{ab} acts as though it were negligible; this in turn means that the two states in question act as if they were resonant.

* Tsao and Curnutte[198] have also considered dipole–quadrupole and quadrupole–quadrupole interaction within the framework of the Anderson theory.

Let us now review the major facets of the Anderson theory. We began with the expression Eq. (6.137) for the intensity distribution in the broadened line. We then obtained from the time development matrix a modified time development matrix, which proves mathematically more convenient, as given by Eq. (6.139a). A few straightforward, mathematical manipulations were carried out to obtain Eq. (6.143b) at which point one of the oft used approximations of the Interruption theory was introduced, namely, (A) THE DURATION OF THE COLLISION IS VERY SHORT COMPARED TO THE INTERVAL BETWEEN COLLISIONS. By this time the intensity integral had been written in Correlation Function Form. The Maxwell–Boltzmann form of the density matrix was introduced and the approximations of (B) BINARY COLLISIONS was introduced. This led to Eqs. (6.146). The fact that collisions may certainly be expected to occur isotropically allows us to obtain relations of the form Eq. (6.148) which yield further simplifications. Additional manipulations yield Eqs. (6.155) and (6.156) for the quantity $F(\tau)$, which, when obtained, may be utilized in Eq. (6.146a) to obtain Eqs. (6.157) and (6.158) for the line shape, width, and shift. We may note here the decrease of the intensity with the inverse square of the frequency separation from line center which is certainly not generally correct.

In evaluating σ an expansion for T was utilized. This required the approximation of a (C) MINIMUM DISTANCE OF APPROACH OF THE COLLIDING PARTICLES.

At any rate, σ was obtained in terms of $S(\varrho)$, an expansion whose third term, S_2, as given by Eq. (6.163) contributes only to the line width and whose second term, S_1, as given by Eqs. (6.163), contributes only to the line shift. Thus, when one has computed P from Eq. (6.162b) and subsequently S_1 and S_2 from Eqs. (6.163) the shift and shape may be computed from Eq. (6.159a) and (6.155c). Also included in the development was the approximation of (D) A CLASSICAL PATH.

6.18. Some applications of Anderson's theory

We will consider the self broadening of the inversion line which we have already considered according to Margenau's statistical theory method. This appears to be a very popular transition. To begin with $S_2(\varrho)_{\text{middle}}$, the "'difficult' term" vanishes in this problem.

We consider the inversion doubled levels of the symmetric top under the interaction potential:

$$H = \left[\mu_1 \cdot \mu_2 - 3\,\frac{(\mu_1 \cdot \mathbf{r})\,(\mu_2 \cdot \mathbf{r})}{r^2}\right] r^{-3} \qquad (6.165)^*$$

the familiar dipole–dipole interaction. The polar axis of coordinates is taken as the optical collision diameter to which the polar coordinates ϑ_1, ϑ_2, φ_1 and φ_2 are referred. We designate $\tan^{-1}((vt)/(\varrho))$ as the angle ψ. Eq. (6.165) may now be expressed in terms of these angular coordinates, and certain of the matrix elements of the symmetric top provide the matrix elements of the resulting equation.

* Here the bold face type refers to vectors *not* matrices.

There are now essentially two types of interaction matrix elements involved in the expression for P, Eq. (6.162b), namely, (1) the first order Stark effect where collisions between molecules having the same J values are considered and (2) the case of rotational resonance where the J values for the colliding molecules differ by $\pm$ 1.

Now let us evaluate k, the resonance region parameter. To begin with we suppose that both molecules are either in the plus inversion state or the minus. Then for both of them to undergo a transition to the other inversion state, an energy change of about 1.6 cm^{-1} is called for. The mean velocity here is about 8×10^4 so that according to Eq. (6.164b)

$$k = \frac{\varrho}{v} = \frac{\varrho(A) \times 10^{-8} \times 3 \times 1.6 \times 10^{18} \times 2}{8 \times 104} = \varrho(A) \times 0.035$$

of which we shall subsequently make use.

Next we wish to evaluate P as given by Eq. (6.162b) or (6.164a). If one molecule is in a plus inversion state and the other in a minus we can have true resonance and neglect the exponential time factor. If the optical collision diameter is less than $15A$ according to the above k calculation the pseudo resonance situation will arise, and we may still neglect the exponential factor. As a consequence of this situation, H_c may be integrated before the matrix elements of it are taken to obtain O. If we let:

$$r^2 = \varrho^2 + v^2 t^2 \text{ so that: } \cos\psi = \varrho/r; \quad \sin\psi = v t/r$$

we may obtain:

$$\int_{-\infty}^{+\infty} \frac{H_1 \, dt}{\hbar} = \frac{2\mu^2}{\varrho^2 v \hbar} [-\cos\vartheta_1 \cos\vartheta_2 - \sin\vartheta_1 \sin\vartheta_2 \sin\varphi_1 \sin\varphi_2] \qquad (6.166)$$

The S_2 sums may now be computed according to Eq. (6.168) with the result:

$$S_2(J_2 K_2) = \frac{8}{9} \frac{\mu^4}{\varrho^4 v^2 \hbar^2} \frac{K_1^2 K_2^2}{J_1(J_1+1) J_2(J_2+1)} \qquad (6.167a)$$

for the case $J_1 = J_2$, and:

$$S_2(J_1-1, K_2) = \frac{8}{9} \frac{\mu^4}{\varrho^4 v^2 \hbar^2} \frac{(J_1^2 - K_1^2)(J_2^2 - K_2^2)}{J_1^2(2J_1+1)(2J_1-1)} \qquad (6.167b)$$

$$S_2(J_1+1, K_2) = \frac{8}{9} \frac{\mu^4}{\varrho^4 v^2 \hbar^2} \frac{[(J_1+1)^2 - K_1^2][(J_1+1)^2 - K_2^2]}{(J_1+1)^2(2J_1+1)(2J_1+3)} \qquad (6.167c)$$

for the case $J_2 = J_1 \pm 1$.

We are evaluating Eq. (6.163d) which we now consider. Now neither in this consideration of ammonia nor in subsequent considerations of HCN and HCl does Anderson consider the Alignment Forces* since, according to this author, the sums required become difficult for the second-order alignment forces.

If the abbreviation:

$$S(J_2, K_2) = A^2/\varrho^4 \qquad (6.168)$$

* See *supra*, Sec. 3.

is used, there results for the three approximate forms of S:

$$\#1.\quad \sigma_1(J_2, K_2) = \varrho_3(J_2, K_2)\, 2\pi\, (1.11 A) \tag{6.169a}$$

$$\#2.\quad \sigma_2(J_2, K_2) = \varrho_3(J_2, K_2)\, 2\pi A \tag{6.169b}$$

$$\#3.\quad \sigma_3(J_2, K_2) = \varrho_3(J_2, K_2)\, 2\pi\, (0.885 A) \tag{6.169c}$$

in the evaluation of which from Eq. (6.169a) the method of Jensen† was used.

Finally one obtains:

$$\sigma = \sum_{J_2 K_2} \sigma\,(J_2\, K_2) \tag{6.170a}$$

$$\delta = \frac{n\, v}{\pi\, c}\, \sigma \text{ cm}^{-1} \tag{6.170b}$$

from Eqs. (6.155c) and (6.158a) respectively.

Anderson presented two types of comparison with experiment, both of which are rather impressive in this case. The first type is a comparison of the absolute width of the line as measured and as computed with this theory. The line used was the inversion line having rotational quantum numbers $J_1 = 3$ and $K_1 = 3$ for which there should be negligible rotational resonance so that Eqs. (6.167b) and (6.167c) may be neglected. Under these conditions the results of Table 6.5 are obtained where the experimental data is that of Bleaney and Penrose.

TABLE 6.5

A COMPARISON OF AN ABSOLUTE WIDTH BY THE ANDERSON THEORY WITH EXPERIMENT

(After Anderson[3])

	Theory			Experiment
Approx.	#1	#2	#3	
$\frac{1}{2}\delta$	0.86	0.77	0.68 cm^{-1} atm^{-1}	0.74

Finally a comparison of the relative line breadths with experiment[13, 14] is made in which the theoretical value for $J = 33$ is made to agree with the experimental for normalization purposes.

Anderson also considered the self broadening of HCN and HCl and the foreign gas broadening by several molecules. The HCN results are given in Fig. 6.18 in comparison with those earlier results of Lindholm and Foley. He reported his agreement as fair in the case of HCl which to the author in question meant 30%. Insofar as his foreign gas broadening results are concerned we shall let the author describe them*. "We do not present this evaluation because the results, in general, do not have any relation to experimental results."[3]

† See *supra*, Chap. 2.

* We recall that here Anderson has utilized only the dipole–dipole interaction. In a later consideration of the broadening of ammonia by foreign gases[198] he included also the interaction between the quadrupole moment of ammonia and the induced dipole moment of the broadening atoms with a resulting agreement which was quite good. Many later authors have, of course, applied this Anderson theory to experimental comparison.

6.19. A Maxwell–Boltzmann distribution of dipole moments

First let us recall the Lorentz assumption to the effect that the electric dipole moment and velocity had a mean value of zero following a collision.* Next we recall touching briefly on the assignment of a Maxwell–Boltzmann distribution of these quantities in Section 1.9. The matter has not been mentioned since, primarily because the distribution has an appreciable effect on the line shape only in the low frequency or microwave region of the spectrum. As a consequence it is chiefly a matter of molecular broadening, a matter to which we now turn our attention.

Van Vleck and Weisskopf defined two types of collisions, strong and, yes, weak ones. Following a strong collision the molecule retains no "'hangover' or memory" of its precollision orientation. Adiabaticity is also assumed here, and in this connection, the adiabatic character assures us that the electric field of the light is a constant during the very short time of the collision, i. e.:

$$E \cos(\omega t) = E \cos(\omega t_0)$$

The difference between this and the Lindholm theory is apparent. Weak collisions, on the other hand, leave the molecule with a hangover. The assumption of strong collision yielded Eq. (1.86) which we write down:

$$\alpha = \frac{\omega}{c} \frac{4\pi N \mu^2}{3kT} \frac{\omega\tau}{1+\omega^2\tau^2} \tag{1.86}$$

This is Debye's equation for the linear absorption coefficient which we obtained in Chapter 1 after the fashion of Van Vleck and Weisskopf. Let us recall that ω appearing in Eq. (1.86) relates only to the frequency of the incident radiation, and that nowhere does any natural frequency of the electronic vibration (in the classical sense) appear. Thus, one might consider Eq. (1.86) to broaden a line of zero frequency through a collision mechanism. In order to determine the quantum mechanical analog of Eq. (1.86) we may recall that, instead of integrating over $\cos \omega t$ as in Eq. (1.86), we sum over the various quantum states. We obtain the same result, however, since here also $\langle \cos^2 \vartheta \rangle = 1/2$. Thus, we need only consider Eq. (1.86).

From the Lorentz theory we obtained in Chapter 1:

$$\alpha = \frac{2\pi N e^2}{mc} \left(\frac{\omega}{\omega_0}\right) \left[\frac{1/\tau}{(\omega-\omega_0)^2+(1/\tau)^2} - \frac{1/\tau}{(\omega+\omega_0)^2+(1/\tau)^2}\right] \tag{1.78a}$$

The equations,

$$s = \frac{4\pi^2}{3} \frac{e^2}{\hbar c} \omega_{ij} |\mathbf{x}_{ij}|^2 I_0(\omega) \tag{6.171a}$$

$$s = \frac{2\pi^2 e^2}{mc} I_0(\omega) \tag{6.171b}$$

* See *supra*, Chap. 1.

tell us that we must needs replace $1/m$ by $(4\pi\nu_{ij}/3\hbar)\,|\mathrm{x}_{ij}|^2$ in Eq. (1.78a) in order to arrive at a quantum analog of this equation. In Eq. (6.171a) $e\,x_{ij}$ is, of course, the matrix element of the electric dipole moment* — if it is electric dipole radiation that we are considering — between the states i and j. In addition, it is apparent that a temperature distribution of the molecules over the various stationary quantum states according to the applicable Maxwell–Boltzmann law must be imposed and a subsequent summing over the possible transitions introduced. These three modifications in Eq. (1.78a) result in:

$$\alpha = \left(\frac{4\pi\nu N e^2}{3hc}\right)\frac{\sum_j\sum_i |\mathrm{x}_{ij}|^2 f(\nu_{ij},\nu)\,\mathrm{e}^{-W_j/kT}}{\sum_j \mathrm{e}^{-W_j/kT}} \tag{6.172a}$$

where the shape factor is given by:

$$f(\nu_{ij},\nu) = \frac{1}{\pi}\left[\frac{\Delta\nu}{(\nu_{ij}-\nu)^2+(\Delta\nu)^2} - \frac{\Delta\nu}{(\nu_{ij}+\nu)^2+(\Delta\nu)^2}\right] \tag{6.172b}$$

$$\Delta\nu = 1/2\pi\tau. \tag{6.172c}$$

Let us write:

$$\sum_j\sum_i |\mathrm{x}_{ij}|^2 f(\nu_{ij},\nu)\,\mathrm{e}^{-W_j/kT} = \sum_{j\neq i}\sum [|\mathrm{x}_{ij}|^2 f(\nu_{ij},\nu)\,\mathrm{e}^{-W_j/kT}$$
$$+ |\mathrm{x}_{ji}|^2 f(\nu_{ij},\nu)\,\mathrm{e}^{-W_j/kT}] = \sum_i\sum_j |\mathrm{x}_{ij}|^2 f(\nu_{ij},\nu)\,\frac{h\nu_{ij}}{2kT}\,\mathrm{e}^{-W_j/kT}$$

since:

$$\nu_{ij} = -\nu_{ji};\quad f(\nu_{ij},\nu) = -f(\nu_{ji},\nu);\quad |\mathrm{x}_{ij}| = |\mathrm{x}_{ji}| \tag{6.173a}$$

and:

$$(\mathrm{e}^{-y} - \mathrm{e}^{-x-y}) \doteq \frac{1}{2}\,x(\mathrm{e}^{-y} + \mathrm{e}^{-x-y}) \text{ for } x \ll 1 \tag{6.173b}$$

so that Eq. (6.172a) becomes:

$$k = \left(\frac{4\pi\nu N}{6\hbar c}\right)\frac{h}{kT}\,\frac{\sum_j\sum_i |\mathrm{x}_{ij}|^2 \nu_{ij}, f(\nu_{ji},\nu)\,\mathrm{e}^{-W_j/kT}}{\sum_j \mathrm{e}^{-W/kT}} \tag{6.174}$$

with the shape factor still given by Eq. (6.172b).

Now if Eqs. (1.86) and (6.172a) are truly equivalent, as it would appear they should be, Eq. (6.172a) should reduce to Eq. (1.86) for $\nu_{ij} = 0$. It is apparent that instead of this reduction we simply obtain zero for Eq. (6.172a). This discrepancy was remedied by Van Vleck and Weisskopf by utilizing a Maxwell–Boltzmann distribution for x and $\dot{x}$ instead of Eq. (1.57) as a basis for obtaining the constants C_1 and C_2 in:

$$x = \frac{Ee}{m(\omega_0^2-\omega^2)}\,\mathrm{e}^{i\omega t} + C_1\mathrm{e}^{i\omega_0 t} + C_2\mathrm{e}^{i\omega_0 t} \tag{1.52}$$

From Eqs. (1.49) the Hamiltonian for our vibrating electron is surely:

$$H(t) = \frac{p^2}{2m} + \frac{1}{2}\,(\omega_0 x)^2 m - e\,x\,E\cos\omega t \tag{6.175}$$

* Van Vleck–Weisskopf allow either magnetic or electric moment.

Then the right side of Eq. (1.58a) instead of being zero will be x averaged over the Maxwell–Boltzmann distribution for Eq. (6.175). The right side of Eq. (1.58b) will still be zero, however, since:

$$\frac{\int\int_{-\infty}^{+\infty} p \exp\left[-H(t_0)/kT\right] dx\, dp}{\int\int_{-\infty}^{+\infty} \exp\left[-H(t_0)/kT\right] dx\, dp} = \frac{\int_{-\infty}^{+\infty} p e^{-\frac{p^2}{2mkT}} dp \int_{-\infty}^{+\infty} f(x)\, dx}{\int\int_{-\infty}^{+\infty} f(x, p)\, dx\, dp}$$

$$= \frac{-mkT e^{-\frac{p^2}{2mkT}} \Big|_{-\infty}^{+\infty} \cdot \int_{-\infty}^{+\infty} f(x)\, dx}{\int\int_{-\infty}^{+\infty} f(x, p)\, dx\, dp} = 0.$$

Thus, Eqs. (1.58) now become:

$$\left.\begin{aligned} &\frac{Ee}{m(\omega_0^2-\omega^2)} e^{i\omega(t-\vartheta)} + C_1' e^{i\omega_0(t-\vartheta)} + C_2' e^{-i\omega_0(t-\vartheta)} \\ &= A = \frac{\int\int_{-\infty}^{+\infty} x \exp\left[-H(t_0)/kT\right] dx\, dp}{\int\int_{-\infty}^{+\infty} \exp\left[-H(t_0)/kT\right] dx\, dp} \end{aligned}\right\} \quad (6.176a)$$

$$\frac{Eei\omega}{m(\omega_0^2-\omega^2)} e^{i\omega(t-\vartheta)} + C_1' i\omega_0 e^{i\omega_0(t-\vartheta)} - i\omega_0 C_2' e^{i\omega_0(t-\vartheta)} = 0 \quad (6.176b)$$

Eq. (6.101) may be solved in a very straightforward manner to yield:

$$C_1' = C_1 + \Delta C_1 = C_1 + \frac{A}{2} e^{i\omega_0 t_0} \quad (6.177a)$$

$$C_2' = C_2 + \Delta C_2 = C_2 + \frac{A}{2} e^{i\omega_0 t_0} \quad (6.177b)$$

where the C are now given by Eqs. (1.59) and $t_0 = t - \vartheta$.

Letting $a = \frac{1}{2mkT}$, $b = \frac{m\omega_0^2}{2kT}$, $c = \frac{eE\cos\omega t}{kT}$, we may integrate A by parts as follows:

$$\left.\begin{aligned} A &= \frac{\int_{-\infty}^{+\infty} x e^{-bx^2+cx}\, dx \int_{-\infty}^{+\infty} e^{-ap^2}\, dp}{\int_{-\infty}^{+\infty} e^{-bx^2+cx}\, dx \int_{-\infty}^{+\infty} e^{-ap^2}\, dp} = \frac{\int_{-\infty}^{+\infty} x e^{-bx^2+cx}\, dx}{\int_{-\infty}^{+\infty} e^{-bx^2+cx}\, dx} \\ &= -\frac{1}{2b} e^{cx} e^{-bx^2} \Big|_{-\infty}^{+\infty} + \frac{C}{2b} \frac{\int_{-\infty}^{+\infty} e^{-bx^2+cx}\, dx}{\int_{-\infty}^{+\infty} e^{-bx^2+cx}\, dx} = \frac{C}{2b} = \frac{2eE\cos\omega t}{m\omega_0^2} \end{aligned}\right\} \quad (6.178)$$

As a result of Eq. (6.176):

$$\Delta C_1 e^{-i\omega_0 t_0} = \Delta C_2 e^{-i\omega_0 t_0} = \frac{eE \cos \omega t_0}{m\omega_0^2}. \tag{6.179}$$

It is now evident that to Eq. (1.60) is next added the term

$$\frac{Ee}{m\omega_0^2} \cos(\omega t_0) \cos \omega_0 (\varepsilon - t_0) = \frac{Ee}{2m\omega_0^2} \Re [e^{i\omega(t-\vartheta)} (e^{i\omega_0\vartheta} + e^{-i\omega_0\vartheta})] \tag{6.180}$$

and the thus amended version of Eq. (1.60) yields, according to Eq. (1.61):

$$\langle x \rangle = \Re \left[\frac{eE e^{i\omega t}}{m(\omega_0^2 - \omega^2)} \left\{ 1 - \frac{(\omega_0 + \omega)\,\omega/\omega_0^2 \tau}{2[1/\tau - i(\omega_0 - \omega)]} + \frac{(\omega_0 - \omega)\,\omega/\omega_0^2 \tau}{2[1/\tau + i(\omega_0 + \omega)]} \right\} \right] \tag{6.181}$$

In analogy to the method of obtaining Eq. (1.78a) we may then obtain:

$$\alpha = \frac{2\pi N e^2}{mc} \left(\frac{\omega}{\omega_0} \right)^2 \left[\frac{1/\tau}{(\omega - \omega_0)^2 + (1/\tau)^2} + \frac{1/\tau}{(\omega + \omega_0)^2 + (1/\tau)^2} \right] \tag{6.182}$$

In analogy to the manner of obtaining Eqs. (6.172b) and (6.174) from Eq. (1.78a) we may now obtain Eq. (6.174) from Eq. (6.182) where here:

$$f(\nu_{ij}, \nu) = \frac{1}{\pi} \frac{\nu}{\nu_{ij}} \left[\frac{\Delta\nu}{(\nu_{ij} - \nu)^2 + (\Delta\nu)^2} + \frac{\Delta\nu}{(\nu_{ij} + \nu)^2 + (\Delta\nu)^2} \right] \tag{6.183}$$

Utilizing Eqs. (6.174) and (6.183), we do indeed obtain Eq. (1.86) for the case of zero resonant frequency, $\nu_{ij} = 0$, and $e^2 \Sigma |\mathbf{x}_{ij}|^2 = \mu^2$. Eq. (6.182) on the other hand, offers little change in the visible region, the $(\omega/\omega_0)^2$ factor yielding slightly more line asymmetry.

6.20. The application of Boltzmann's equation to oscillator distributions

In our considerations of the past section we treated the Maxwell–Boltzmann distribution of the coordinates and velocities associated with the classical oscillators which we may take as our radiator models. This was a relatively straightforward next step after the Lorentz assumption of zero averages for these quantities. Now if equilibrium exists in the neighborhood of the oscillator the Maxwell–Boltzmann distribution would appear the logical choice. If the possibility of non-equilibrium exists, one would expect someone to appeal directly to the Boltzmann equation for the proper manner in which the coordinates and velocities are to be distributed. The equation is, of course:

$$\frac{\partial f}{\partial t} + v_x \frac{\partial f}{\partial x} + v_y \frac{\partial f}{\partial y} + v_z \frac{\partial f}{\partial z} + a_x \frac{\partial f}{\partial v_x} + a_y \frac{\partial f}{\partial v_y} + a_z \frac{\partial f}{\partial v_z} = \left. \frac{\Delta f}{\Delta t} \right|_c \tag{6.184}$$

wherein $f = f(x, y, z, v_x, v_y, v_z, t)$ is our distribution function while the a_x, a_y, and a_z are the components of the acceleration attributable to the external forces. The quantity on the right is, of course, the collision integral.

Fröhlich[50] apparently first appealed to the Boltzmann equation while Gross[67] treated the problem more extensively. Finally, Gora[61] has justified the replacement of the collision integral by a quantity more amenable to manipulation and computation and carried through the distribution treatment to a result which appears quite convincing in its limit behavior.

In general the collision integral is given by:

$$\left.\frac{\Delta f}{\Delta t}\right|_c \equiv \int\int\int \mathrm{d}v_{x1}\,\mathrm{d}v_{y1}\,\mathrm{d}v_{z1} \int b\,\mathrm{d}b \int \mathrm{d}\varphi\, v\,(f' f_1' - f f_1) \tag{6.185a}$$

where f and f' are distribution functions for the particles undergoing collision, the unprimed function to be associated with those particles before collision while the primed function is to be associated with them after collision. The subscript "1" refers to the particles causing the collision. The parameter "b" is the collision parameter.

We remark that the replacement effected by these authors is:

$$\left.\frac{\Delta f}{\Delta t}\right|_c = -\frac{f - f'_{av}}{\tau} \tag{6.185b}$$

It is this replacement which Gora justified.

If we confine the oscillator to x-axis motion, we may recall its Hamiltonian as:

$$H(t) = \frac{p^2}{2m} + \frac{m\,\omega_0^2\,x^2}{2} - e\,x\,E_0\,\cos\omega t$$

where the oscillator exists in the external electric field $E_0 \cos\omega t$. The acceleration,

$$a = -\frac{1}{m}\frac{\partial H}{\partial t} = -\omega_0^2\,x + e\,E_0\frac{\cos\omega t}{m}$$

leads to the Boltzmann equation:

$$\frac{\partial f}{\partial t} + v\frac{\partial f}{\partial x} - \omega_0^2\,x\frac{\partial f}{\partial v} + \frac{e\,E_0\cos\omega t}{m}\frac{\partial f}{\partial v} = \left.\frac{\Delta f}{\Delta t}\right|_c \tag{6.186}$$

With no external electric field present the equilibrium distribution is:

$$f_0 = \frac{N\,\omega_0\,m}{2\pi\,kT}\exp\left[-(m\,v^2 + m\,\omega_0^2\,x^2)/2\,kT\right] \tag{6.187a}$$

while in the presence of the electric field:

$$\left.\begin{aligned} f_{\mathrm{eq}} &= \frac{N\,m\,\omega_0}{2\pi\,kT}\exp\left[-(m\,v^2 + m\omega_0^2 x^2 - 2\,e\,E_0\,x\cos\omega\,t)/2\,kT\right] \\ &= f_0\left(1 + \frac{e\,x\,E_0}{kT}\cos\omega\,t\right) \end{aligned}\right\} \tag{6.187b}$$

Now Gora supposed that f' could be replaced by:

$$f' = \frac{1}{\tau_0(\alpha+\beta)}\int_{-\alpha\tau_0}^{\beta\tau_0} f_{\mathrm{eq}}(t + t')\,\mathrm{d}t \tag{6.188}$$

where:

(Case I) $\alpha = \beta = \frac{1}{2}$ (symmetric)

(Case II) $\alpha = \frac{1}{2}, \quad \beta = 0$ (interaction during perturber approach)

(Case III) $\alpha = 0, \quad \beta = \frac{1}{2}$ (perturber recession)

total duration of collision

and where f_{eq} is given by Eq. (6.187b).

The assumption that only the incident field need be time averaged leads to:

$$\left.\begin{aligned}\langle\cos\omega t\rangle_t &= \frac{1}{\tau_0(\alpha+\beta)}\int_{-\tau_0\alpha}^{\tau_0\beta}\cos\left[\omega(t+t')\right]dt' \\ &= \frac{1}{\tau_0(\alpha+\beta)}\left[\sin\omega(t+\beta\tau_0)-\sin\omega(t-\alpha\tau_0)\right]\end{aligned}\right\} \quad (6.189)$$

Gora now chooses $\tau_0 = \gamma b/v$ where γ is the factor introduced in Section 2.9. Next it is assumed that some convergence factor of the form $1/2\exp(-b/R)$ will be introduced into Eq. (6.185a) in a more specific quantum treatment. As a consequence Eq. (6.185a) becomes:

$$\frac{1}{2}f_0\int_0^\infty e^{-b/2R}\,b\,db\int_0^{2\pi}d\varphi\left[1+\frac{e\,E_0\,x}{kT}\langle\cos\omega t\rangle\right] \quad (6.190)$$

which may be evaluated for the three cases and divided by the effective collision cross section πR^2 with the result:

$$f_{av}' = f_0\left[1+\frac{e\,E_0\,x}{kT}\,\frac{\exp(i\,\omega\,t)}{1+\omega^2\tau_c^2/4}\right] \quad (6.191\text{a})$$

$$f_{av}' = f_0\left[1+\frac{e\,E_0\,x}{kT}\,\frac{\exp(i\,\omega\,t)}{1\pm\omega\,\tau_c/2}\right] \quad (6.191\text{b})$$

In Eqs. (6.191) we have replaced the cosines by the imaginary exponential with the understanding that the real part is to be taken. These then are our distribution functions to be anticipated after a collision.

The application of f_{av}' to the line shape problem is begun by supposing that:

$$f = f_0(1+\varphi) \quad (6.192)$$

where now we shall neglect all terms not linear in E and Eq. (6.185a) becomes:

$$\frac{\partial\varphi}{\partial t}+\frac{\varphi}{\tau}+v\frac{\partial\varphi}{\partial x}-\omega_0^2\,x\frac{\partial\varphi}{\partial v} = \frac{e\,E_0\,x}{kT}\,e^{i\omega t}+g_{av}' \quad (6.193\text{a})$$

with:

$$g_{av}' = \frac{f_{av}'-f_0}{f_0} \quad (6.193\text{b})$$

To the level of our approximation here φ may be assumed linear in x and v, so we take:

$$\varphi = (Ax + Bv) \frac{e\,E\,e^{i\omega t}}{kT} \tag{6.194}$$

From Eqs. (6.193) and (6.194) we may obtain:

$$A = \frac{h_{av}'(i\,\omega - 1/\tau)/\tau + \omega_0^2}{(i\,\omega + 1/\tau)^2 + \omega_0^2} \tag{6.195a}$$

with:

$$h_{av}' = g_{av}' \left(\frac{kT}{e\,E}\right) e^{-i\omega t} \tag{6.195b}$$

The polarization and particle density are given by:

$$P = e \iint_{-\infty}^{+\infty} x\,f\,dx\,dv = e \iint_{-\infty}^{+\infty} x\,f_0\,\varphi\,dx\,dv \tag{6.196a}$$

$$N = \iint_{-\infty}^{+\infty} f_0\,dx\,dv \tag{6.196b}$$

From Eqs. (6.196) and (6.194) we may obtain:

$$P = \frac{Ne^2\,E_0}{m\,\omega_0^2} e^{i\omega t} A \tag{6.197}$$

The polarization is related to the complex dielectric constant by the familiar expression:

$$\varepsilon = \varepsilon_1 - i\varepsilon_2 = 1 + \frac{4\pi\,P}{E}$$

with the absorption coefficient, of course, given by $\varepsilon_2\,\omega/c$. As a consequence:

$$\alpha = \frac{4\pi\,Ne^2}{m\,\omega_0^2\,c}\,\omega\,\mathfrak{J}(A) \tag{6.198}$$

We now use Eqs. (6.191) in Eq. (6.195b) to obtain:

$$h_{av}' = \frac{1}{1 + \omega^2\,\tau_c^2/4} \qquad \text{(Case I)} \tag{6.199a}$$

$$h_{av}' = \frac{1}{1 \pm i\,\omega\,\tau_c/2} \qquad \text{(Case II, III)} \tag{6.199b}$$

or, if $S = 0$ (Case I), $+1$ (Case II), -1 (Case III),

$$h_{av}' = (1 - i\,S\,\omega\,\tau_c/2)\,\Delta \tag{6.200a}$$

$$\Delta = \frac{1}{1 + \omega^2\,\tau_c^2/4} \tag{6.200b}$$

Eq. (6.198) now yields the absorption coefficient:

$$\alpha = \frac{4\pi\,Ne^2}{m\,c\,\tau} \frac{\omega^2}{\omega_0^2} \frac{1}{D} \left[2\,\omega_0^2(1 - \Delta) + \Delta\left(\omega^2{}_0 + \omega^2 + \frac{1}{\tau^2}\right)\left(1 + \frac{S\tau_c}{2\,\tau}\right)\right] \tag{6.201}$$

For $\Delta = 1$ or low frequencies this reduces to the Van Vleck Weisskopf result. For high frequencies ($\Delta = 0$) this reduces to the Lorentz result. Therefore, from the point of view of the microwave, this work of Gora has borne out the earlier work of Van Vleck and Weisskopf. However, the application of the Boltzmann equation to the study of these phenomena peculiar to the microwave region still affords a rather large field of endeavor.

6.21. Review and summary

The broadening of polyatomic spectral lines has been considered as a unique phenomenon. The uniqueness does not arise from any different theories as to the actual broadening of the lines in question. Rather, it is due to the uniqueness of the forces which may exist between polyatomic molecules and to the very low frequency or microwave radiation which these molecules may emit.

In approaching such a broadening problem then one would begin by deciding whether a classical treatment is applicable. If it is, then the Interruption or Statistical Theory may certainly be applied. Within the framework of these theories, however, one would use the particular molecular interaction force which arises and which we have discussed in this chapter. Of course, it still might transpire that the force in question was a familiar van der Waals force which we have encountered so often in atomic broadening considerations. Some examples such as the Lindholm treatment of HCl have been given as to how we could go about developing the theory for a particular broadening case of our interest.

If one found that the molecular broadening was not to be treated by a classical consideration then one would turn to a quantum theory of the phenomenon. The most familiar and most widely used of these theories is, of course, the Anderson treatment. We remark the basis of this in the expression for the quantum analog of the radiation from a classical oscillator which we have discussed in an earlier chapter. Further, we recall that the assumption as to very short collision durations actually renders this an Interruption type treatment.

Finally, we saw how, for the microwave region of the spectrum, certain considerations as to distribution of oscillator coordinate and velocity had to be brought into our Interruption thinking. The van Vleck Weisskopf treatment supposed the equilibrium Maxwell–Boltzmann distribution of these quantities while the Gross and, later, Gora considerations utilized a direct appeal to the Boltzmann equation. For these very low frequencies too, diabaticity can be anticipated in greater plenitude than for the optical region of the spectrum, primarily because of the ease with which transitions may be collision induced between the close lying levels required for the production of these low frequencies. Thus, for precise treatment every problem becomes a problem requiring special attention to the particular diabatic effects appropriate to the physical system.

CHAPTER 7

THE BROADENING AND SHIFT OF THE HIGH SERIES MEMBERS

THE FOREIGN gas broadening theories which we have considered previously have all been amenable to a more or less tenuous association with the labels Interruption or Statistical. Although we have devoted a separate chapter to it, this association also holds for the Stark theory. The theory which we are now to consider might, by rather stretching a point, be labelled pseudo-Statistical, but, since the general principles of it are sufficiently unique, let us not attempt to so classify it. In all the theories considered we have been able to suppose the molecule, whose spectral lines were being broadened, as an entity of itself which was affected from without by one or a combination of broadening molecules. The theory which we are about to examine proves quite different from this, however.

7.1. A qualitative explanation of the high series shift

The indication that the higher series members should be treated any differently from any other spectral lines was quite clearly given in the results which were obtained by Amaldi and Segre[1,2] for the shift and broadening of the spectral lines of the two alkali atoms, sodium and potassium, by H_2, N_2, He and A.

Let us recall that when we speak of the higher series members we refer to spectral lines which arise from transitions from photoelectron states of high principal quantum number n. In the conceptual language of the Bohr theory this means that the orbit of the photoelectron is at a great distance from the atomic nucleus, in fact, for principal quantum number 30 the radius of the Bohr orbit is around 500 A. The size of this orbit — or any other high n orbit — is the basis which Fermi[45] used for the theory by means of which he sought to explain the anomalies which Amaldi and Segre found in their high series spectra, namely: (1) The spectral line shift increases as the series member increases, that is, the shift increases for upper states with higher n. (2) As the order of the upper state continues to increase the line shift appears to converge toward a value which is approximately proportional to the pressure. (3) The magnitude of the shift was found to be the same for the spectral lines of sodium and potassium. (4) The amount and direction (violet or red) appear to depend on the nature of the foreign broadening gas. Qualitatively Fermi explained these results, or some of them, as follows:

To begin with only those electron orbits will be treated which are of sufficient radius to include within the spheres corresponding to them several thousand foreign gas atoms. It is precisely this orbit size which forms the basis for the present treatment. In the case $n = 30$ of the last paragraph there are, at atmospheric pressure, 30,000 atoms within the orbit.* In addition the velocity of the electron at these distances from its nucleus is low. The picture that then presents itself, and which is all important to the theory, is that of the photoelectron moving slowly through the atoms of the foreign gas. From this picture Fermi evolved two co-acting methods by which the lines are broadened.

(a) *The Polarization Effect.* As we are well aware, the electrons in an atom may be considered as arranged in shells around the nucleus with a sufficient number of shells filled so that the positive charge of the nucleus is screened or counteracted in such a manner that no electric field will arise due to the nuclear charge. Now when we move the photoelectron out to such a distance that one or more atoms are either at the same or lesser distance from the nucleus, the shielding effect of this eleetron will be absent insofar as these atoms are concerned. As a consequence these atoms will be under the influence of a field of an ion of charge e. We shall assume that they are not possessed of a permanent dipole, but an interaction between the dipole induced on these atoms and the electric field will occur. The energy of this interaction can only arise from one place, and that is the emitting atom. The result is that the energies of the energy levels of the atom are lowered by the amount contributed to this interaction. *In toto* then, the emitted spectral line is shifted to the red as a result of the Polarization Effect.

(b) *The Potential Valley Effect.* We have considered an interaction between the partially unscreened nucleus and the broadeners, and now we consider the interaction between the photoelectron and the broadeners. Fermi considered the broadeners as effectively scattering potential valleys along the flight path of the photoelectron. These valleys would not be large in extent, however, since the atoms causing them are neutral. Quite obviously these valleys will have some effect on the energies of the electron, and we shall see that (1) the line shift resulting may be either to the red or the violet and (2) the magnitude is always greater than that of the Polarization Effect.

7.2. The polarization effect

Let us first consider one of the many foreign gas atoms contained within the orbit of the photoelectron and hence acted on by the field of a charge e. If the separation of this atom from the center of the emitting atom is r_i, then the field in which this atom finds itself is e/r_i^2.

* Quite often nannies and kids are pastured separately for reasons which need not concern us. In the evening when the kids and their maternal ancestors are again placed in the same corral it is most impressive to watch a nanny rapidly and unerringly pick her offspring from among hundreds of other and apparently identical kids while refusing to be cozened by the younger members of other families. Although impressive, we must admit that this cannot hold a candle to the selectivity of the electron and nucleus to which we refer.

We now make our first assumption, namely, (A) THE BROADENERS POSSESS NO PERMANENT DIPOLES. If, as usual, we take the polarizability as α, then the interaction energy will be given by $-(1/2)\,\alpha E^2 = -\alpha e^2\, 2/r_i^4$ so that the total energy shift resulting may be found by summing over all atoms within the orbit as:

$$\Delta E = -\frac{\alpha e^2}{2} \sum_i \frac{1}{r_i^4} \tag{7.1}$$

Very little error is introduced by extending this sum over all perturbers (this arises from the rapid convergence of r_i^4) so we do precisely this.

Now the spectral line may be supposed shifted by the same amount as the upper level since we may imagine this effect as not present for the lower level involved in the transition. The shift of the line intensity maximum due to the effect under consideration may then be taken as the energy perturbation resulting from the most probable perturber configuration. This most probable configuration is a uniform distribution of the perturbers, with the nearest one a distance $(4/3)\,\pi R_1^3 = 1/N$, where now the volumes of the individual perturbers of size $1/N$ are distributed in spherical shells. In order to find the value of the summation over r_i then, we simply replace the summation in Eq. (7.1) by an integration of lower limit r_1 with the result:

$$\sum_i \frac{1}{r_i^4} = 4\pi N \int_{R_1}^{\infty} \frac{r^2\, dr}{r^4} = \frac{4\pi N}{R_1} \tag{7.2}$$

Eq. (7.2) may now be substituted into Eq. (7.1) to obtain:

$$\left.\begin{aligned} \Delta E &= -10\, e^2 \alpha N^{4/3} \text{ ergs} = -2.8 \times 10^{n} (\varepsilon - 1)\, N^{1/2} \sec^{-1} \\ &= +0.00093\, (\varepsilon - 1)\, N^{1/3} \text{ cm}^{-1} \end{aligned}\right\} \tag{7.3}$$

according to Eq. (7.41b) under the standard assumption $\varepsilon \doteq 1$. The symbol ε represents the dielectric constant of the foreign gas.

Eq. (7.3) yields a shift of about a reciprocal centimeter for a foreign gas pressure of one atmosphere.

7.3. The potential valley effect

In taking up this second effect let us first introduce the assumption (B) THE MOLECULES ARE MONATOMIC. This will tend to restrict the theory, but it will be important in the proof.

We wish to set up the Schrödinger equation for the photoelectron. First then we are in need of a potential function which conveniently presents itself as a sum of the potential due to the remainder of the emitter (excluding the photoelectron) U, and the potential due to all broadeners present, $\sum_i V_i$. Now the de Broglie wavelength* of our earlier example — the atom with an n of 30

* It may be remembered that the de Broglie wavelength of a particle is related to the linear momentum of the particle by the relation: $h/\lambda = p$.

— is in the neighborhood of 100 A. The perturbing atoms' potential V_i will be of limited extent — and spherical symmetry — since the atoms are neutral, however, so that the de Broglie wavelength will be large compared to the spatial extent of the V_i.

Our Schrödinger equation will then be:

$$\nabla^2 \psi + \frac{2m}{\hbar^2}\left[E - U - \sum_i V_i\right]\psi = 0 \tag{7.4}$$

Next a function, $\langle\psi\rangle$, is defined which is the space average of the eigenfunction of Eq. (7.4) in a domain of small extent compared to the de Broglie wavelength but sufficiently large to contain a goodly number of perturbers. The potential U will surely be about constant in such a region, and we shall so consider it. The function which we have defined satisfies the equation:

$$\nabla^2 \langle\psi\rangle + \frac{2m}{\hbar^2}[E - U]\langle\psi\rangle - \frac{2m}{\hbar^2}\langle\sum_i V_i \psi\rangle = 0 \tag{7.5}$$

We are now faced with the problem of computing $\langle\sum_i V_i \psi\rangle$ which may be accomplished in the following approximate manner: Since the remainder of the atom is some hundreds of angstroms away, we may expect $E - U$ to be sufficiently small so that it may be neglected in comparison to V_i in a small region about the i-th perturber. In addition we have supposed our broadeners in a spherically symmetric state so that the potential will be simply a function of r, measured, of course, from the perturber center. Consequently $V_i = V_i(r)$, and the Schrödinger equation takes the approximate form:

$$\nabla^2 \psi = \frac{2m}{\hbar^2} V_i \psi \tag{7.6a}$$

When we set $\psi = u(r)/r$, this becomes:

$$u'' = \frac{2m}{\hbar^2} V_i u \tag{7.6b}$$

In that part of space sufficiently far removed from the perturber, ψ will tend toward the average value $\langle\psi\rangle$ which we have already defined. On the other hand, the solution to Eq. (7.6b) at a position far from the potential valley of the perturber is:

$$u = c_1 + c_2 r \tag{7.7a}$$

Since ψ is tending toward $\langle\psi\rangle$ at great distance, c_2 is given by $\langle\psi\rangle$ since $u(r) = r\psi$. There can be no objection to replacing the constant c_1 by the constant $a\langle\psi\rangle$, so that Eq. (7.7) becomes:

$$u = (a + r)\langle\psi\rangle \tag{7.7b}$$

and the significance of a is apparent from Fig. 7.1.

We now integrate Eqs. (7.6) with the result:

$$\left.\begin{aligned} \frac{2m}{\hbar^2}\int V\psi \,\mathrm{d}\tau &= 4\pi\frac{2m}{\hbar^2}\int V u r \,\mathrm{d}r = 4\pi u'' r \,\mathrm{d}r \\ &= 4\pi \left| u' r - u \right|_0^r = -4\pi a \langle\psi\rangle \end{aligned}\right\} \tag{7.8}$$

where the final evaluation is obtained directly from Eq. (7.7b).

Eq. (7.8) leads directly to:

$$\frac{2m}{\hbar^2}\left\langle \sum_i V_i \psi \right\rangle = -4\pi a N \langle\psi\rangle \tag{7.9}$$

Eq. (7.9) may now be substituted into Eq. (7.5) to obtain:

$$\Delta^2 \langle\psi\rangle + \frac{2m}{\hbar^2}[E_0 - U]\langle\psi\rangle = 0 \tag{7.10}$$

where:

$$E_0 = E + \frac{h^2 a N}{2\pi m} \tag{7.11}$$

Eq. (7.10) is the Schrödinger equation for the emitting atom — considered as central core plus photoelectron — when the perturbing foreign gas atoms are not present. Thus, E_0 will take on the various values of the unperturbed atom energies. Since E is the energy of the perturbed atom, it will, according to Eq. (7.11), differ from the unperturbed energy by amount:

$$\Delta E = -\frac{h^2 a N}{2\pi m} \tag{7.12}$$

It is apparent from Eq. (7.12) that the direction of level and line shift will depend on the sign of the quantity a. The problem is essentially solved then if some further information about the constant a can be obtained. Fermi was able to obtain an approximate value for the magnitude of this factor — not the sign — by beginning with Wentzel's expression[210] for the collision cross section for very slow electrons on the foreign gas atoms:

$$\sigma = \frac{h^2}{\pi p^2}\sin^2 \Upsilon_0 \tag{7.13}$$

where p is the electronic linear momentum, and Υ_0 is the phase change occurring in the matter waves associated with the electron due to the perturber valley. Since the sign is equal to its argument for very small p:

$$\Upsilon = \frac{p}{h}\sqrt{\pi\sigma} \tag{7.14}$$

We may obtain an alternate expression for the phase change by a study of Fig. 7.1. A consideration of this figure is sufficient to tell us that the matter wave of the electron has suffered a phase change of a/λ wavelengths after having undergone the influence of one of the potential valleys. It follows then that:

$$a/\lambda = \frac{\Upsilon}{2\pi} \tag{7.15}$$

We substitute Eq. (7.15) into Eq. (7.14) to obtain:

$$\sigma = 4\pi a^2 \tag{7.16}$$

which may be utilized for the magnitude of a in Eq. (7.12), but, as we have noted earlier, which tells us nothing about the sign and, hence, nothing about the direction of the shift.

Finally, Fermi added the effect of Section 2 as given by Eq. (7.3) to Eq. (7.12) to obtain, as the overall shift due to the interactions of this and the preceding sections:

$$\Delta = -2.8 \times 10^7 (\varepsilon - 1) N^{1/3} \pm 0.33\, \sigma^{1/2} N \tag{7.17}$$

Eq. (7.17) is valid then (C) FOR PRESSURES HIGH ENOUGH SO THAT SEVERAL ATOMS WILL BE FOUND IN A CUBE OF SIDE THE DE BROGLIE WAVELEGTH OF THE ELECTRON AND LOW ENOUGH SO THEIR SEPARATION IS MUCH GREATER THAN THEIR RADIUS OR THE RADIUS OF THE COLLISIONAL CROSS SECTION. On this basis Fermi set the minimum pressure at about one atmosphere.

Fermi considered the broadening of the high series members very briefly, although it is directly apparent that the same phenomena will lead to broadening as led to shift. By a rather qualitative line of reasoning he gave as the width due to the Polarization Effect Eq. (7.3). It might also be of some interest to obtain his result for the highest order line appearing. The premise here is that, in order for a line to be distinct, the free path of the photoelectron must be sufficiently long that said electron is given the opportunity of completing several circuits of its orbit. If σ is the collisional cross section for the electron on the foreign gas atoms, we will readily concede that the free path is given by:

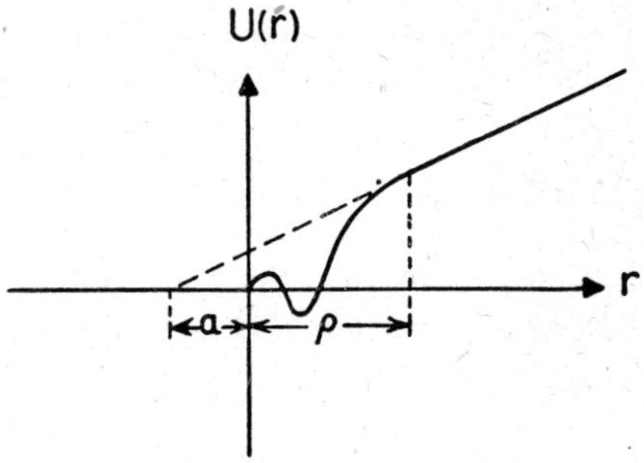

FIG. 7.1. A plot of Eq. (9.7b). (After Fermi[45])

$$l = 1/\sigma N \tag{7.18}$$

and the orbit circumference may be approximated by:

$$4 r_0 n^2 \tag{7.19}$$

where r_0 is the radius of the first Bohr orbit.

Then our condition for distinctness of lines leads to the relation:

$$4 r_0 n^2 < \frac{1}{\sigma N}$$

from which we conclude that the highest principal quantum number specifying the upper state of a spectral line is given by:

$$n_0 = \frac{1}{2\sqrt{r_0 \sigma N}} \tag{7.20}$$

As Fermi noted, however, one may not be able to observe quite this high order a line since other causes of broadening will further affect the line distinctness.

7.4. Axially symmetric broadeners and the shift direction

Reinsberg next considered the high series shift problem[176] under the same basic assumptions as Fermi had first stated. Insofar as the actual treatment which he gave the problem was concerned, his work was about the equivalent of Fermi's for the spherically symmetric broadeners. This, of course, resulted in his obtaining Fermi's earlier results for this case with one magnitudewise minor but otherwise important difference which we shall note in due course.

From Eqs. (7.5) and (7.6) Reinsberg obtained instead of Eq. (7.6):

$$\nabla^2 \psi + \frac{2m}{\hbar^2}[E - V_i]\psi = 0 \tag{7.20a}$$

$$\nabla^2 \langle\psi\rangle + \frac{2m}{\hbar^2} E\langle\psi\rangle - \frac{2m}{\hbar^2}\langle V_i\psi\rangle = 0 \tag{7.20b}$$

where all symbols continue to have the meaning previously ascribed to them.

The undisturbed solution of:

$$\nabla^2 \psi^0 + \frac{2m}{\hbar^2} E\psi^0 = 0 \tag{7.21}$$

will also be useful. For the same reasons leading to the choice in Eqs. (7.6) a first approximation to ψ and ψ^0 may be taken as $u(r)/r$ and $u^0(r)/r$ respectively. There then results for Eq. (7.20a) and (7.21):

$$u'' + \frac{2m}{\hbar^2}(E - V_i)u = 0 \tag{7.22a}$$

$$u^{0\prime\prime} + \frac{2m}{\hbar^2} E u^0 = 0 \tag{7.22b}$$

Clearly, the solution to Eq. (7.22b) is:

$$u^0 = A \sin \frac{\sqrt{2mE}}{\hbar} r \tag{7.23}$$

Now for very large r, $V_i(r)$ goes to zero, and, since this potential constitutes the difference only between Eqs. (7.22), we take the solutions to these two equations as differing only by an amplitude factor and a phase factor for large r. Thus, there results from Eq. (7.23):

$$u = \alpha_0 A \sin\left(\frac{\sqrt{2mE}}{\hbar} r + \Upsilon_0\right) \tag{7.24a}$$

or, since $\sin(a + b) = \cos a \cdot \cos b(\tan a + \tan b)$ and p is presumed very small, Eq. (7.24a) may be approximated as:

$$u = \alpha_0 A \frac{\sqrt{2mE}}{\hbar} \cos \Upsilon_0 \left[r + \frac{\hbar}{\sqrt{2mE}} \tan \Upsilon_0\right] \tag{7.24b}$$

since in this case $\cos a = 1$ and $\tan a = \sin a = a$ for a quite small.

Now in place of Eq. (7.7b) we obtain, after a solution exactly coinciding to Eqs. (7.6):

$$u = \left(\frac{\hbar}{\sqrt{2mE}} \tan \Upsilon_0 + r \right) \langle \psi \rangle \tag{7.25}$$

A comparison of Eq. (7.25) with Eq. (7.7b) tells us the most important relation to be obtained here:

$$a = \frac{\hbar}{\sqrt{2mE}} \tan \Upsilon_0 \tag{7.26}$$

The reason for the importance of this expression as opposed to an analogous expression obtained from Eqs. (7.13) and (7.16) is that here a possibility exists for obtaining different signs for a for different values of the phase shift Υ_0. One further point should be made. The equation obtained for a from Eqs. (7.13) and (7.16) must be expected to agree with Eq. (7.26), and let us remark that they will so agree for small linear momentum (slow electrons). Υ_0 can be expected to differ but little from some integral multiple of π. In consequence the sine and tangent are about equal.

Reinsberg treated the case of the axially symmetric broadener by setting the problem up in ellipsoidal coordinates and solving the resulting Schrödinger equation by a method quite similar to that which we have used in this and the previous section for the spherically symmetric perturber. The shift is found to be that of the spherically symmetric molecule, namely, Eq. (7.12) with a given by Eq. (7.26). Now the matter of shift direction may more conveniently be considered for application to either type broadener.

The particular phenomenon of slow electron deflection by the noble gases had been investigated in considerable detail by Ramsauer.* Faxen and Holtsmark[44] had treated the situation theoretically from the quantum mechanical point of view, and, finally, Holtsmark[83, 84] and Heuneberg[77] had furnished the results of which we shall make specific use.

Firstly, Heuneberg had shown that for medium and high velocities the phase constant increases monotonically with decreasing electron velocities. Holtsmark carried his considerations of the problem to much lower velocities and determined that the phase shift must needs proceed to multiples of π unless, as may possibly occur, Υ_0 has already attained the limiting value for infinitely small velocities for some small value of the electronic energy. This means that the phase shift has a maximum and the Ramsauer cross section has a maximum as indicated by Fig. 7.2. Now for this case, since in this limiting case Υ_0 approaches some multiple of π, the phase change becomes equal to $k\pi + \beta$ where $0 \leq \beta \leq \pi/2$. This means that $\tan \Upsilon_0$ occurring in Eq. (7.26) must be positive so that a is positive, and the line shift as given by Eq. (7.11) is to the red.

Thus, IF THE RAMSAUER CROSS SECTION FOR THE BROADENING GAS FOR LOW ELECTRON VELOCITIES HAS A MINIMUM THE SPECTRAL LINE IS SHIFTED TOWARD THE RED, AND IF THIS CROSS SECTION HAS NO MINIMUM THE LINE IS SHIFTED TOWARD THE VIOLET.

* References 142, 143, and 144.

A and *Xe* have minimums while *He* and *Ne* do not, so that one would expect from the theory that the shift direction would be red for the two former and violet for the two latter. This has been found to be indeed the case.[3,52,56]

Hg has been experimentally determined as displaced to the red which to a cursory inspection might indicate a minimum. If, however, the Polarization

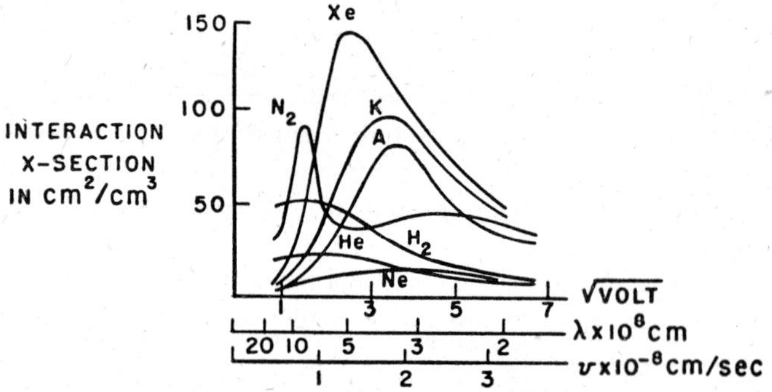

FIG. 7.2. Ramsauer cross sections for slow electrons. $T = 0\,°C$. $p = 1$ mm. (After Faxen and Holtsmark[44])

Effect† is large enough, the Potential Valley Effect of this and the last section might not be large enough to overcome it so that even though the aggregate shift is toward the red, the Valley shift might be toward the violet. This is Reinsberg's assumption in this case at any rate.

Reinsberg calculated the magnitudes of some shifts with quite good results. In doing so he took the values of the phase shift from the work of Faxen and Holtsmark[44].

7.5. The limiting breadths of the high series lines

As our last consideration of this specialized effect, we shall consider Reinsberg's treatment[177] of the breadth of the spectral lines in the limiting case of very high principal quantum number. Let us first remark the results of Füchtbauer on the breadths of spectral lines as a function of principal quantum number[54]. Quite concisely, Fig. 7.3 gives the manner in which, as an example, the $(1s - np)$ *Na* line broadened by *A* varies with principal quantum number, n. It is very apparent from this figure that, for sufficiently high n, the line width approaches a limiting value. It is this limiting line width which Reinsberg attempted to obtain.

In carrying through this calculation the general method will be an incorporation of the interaction forces of Sections 2 and 3 into the framework of the Lorentz collision theory through the medium of the collision cross section.

† See *supra*, Sec. 7.2.

To begin with then the level* width is given by:

$$\delta = \frac{1}{\pi \tau} = \frac{2\, T_0 N_0}{\pi} \sqrt{\frac{2\,k}{\pi}}\, \frac{p\,\sigma}{\sqrt{m\,T}} \tag{7.27}$$

where now T_0 and N_0 are a temperature of 273.3°K and the number of atoms per unit of volume respectively. σ is the cross section through which the special effects of this chapter will enter.

Now we begin the problem by choosing as the absorber the hydrogen atom. Next the Bohr assumption is utilized to the effect that we take as the radius of an orbit corresponding to principal quantum number n the value $r_0\, n^2$. As a consequence, when we suppose the electron to be a unit of charge uniformly smeared out** through the sphere corresponding to this orbit, the constant electron density (or charge density) is:

$$d = \frac{3}{4\,\pi\, r_0^3\, n^6} \tag{7.28}$$

Now Eqs. (7.12) and (7.26) tell us that *one* noble gas atom in interaction with one electron has the interaction energy:

$$E_1 = \frac{h^2 \tan \Upsilon_0}{4\, m\, \pi^2 \sqrt{2\, m\, E'}} \tag{7.29}$$

so that this atom will now be affected by a potential:

$$E_1\, d \tag{7.30}$$

as given by Eqs. (7.28) and (7.29).

Eq. (7.30) gives the main contribution to the potential curve shown in Fig. 7.4 between the points r_1 and r_2. For convenience one might consider it as the only contribution in this region to a fair approximation. Thus, the Potential Valley Effect builds the potential curve in this portion of space.

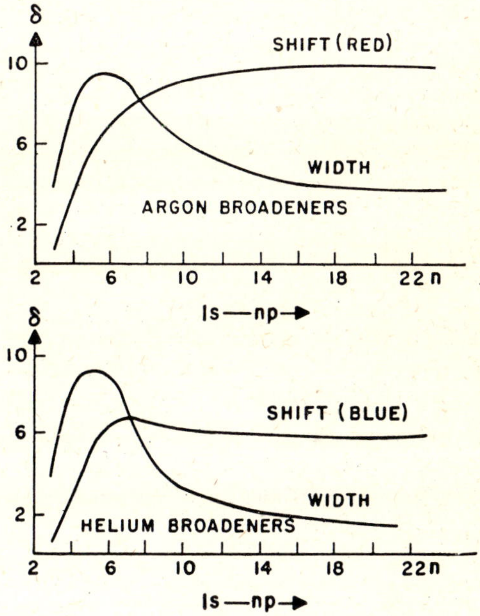

FIG. 7.3. Line shift and halfwidth as a function of order. (After Füchtbauer and Schulz[54])

To build the remainder of the potential curve we utilize, as one might suspect, the Polarization Effect so that the interior portion of the curve is given by the equation:

$$\frac{\alpha\, e^2}{2\, r^4} \tag{7.31}$$

* Reinsberg simply used the Lorentz result for the levels giving rise to a line rather than for the line itself.

** This may better be considered as smeared out location probability and hence timewise charge smearing.

Broadeners of thermal $E = (3/2)kT$ are now to be considered, and one of two rather obvious situations may arise, either $E < V_0$ or $E > V_0$. Although these "things" which the energy may do are rather obvious, they lead to some rather important consequences.

First the case $E < V_0$. In this case the foreign gas atom cannot climb the wall at r_2, and, as a result, all within a sphere of this radius is forbidden to it. This means that the collisions cross section for such an atom is:

$$\pi r_2^2 = \pi r_0^2 n^4 \tag{7.32}$$

It also hardly needs remarking that this will form a lower limit on cross sections for atoms of higher energies. We shall return to a consideration of the more complex problem of cross sections for these higher energies after a few preliminary remarks.

Through Eq. (7.27) the level widths are directly proportional to the cross sections with which we have been concerning ourselves. Now Reinsberg supposed that the width of the np level might simply be added to the width of the $1s$ (only transitions $(1s - np)$ are being considered) level in order to obtain the width of spectral line. Further the $1s$ level width is negligible compared to the width

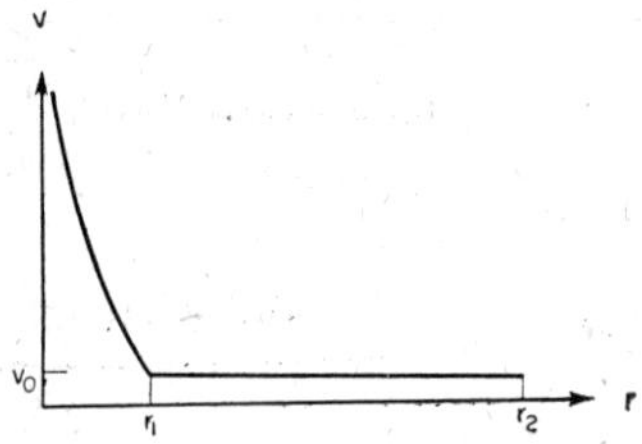

Fig. 7.4. The potential curve due to the high series interactions. (After Reinsberg[177])

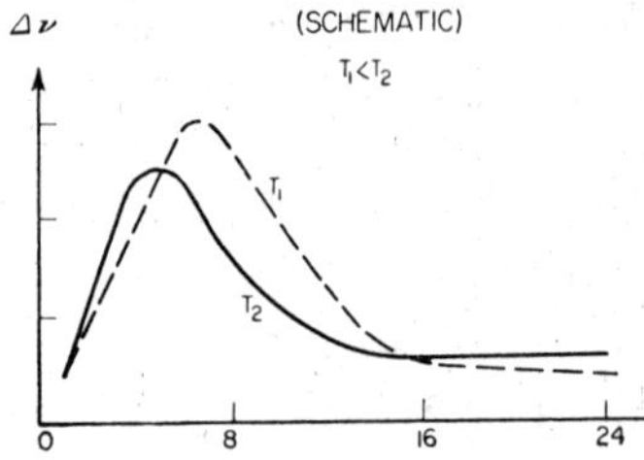

Fig. 7.5. Displacement of evel lwidth curve with temperature. (After Reinsberg[177])

of the other combining levels so that the course of the level width with principal quantum number may be inferred directly from Fig. 7.5. Another interesting point is illustrated by Fig. 7.5. This figure shows the displacement of the curve of the level widths with temperature. This latter effect may be directly traced to the increase of the cross section (and consequently the level width) with a decrease in temperature (accompanied, of course, by a decrease in energy) according to Eq. (7.32).

It now appears that the obtention of the cross section for $E > V_0$ will mean that the problem as set up has been solved. The solution is furnished us by the work of Massey and Mohr[199]. These two authors[148, 149], in quantizing the earlier work of Chapman[21, 22] on free paths and various transport phenomena, obtained the expression for the cross section which will prove applicable to the present situation.

We begin by defining the quantity $I(\vartheta) \sin\vartheta \, d\vartheta$ as the probability that a particle will be scattered on collision through an angle between ϑ and $\vartheta + d\vartheta$. From this definition, it follows logically that we may find the collision cross section — the total probability that the particle will be scattered in any direction whatsoever — as:

$$\sigma = 2\pi \int I(\vartheta) \sin\vartheta \, d\vartheta \tag{7.33}$$

Quite evidently then, the determination of the collision cross section demands the prior determination of $I(\vartheta)$. This we proceed to do.

We initially assume that a solution of the equation

$$\Delta^2 \psi + \frac{2m}{\hbar^2}[E - V(r)]\psi = 0 \tag{7.34}$$

at some distance from the scattering atom, the absorber, is desired. It will be of the form:

$$e^{ikz} + \frac{1}{r} e^{ikr} f(\vartheta) \text{ where: } k^2 = \frac{2mE}{\hbar^2} \tag{7.35}$$

where the first term represents the incident particle in the form of a plane wave and the second term represents the scattered particle. Since the solution $P_n(\cos\vartheta) f_n(r)$ is a quite satisfactory one, the solution

$$\sum_n A_n P_n(\cos\vartheta) f_n(r) \tag{7.36}$$

is equally so. From this equation one may, by integration and requiring that the asymptotic form of $f_n(r)$ be:

$$f_n(r) \to \frac{1}{kr} \sin\left(kr - \frac{n\pi}{2}\right) \tag{7.37}$$

determine the A_n so that:

$$e^{ikz} = \sum_n (2n+1) i^n P_n(\cos\vartheta) f_n(r) \tag{7.38}$$

We take the solution, Eq. (3.55), where, we may recall, the $rL_n(r)$ is the solution to Eq. (3.56a). Now Eq. (7.38) is subtracted from Eq. (3.55) with the intention of obtaining the expression for the scattered wave in this fashion:

$$A_n L_n(r) - (2n+1) i^u f_n(r) \doteq C_n r^{-1} e^{ikr} \tag{7.39}$$

When the asymptotic expressions for the $f_n(r)$ from Eq. (7.37) and $L_n(r)$ from Eq. (3.59) are utilized, there results:

$$\left.\begin{aligned} &\frac{e^{ik\varrho}}{2ikr}[A_n e^{i\Upsilon_n} - (2n+1) i^n] \\ &- \frac{e^{ik\varrho}}{2ikr}[A_n e^{-i\Upsilon_n} - (2n+1) i^n] \text{ where: } k\varrho = kr - \frac{n\pi}{2} \end{aligned}\right\} \tag{7.40}$$

It is apparent from Eq. (7.35) that, since this is to be the expression for the scattered wave, A_n must be given by:

$$A_n = (2n+1) i^n e^{i\Upsilon_n} \tag{7.41}$$

Thus, the wave function representing the incident and the scattered wave is:

$$\psi = \sum_n (2n+1) i^n e^{i\Upsilon_n} L_n(r) P_n(\cos\vartheta) \tag{7.42}$$

of asymptotic form:

$$r^{-1} e^{ikr} f(\vartheta)$$

so that:

$$f(\vartheta) = \frac{1}{2ik} \sum_n (2n+1) [e^{2i\Upsilon_n} - 1] P(\cos\vartheta) \tag{7.43}$$

This equation is precisely the one for which we have been looking since, by quantum mechanical definition, the absolute square of $f(\vartheta)$ multiplied by the angular volume element gives us the probability that the particle will be in the angular volume element. Thus, from Eq. (7.43):

$$I(\vartheta) = \frac{1}{4k^2} \sum_n |(2n+1) [e^{2i\Upsilon_n} - 1] P_n(\cos\vartheta)|^2 \tag{7.44}$$

The substitution of this expression into Eq. (7.33) leads to:

$$\sigma = \frac{4\pi}{k^2} \sum_n (2n+1) \sin^2 \Upsilon_n \tag{7.45}$$

Eq. (7.45) is the one which Massey and Mohr treated under the assumption of a potential of the form Cr^{-s}, exactly the type with which we are concerned. Beginning with Jeffery's form[97] for the phase shifts in Eq. (7.45), these authors carried out an asymptotic solution for large n to obtain:

$$\sigma = \pi \frac{2s-3}{s-2} \left[\frac{(s-3)!!}{(s-2)!!} \frac{\pi}{2} \right]^{2/s-1} \left(\frac{C}{p} \right)^{2/s-1} \tag{7.46}$$

Eq. (7.31) tells us that $s = 4$ and $C = (1/2)\, \alpha\, e^2$, and, when we recall that $\alpha = \dfrac{\varepsilon - 1}{4\pi N_0}$:

$$\sigma = \frac{5}{2} \pi^3 \left(\frac{e^2}{4h} \right)^{2/3} \left(\frac{\varepsilon - 1}{4\pi N_0} \right)^{2/3} \left(\frac{1}{2\pi k} \right)^{1/3} \left(\frac{m}{T} \right)^{1/3} \tag{7.47}$$

The limiting value for the line width may then be found by substituting Eq. (7.47) into Eq. (7.27):

$$\delta_\infty = \frac{5}{2} \frac{T_0}{c} \left(\frac{N_0 \pi}{2} \right)^{1/3} \left(\frac{e^2}{4h} \right)^{2/3} (2\pi k)^{1/6} (\varepsilon - 1)^{2/3} p (m T^5)^{-1/6} \tag{7.48}$$

To this widths is added, for the line width, the width due to the Valley Effect.

BIBLIOGRAPHY

Periodicals

1. Amaldi, E. and E. Segré, *Nature* **133**, 141 (1934).
2. Amaldi, E. and E. Segré, *Nuovo Cimento* **11**, 145 (1934).
3. Anderson, P. W., *Phys. Rev.* **76**, 647 (1949).
4. Anderson, P. W., *Phys. Rev.* **80**, 511 (1950).
5. Anderson, P. W., *Phys. Rev.* **86**, 809 (1952).
6. Baranger, M. and B. Mozer, *Phys. Rev.* **115**, 521 (1959).
7. Baranger, M. and B. Mozer: *Tech. Rept.* No. 2, Contract Nonr-760(15).
8. Bassett, A. B., *Proc. Camb. Phil. Soc.* **6**, 11 (1889).
9. Bethe, H., *Handbuch der Physik* XXIV/1, 2. Aufl. Berlin (1933).
10. Bleaney, B., *Physica* **12**, 595 (1946).
11. Bleaney, B. and J. H. N. Laubser, *Nature* **161**, 522 (1948).
12. Bleaney, B. and R. P. Penrose, *Proc. Roy. Soc.* A **189**, 358 (1947).
13. Bleaney, B. and R. P. Penrose, *Proc. Phys. Soc. (London)* **59**, 424 (1947).
14. Bleaney, B. and R. P. Penrose, *Proc. Phys. Soc. (London)* **60**, 540 (1948).
15. Bloom, S. and H. Margenau, *Phys. Rev.* **90**, 791 (1953).
16. Born, M. and P. Jordan, *Z. Physik* **34**, 858 (1925).
17. Bothe, W., *Ann. Physik* **64**, 693 (1921).
18. Broyles, A. A., *Phys. Rev.* **100**, 1181 (1955).
19. Burkhardt, G., *Z. Physik* **115**, 592 (1940).
20. Carroll, K. G., *Phys. Rev.* **53**, 310 (1938).
21. Chapman, S., *Roy. Soc. (London) Phil. Trans.* **216**, 279 (1916).
22. Chapman, S., *Roy. Soc. (London) Phil. Trans.* **217**, 115 (1918).
23. Ch'en, S. Y., *Phys. Rev.* **58**, 884 (1940).
24. Childs, W. H. J., *Proc. Roy. Soc.* **153**, 555 (1936).
25. Condon, E., *Phys. Rev.* **28**, 1182 (1926).
26. Condon, E., *Phys. Rev.* **32**, 858 (1928).
27. Cornell, S. D., *Phys. Rev.* **51**, 739 (1937).
28. Cornell, S. D. and W. W. Watson, *Phys. Rev.* **50**, 279 (1936).
29. Debye, P., *Physik. Z.* **20**, 160 (1919).
30. Debye, P., *Physik. Z.* **21**, 178 (1920).
31. Debye, P., *Physik. Z.* **22**, 302 (1921).
32. Dennison, D. M., *Phys. Rev.* **28**, 318 (1926).
33. Dennison, D. M. and S. B. Ingram, *Phys. Rev.* **36**, 1451 (1930).
34. Dieter-Henkel, W., *Z. Physik* **137**, 295 (1954).
35. Dirac, P. A. M., *Proc. Roy. Soc.* A **114**, 263 (1928).
36. Ebert, H., *Ann. Physik* **36**, 466 (1889).
37. Eckart, C., *Phys. Rev.* **47**, 455 (1935).
38. Ecker, G., *Z. Physik* **148**, 593 (1957).
39. Ecker, G., *Z. Physik* **149**, 254 (1957).
40. Ecker, G., *Z. Naturforsch.* **12**, 346, 517 (1957).
41. Einstein, A., *Verh. d. deut. phys. Ges.* **18**, 318 (1916).
42. Epstein, P. S., *Phys. Rev.* **28**, 695 (1926).
43. Ewald, P. P., *Ann. Physik* **49**, 117 (1916).

44. Faxen, H. and J. Holtsmark, *Z. Physik* **45**, 307 (1927).
45. Fermi, E., *Nuovo Cimento* **11**, 157 (1934).
46. Foley, H. M., *Phys. Rev.* **69**, 616 (1946).
47. Foley, H. M., *Phys. Rev.* **73**, 259 (1948).
48. Franck, J., *Faraday Soc.* **21**, 536 (1926).
49. Frenkel, J., *Z. Physik* **59**, 198 (1930).
50. Fröhlich, H., *Nature* **157**, 478 (1946).
51. Füchtbauer, C., *Physik Z.* **55**, 775 (1934).
52. Füchtbauer, C. and F. Gossler, *Z. Physik* **87**, 89 (1933).
53. Füchtbauer, C. and U. W. Hoffmann, *Ann. Physik* **43**, 96 (1914).
54. Füchtbauer, C. and P. Schultz, *Z. Physik* **97**, 699 (1935).
55. Füchtbauer, C., G. Joos, and O. Dinkelacker, *Ann. Physik* **71**, 204 (1923).
56. Füchtbauer, C., P. Schulz, and F. Brandt, *Z. Physik* **90**, 403 (1934).
57. Furssow, W. and A. Wlassow, *Physik Z. Sowjetunion* **10**, 378 (1936).
58. Furssow, W. and A. Wlassow, *J. Phys. (USSR)* **1**, 335 (1939).
59. Gans, R., *Ann. Physik* **66**, 396 (1921).
60. Godfrey, C., *Roy. Soc. (London) Phil. Trans.* A **195**, 329 (1901).
61. Gora, E. K., AFCRC Report AF 19(604)-831, 21 June 1956.
62. Gorter, C. J. and R. de L. Kronig, *Physica* **3**, 1009 (1936).
63. Grasse, W., *Z. Physik* **89**, 261 (1934).
64. Gregory, C., *Phys. Rev.* **61**, 465 (1942).
65. Griem, H., *Z. Physik* **137**, 280 (1954).
66. Griem, H. R., A. C. Kolb, and K. Y. Shen, *Phys. Rev.* **116**, 4 (1959).
67. Gross, E. P., *Phys. Rev.* **97**, 395 (1955).
68. Guttinger, P., *Z. Physik* **73**, 169 (1931).
69. Hamilton, W. R., *Roy. Soc. (London) Phil. Trans.* **247** (1834).
70. Hamilton, W. R., *Roy. Soc. (London) Phil. Trans.* **95** (1835).
71. Hedfield, K. and R. Mecke, *Z. Physik* **64**, 151 (1930).
72. Heisenberg, W., *Z. Physik* **33**, 879 (1925).
73. Herschberger, W. D. and L. E. Norton, *RCA Rev.* **9**, 38 (1948).
74. Herzberg, G. and J. W. T. Spinks, *Z. Physik* **91**, 386 (1934).
75. Herzberg, G. and J. W. T. Spinks, *Proc. Roy. Soc.* A **147**, 434 (1934).
76. Herzberg, G. and J. W. T. Spinks, and W. W. Watson, *Phys. Rev.* **50**, 1186 (1936).
77. Heuneberg, W., *Z. Physik* **83**, 555 (1933).
78. Hoffman, H. and O. Theimer, *Astrophys. J.* **127**, 477 (1958).
79. Holtsmark, J., *Ann. Physik* **58**, 577 (1919).
80. Holtsmark, J., *Physik. Z* **20**, 162 (1919).
81. Holtsmark, J., *Physik. Z.* **25**, 73 (1924).
82. Holtsmark, J., *Z. Physik* **34**, 722 (1925).
83. Holtsmark, J., *Z. Physik* **55**, 437 (1929).
84. Holtsmark, J., *Z. Physik* **66**, 49 (1930).
85. Holtsmark, J. and B. Trumpy, *Z. Physik* **31**, 803 (1925).
86. Houston, W. V., *Phys. Rev.* **54**, 884 (1938).
87. Hughes, D. S. and P. E. Lloyd, *Phys. Rev.* **52**, 1215 (1937).
88. Hull, G. F., *Phys. Rev.* **50**, 1148 (1936).
89. Inglis, D. R. and E. Teller, *Astrophys. J.* **90**, 439 (1939).
90. Jablonski, A., *Z. Physik* **70**, 723 (1931).
91. Jablonski, A., *Acta Phys. Polon.* **6**, 371 (1937).
92. Jablonski, A., *Acta Phys. Polon.* **7**, 196 (1938).
93. Jablonski, A., *Acta Phys. Polon.* **8**, 71 (1939).
94. Jablonski, A., *Physica* **7**, 541 (1940).
95. Jablonski, A., *Phys. Rev.* **68**, 78 (1945).
96. Jablonski, A., *Phys. Rev.* **73**, 258 (1948).
97. Jeffery, H. M., *Proc. London Math. Soc.* **23** (1892).
98. Jensen, H., *Z. Physik* **80**, 448 (1933).

99. Kallmann, H. and F. London, *Z. f. phys. Chemie* B **2**, 207 (1929).
100. Karplus, R. and J. Schwinger, *Phys. Rev.* **73**, 1020 (1948).
101. Kauzmann, W., *Rev. Mod. Phys.* **14**, 12 (1942).
102. Keesom, W. H., *Physik Z.* **22**, 129 (1921).
103. Kivel, B., *Phys. Rev.* **98**, 1055 (1955).
104. Kivel, B., S. Bloom, and H. Margenau, *Phys. Rev.* **98**, 495 (1955).
105. Kleman, A. and E. Lindholm, *Ark. Mat. Astron. Fys.* A **32**, No. 17 (1946).
106. Kolb, A. C., AFOSR-TN-57-8, Astia Document No. AD 115-400.
107. Kolb, A., Thesis, University of Michigan, 1957.
108. Kramers, H. A., *Z. Physik* **39**, 828 (1926).
109. Krogdahl, M., *Astrophys. J.* **110**, 355 (1949).
110. Kronig, R. De L., *Physica* **5**, 65 (1938).
111. Kronig, R. De L. and I. I. Rabi, *Phys. Rev.* **29**, 262 (1927).
112. Kuhn, H., Phil. Mag. **18**, 987 (1934).
113. Kuhn, H., *Proc. Roy. Soc.* A **158**, 212 (1937).
114. Kuhn, H. and F. London, *Phil. Mag.* **18**, 983 (1934).
115. Kulp, M., *Z. Physik* **79**, 495 (1932).
116. Kulp, M., *Z. Physik* **87**, 245 (1933).
117. Kussman, H. W., *Z. Physik* **48**, 831 (1928).
118. Landwehr, G., Thesis, Yale University, 1956 (unpublished).
119. Lasareff, W., *Z. Physik* **64**, 598 (1930).
120. Lenz, W., *Z. Physik* **25**, 299 (1924).
121. Lenz, W., *Z. Physik* **80**, 423 (1933).
122. Lewis, M. B. and H. Margenau, *Phys. Rev.* **109**, 842 (1958).
123. Lindholm, E., *Z. Physik* **109**, 223 (1938).
124. Lindholm, E., *Z. Physik* **113**, 596 (1939).
125. Lindholm, E., *Ark. Mat. Astron. Fys.* **28** A, No. 3 (1942).
126. Lindholm, E., *Ark. Mat. Astron. Fys.* **32** A, No. 17 (1946).
127. Lochte-Holtgreven, W. and E. Eastwood, *Z. Physik* **79**, 450 (1932).
128. London, F., *Z. Physik* **63**, 245 (1930).
129. London, F., *A. f. phys. Chemie* **11**, 222 (1930).
130. Lorentz, H. A., *Proc. Roy. Acad. (Amsterdam)* **8**, 591 (1906).
131. Lussen, M., *Ann. Physik* **49**, 865 (1916).
132. Malmsten, C. J. and K. Svenska, *V. Akad. Handl.* **62**, 65 (1841).
133. Margenau, H., *Phys. Rev.* **38**, 747 (1931).
134. Margenau, H., *Phys. Rev.* **40**, 387 (1932).
135. Margenau, H., *Phys. Rev.* **43**, 129 (1933).
136. Margenau, H., *Phys. Rev.* **44**, 931 (1933).
137. Margenau, H., *Phys. Rev.* **48**, 755 (1935).
138. Margenau, H., *Phys. Rev.* **49**, 596 (1936).
139. Margenau, H., *J. Chem. Phys.* **6**, 896 (1938).
140. Margenau, H., *Phys. Rev.* **76**, 121 (1949).
141. Margenau, H., *Phys. Rev.* **76**, 1423 (1949).
142. Margenau, H., *Phys. Rev.* **82**, 156 (1951).
143. Margenau, H. and B. Kivel, *Phys. Rev.* **98**, 1822 (1955).
144. Margenau, H. and R. Meyerott, *Astrophys. J.* **121**, 194 (1955).
145. Margenau, H. and D. T. Warren, *Phys. Rev.* **51**, 748 (1937).
146. Margenau, H. and W. W. Watson, *Phys. Rev.* **44**, 92 (1933).
147. Margenau, H. and W. W. Watson, *Phys. Rev.* **44**, 748 (1933).
148. Margenau, H. and W. W. Watson, *Phys. Rev.* **51**, 48 (1937).
149. Massey, H. S. W. and C. B. O. Mohr, *Proc. Roy. Soc.* A **141**, 434 (1933).
150. Massey, H. S. W. and C. B. O. Mohr, *Proc. Roy. Soc.* A **144**, 188 (1934).
151. Mayer, H., Los Alamos Scientific Laboratory Report LA-647, 1947.
152. McKellar, A. and C. A. Bradley, *Phys. Rev.* **46**, 664 (1934).
153. Mensing, L., *Z. Physik* **34**, 611 (1925).

154. Meyerott, R. E. and H. Margenau, *Phys. Rev.* **99**, 1851 (1955).
155. Michelson, A. A., *Astrophys. J.* **2**, 251 (1895).
156. Michelson, A. A., *Phil. Mag.* **31**, 338 (1891); **34**, 280 (1892).
157. Minkowski, R., *Z. Physik* **55**, 16 (1929).
158. Minkowski, R., *Z. Physik* **93**, 731 (1935).
159. Mizushima, M., *Phys. Rev.* **83**, 1061 (1951).
160. Ny, T-Z., *Phys. Rev.* **52**, 1158 (1937).
161. Oldenberg, O., *Z. Physik* **47**, 984 (1928).
162. Oldenberg, O., *Z. Physik* **51**, 605 (1928).
163. Oltramare, G., *Comptes Rendus de l'Assoc. France* **24**, II, 167 (1895).
164. Oseen, C. W., *Ann. Physik* **48**, 1 (1915).
165. Podolsky, B., *Phys. Rev.* **32**, 812 (1928).
166. Ramsauer, C., *Ann. Physik* **54**, 513 (1921).
167. Ramsauer, C., *Ann. Physik* **66**, 546 (1921).
168. Ramsauer, C., *Ann. Physik* **72**, 346 (1923).
169. Ray, B. S., *Z. Physik* **75**, 74 (1932).
170. Rayleigh, Lord, *Phil. Mag.* **27**, 298 (1889).
171. Rayleigh, Lord, *Proc. Roy. Soc.* A **76**, 440 (1905).
172. Rayleigh, Lord, *Phil. Mag.* **26**, 274 (1915).
173. Reiche, F., *Z. Physik* **39**, 444 (1926).
174. Reiche, F. and H. Rademacher, *Z. Physik* **39**, 444 (1926).
175. Reiche, F. and H. Rademacher, *Z. Physik* **41**, 453 (1927).
176. Reinsberg, C., *Z. Physik* **93**, 416 (1935).
177. Reinsberg, C., *Z. Physik* **105**, 460 (1937).
178. Rosenfeld, L. and J. Solomon, *J. de Phys.* **2**, 189 (1931).
179. Rudkjobing, M., *Ann. Astrophys.* **12**, 229 (1949).
180. Schmalljohann, P., Staatsexam. Kiel, 1936 (ungedruckt).
181. Schonrock, O., *Ann. Physik* **20**, 995 (1906).
182. Schonrock, O., *Ann. Physik* **22**, 209 (1907).
183. Schrödinger, E., *Ann. Physik* **79**, 361 (1926).
Schrödinger, E., *Ann. Physik* **79**, 489 (1926).
Schrödinger, E., *Ann. Physik* **79**, 734 (1926).
184. Schulz, P. and R. Rompe, *Z. Physik* **108**, 654 (1938).
185. Schulz, P. and R. Rompe, *Z. Physik* **110**, 223 (1938).
186. Schutz-Mensing, L., *Z. Physik* **61**, 655 (1930).
187. Schwinger, J., *Phys. Rev.* **51**, 648 (1937).
188. Smith, D. F., *Phys. Rev.* **74**, 506 (1948).
189. Spitzer, L., Jr., *Phys. Rev.* **55**, 699 (1939).
190. Spitzer, L., Jr., *Phys. Rev.* **56**, 39 (1939).
191. Spitzer, L., Jr., *Phys. Rev.* **58**, 348 (1940).
192. Stark, J., *Ann. Physik* **48**, 215 (1915).
193. Stark, J. and E. Kirschbaum, *Ann. Physik* **43**, 1029 (1914).
194. Swaim, V. R., *Asrophys. J.* **40**, 137 (1917).
195. Tait, P. G., *Roy. Soc. (Edinb.) Trans.* **33**, 72 (1886).
196. Takamine, T., *Astrophys. J.* **50**, 1 (1919).
197. Takamine, T. and N. Kokubu, *Mem. Coll. Sci. Kyoto Imp. Univ.* **3**, 275 (1919).
198. Tsao, C. J. and B. Curnutte, Sci. Rept. 1A-8. Contr. No. AF 19(122)65. 1954.
199. Vedder, H. and R. Mecke, *Z. Physik* **86**, 137 (1933).
200. Verweij, S., Diss. Amsterdam. 1936.
201. Verweij, S., *Publ. Astron. Inst. Univ. Amsterdam*, No. 5 (1936).
202. Van Vleck, J. H. and H. Margenau, *Phys. Rev.* **76**, 1211 (1949).
203. Van Vleck, J. H. and V. F. Weisskopf, *Rev. Mod. Phys.* **17**, 227 (1945).
204. Watanabe, K., *Phys. Rev.* **59**, 151 (1941).
205. Watson, W. W., *J. Phys. Chem.* **41**, 61 (1937).
206. Watson, W. W. and G. F. Hull, Jr., *Phys. Rev.* **49**, 592 (1936).

207. WATSON, W. W. and H. MARGENAU, *Phys. Rev.* **44**, 748 (1933).
208. WEISSKOPF, V. F., *Z. Physik* **75**, 287 (1932).
209. WEISSKOPF, V. F., *Z. Physik* **77**, 398 (1932).
210. WENTZEL, G., *Handbuch der Physik* XXIV/1, 2 Auflage. Berlin (1933).
211. WLASSOW, A., *Z. f. exp. u. theor. Physik* **4**, 24 (1934).

Reviews

212. "Line Shape", R. G. BREENE, Jr., *Rev. Mod. Phys.* **29**, 94 (1957).
213. "Line Width", R. G. BREENE, Jr., *Handbuch der Physik* XXVII, Springer Verlag, Berlin.
214. "Broadening and Shift of Spectral Lines Due to the Presence of Foreign Gases", SHANG-YI CH'EN and MAKOTO TAKEO, *Rev. Mod. Phys.* **29**, 20 (1957).
215. "Quantum Theory of Radiation", E. FERMI, *Rev. Mod. Phys.* **4**, 87 (1932).
216. "Van der Waals Forces", H. MARGENAU, *Rev. Mod. Phys.* **11**, 1 (1939).
217. "Structure of Spectral Lines from Plasmas", H. MARGENAU and M. LEWIS, *Rev. Mod. Phys.* **31**, 569 (1959).
218. "Pressure Effects on Spectral Lines", H. MARGENAU and W. W. WATSON, *Rev. Mod. Phys.* **8**, 22 (1936).
219. „Beeinflussung der Breite und Lage von Spektrallinien durch Gase", P. SCHULZ, *Physik Z.* **39**, 412 (1938).
220. „Über die Theorie der Linienbreite von Atomen", I. I. SOBELMAN, *Fort. Physik* **5**, 175 (1957).
221. „Über die Theorie der Druckverbreiterung und -verschiebung von Spektrallinien", A. UNSOLD, *Vierteljahrsschriften der Astronomischen Gesellschaft* **78**, 213 (1943).
222. „Die Breite der Spektrallinien in Gasen", V. F. WEISSKOPF, *Physik. Z.* **34**, 1 (1933).

Books

223. *The Mathematical Theory of Non-Uniform Gases*, S. CHAPMAN and T. G. COWLING, Cambridge University Press, England 1953.
224. *Theory of Atomic Spectra*, E. U. CONDON and G. H. SHORTLEY, Cambridge University Press, 1951.
225. *Methods of Mathematical Physics*, Volume I, R. COURANT and D. HILBERT, Interscience, New York 1953.
226. *Polar Molecules*, P. DEBYE, Chemical Catalog Co., New York, 1929.
227. *Theory of Dielectrics*, H. FRÖHLICH, Clarendon Press, Oxford, England, 1949.
228. *Handbuch der Kugelfunction*, H. E. HEINE, G. Reiner, Berlin, 1881.
229. *The Quantum Theory of Radiation*, W. HEITLER, Oxford, University Press, 1954.
230. *Molecular Theory of Gases and Liquids*, J. C. HIRSCHFELDER, C. F. CURTIS, and R. B. BIRD, John Wiley, New York, 1954.
231. *Methods of Mathematical Physics*, H. JEFFREYS and B. S. JEFFREYS, Cambridge University Press, 1950.
232. *Mécanique Analytique*, J. L. DE LAGRANGE, Chez la Veuve Desaint, 1788.
233. *Über die Verbreiterung und Verschiebung von Spektrallinien*, E. LINDHOLM, Almquist and Wiksells, Uppsala, 1942.
234. *Wahrscheinlichkeitsrechnung*, A. A. MARKOFF, Deutsch von Liebmann, B. G. Teubner, Leipzig u. Berlin, 1912.
235. *A Treatise on Astronomical Spectroscopy*, J. SCHEINER, Ginn & Co. Boston and London, 1894.
236. *The Physics of Fully Ionized Gases*, L. SPITZER, Jr., Interscience Publishers, New York, 1956.
237. *Spektralanalyse chemische Atome*, J. STARK, S. Hirzel, Leipzig, 1914.
238. *History of Chemistry*, Sir T. E. THORPE, Watts & Co., London, 1909.
239. *Physik der Sternatmosphaeren*, A. UNSOLD, Springer-Verlag, Berlin, 1955.
240. *Theory of Bessel Functions*, G. N. WATSON, Cambridge University Press, 1944.
241. *Group Theory*, E. P. WIGNER, Academic Press, New York, 1959.

APPENDIX I

DETAILED BALANCING

We here detail certain considerations of Van Vleck and Margenau[202]. The ruminations of these authors did not serve actually to change the shape of the proposed absorption line, but they did finally show that the shapes of the spectral lines are the same in emission as in absorption, a point of no mean import. At any rate, since we accept the thesis that the integrated absorption and emission balance each other — in a Rayleigh–Jeans radiation field, which we shall mention in a moment — detailed balance of emission and absorption results from this equivalence of absorption and emission line shapes. That is, a given frequency interval absorbs as much as it emits. The balance was obtained by a hitherto untried technique, namely, the summing of the work done on the oscillator by the electric field at collision as well as during the intercollision intervals. This yielded the same frequency by frequency power as that emitted between two collisions.

If our oscillator motion is described by $x(t) = x_0 \cos(\omega_0 t + \varphi)$ where φ is, of course, the phase constant, then we may write down the Fourier analysis of $x(t)$. We may then integrate the expression for $x(\omega)$ and average $|x(\omega)|^2$ over a random distribution of φ to obtain a result which when utilized for x in Eq. (1.61) yields:

$$|x(\omega)|^2 = \frac{x_0^2}{4\pi}\left[\frac{1}{a^2 + (\omega_0 - \omega)^2} + \frac{1}{a^2 + (\omega_0 + \omega)^2}\right] \tag{I.1}$$

where now $a = 1/\tau$.

A point charge e which is oscillating in one dimension radiates power of amount:

$$\frac{\mathrm{d}W}{\mathrm{d}t} = \frac{2}{3}\frac{e^2}{c^3}\ddot{x}^2$$

From the equation for simple harmonic motion $\ddot{x} = -\omega^2 x$. In addition, normalization of the Fourier components requires:

$$\int_{-\infty}^{+\infty} [x(t)]^2\,\mathrm{d}t = \int_{-\infty}^{+\infty} |x(\omega)|^2\,\mathrm{d}\omega = 2\int_0^{\infty} |x(\omega)|^2\,\mathrm{d}\omega$$

so that:

$$\ddot{x}^2 = -\omega^4\,|x(t)|^2 = -\omega^4\,|x(\omega)|^2$$

We thus obtain for the power emitted by the oscillator in the frequency interval between ω and $\omega + d\omega$:

$$P_E'(\omega)\, d\omega = \frac{2\, e^2}{3\, c^3}\, 2\, \omega^4\, |x(\omega)|^2\, d\omega \tag{I.2a}$$

This must be modified to include the intervals between all collisions over a long period T, however, since the Fourier analysis has only been carried out over one such period. As a result, Eq. (I.2a) becomes:

$$P_E(\omega)\, d\omega = \frac{a\, T\, P_E'(\omega)\, d\omega}{T} = \frac{2\, e^2}{3\, c^3}\, 2\, \omega^4\, |x(\omega)|^2\, a\, d\omega \tag{I.2b}$$

Substitution of Eq. (I.1) into Eq. (I.2b) then yields:

$$P_E(\omega) = \frac{e^2\, x_0^2}{3\, \pi\, c^3}\, \omega^4 \left[\frac{a}{a^2 + (\omega_0 - \omega)^2} + \frac{a}{a^2 + (\omega_0 + \omega)^2}\right] \tag{I.3}$$

Now in order to obtain detailed balancing, we must needs show that $P_A(\omega)$, the power absorbed at the frequency ω, corresponds to this.

Let us suppose the electric vector of the incident radiation to be given by $E \cos(\omega t + \varphi)$, and, in addition, collisions to occur at $t = t_1, t_2, \ldots$ If the velocity proportional viscous drag force $g\dot{x}$ is dropped from Eq. (1.49b) the van Vleck–Margenau equation is obtained as:

$$\ddot{x} + \omega_0^2\, x = \frac{e\, E_x}{m}\, x \cos(\omega t + \varphi)$$

and a solution of the form:

$$x = \begin{cases} x_1 \text{ for } t_1 \leq t \leq t_2 \\ x_2 \text{ for } t_2 \leq t \leq t \\ \text{etc.} \end{cases}$$

is sought under the Lorentz boundary conditions $x_j(t_j) = \dot{x}_j(t_j) = 0$. The following integral forms satisfy the boundary conditions and the equations of motion:

$$x_j = \frac{e\, E_x}{m\, \omega_0} \int_0^{t-t_j} \cos(\omega\, t - \omega\, \Lambda + \varphi) \sin \omega_0 \Lambda\, d\Lambda \tag{I.4a}$$

$$\dot{x}_j = \frac{e\, E_x}{m} \int_0^{t-t_j} \cos(\omega\, t - \omega\, \Lambda + \varphi) \cos \omega_0 \Lambda\, d\Lambda \tag{I.4b}$$

It appears rather obvious that the work done by the field on the oscillator of charge e *between* collisions is

$$W = \sum_j \int_{t_j}^{t_j+1} e\, E_x \cos(\omega\, t + \varphi)\, \dot{x}_j\, dt \tag{I.5a}$$

or, by Eq. (I.4b)

$$W = \frac{e^2\, E_x^2}{m} \sum_j \int_0^{\vartheta_j} \cos(\omega\, t + \varphi_j)\, dt \int_0^t \cos[\omega(t - \Lambda + \varphi_j)] \cos \omega_0 \Lambda\, d\Lambda \tag{I.5b}$$

where:

$$\vartheta_j = t_{j+1} - t_j \text{ and } \varphi_j = \omega\, t_j + \varphi$$

The total time of observation can be chosen as, say, T, and the distribution function for the intercollision times may again be taken as $c\,e^{-a\vartheta}\,d\vartheta$ in analogy to Eq. (1.61). We may evaluate c from:

$$T = c \int_0^\infty e^{-a\vartheta}\,\vartheta\,d\vartheta = \frac{c}{a^2}$$

and subsequently substitute $a^2\,T \int_0^\infty e^{-a\vartheta} d\vartheta$ for the sum in Eq. (I.5b). Then, using the well known trigonometric relations for $\cos(a \pm b)$ and $\sin(a \pm b)$ and a large sheet of paper, one may reduce the resulting equation to one which, when integrated over a random distribution of the φ_f, yields:

$$W = \frac{e^2 E_x^2}{2m}\,a^2\,T \int_0^\infty e^{-a\vartheta} d\vartheta \int_0^\vartheta dt \int_0^t \cos\omega\Lambda \cos\omega_0\Lambda\,d\Lambda \tag{I.6}$$

We shall accept the formula:

$$a \int_0^\infty e^{-a\vartheta} d\vartheta \int_0^\vartheta f(\Lambda)\,d\Lambda = \int_0^\infty e^{-a\vartheta} f(\Lambda)\,d\Lambda$$

by means of which Eq. (I.6) may be integrated to obtain the work done by the field between collisions. As a consequence, the average power or work per unit time is:

$$\frac{W}{T} = \frac{e^2 E_x^2}{4m}\left[\frac{a}{a^2 + (\omega - \omega_0)^2} + \frac{a}{a^2 + (\omega + \omega_0)^2}\right] \tag{I.7}$$

The work done by the field *at* collision will be obtained. The Lorentz boundary conditions require that $x_j(t_{j+1})$ be zero. Since the oscillator will have some displacement $x_j(t_{j+1})$ probably not zero, an instantaneously infinite velocity of the oscillator is required so that the displacement change may occur in zero time. Since the authors do not appear to claim that this corresponds closely to physical reality, it would not appear that too much complaint against it as a mathematical device is in order. At any rate, in complete analogy to Eq. (I.5a) we obtain:

$$\left.\begin{aligned}
W_c &= \lim_{\Upsilon\to 0} \int_{t_{j+1}-\Upsilon}^{t_{j+1}+\Upsilon} e\,E_x \cos(\omega t + \varphi)\,x_f\,dt \\
&= \lim_{\Upsilon\to 0} \sum_j \Big[e\,E_x \cos(\omega t + \varphi)\,x_f\Big]_{t_{j+1}-\Upsilon}^{j_{j+1}+\Upsilon} \\
&\quad + \lim_{\Upsilon\to 0} \sum_j \int_{t_{j+1}-\Upsilon}^{t_{j+1}+\Upsilon} \omega\,e\,E_x \sin[\omega t + \varphi]\,x_j\,dt \\
&= \lim_{\Upsilon\to 0} \sum_j [e\,E_x \cos(\omega t + \varphi)\,x_j(t_{j+1} + \Upsilon) \\
&\quad - e\,E_x \cos(\omega t + \varphi)\,x_j(t_{j+1} - \Upsilon)] + 0 \\
&= -\sum_j e\,E_x \cos(\omega t + \varphi)\,x_j(t_{j+1})
\end{aligned}\right\} \tag{I.8}$$

by virtue of integration by parts and the fact that as $t \to t_{j+1}$ minus, $x \to x_j(t_{j+1})$ while as $t \to t_{j+1}$ plus, $x \to 0$. Eq. (I.4a) then tells us that we must add the term

$$-\frac{e^2 E_x^2}{m\,\omega} \sum_j \cos(\omega\,\vartheta_j + \varphi_j) \int_0^{\vartheta_j} \cos[\omega(\vartheta_j - \Lambda) + \varphi_j] \sin \omega_0 \Lambda \, d\Lambda \qquad \text{(I.9)}$$

to the right side of Eq. (I.5b). This equation may now be dealt with as was Eq. (I.5b) to obtain the power term:

$$\frac{e^2 E_x^2 a}{4\,m\,\omega} \left[\frac{\omega - \omega_0}{a^2 + (\omega - \omega_0)^2} - \frac{\omega + \omega_0}{a^2 + (\omega + \omega_0)^2} \right] \qquad \text{(I.9)}$$

which when added to Eq. (I.7) yields:

$$P_A(\omega) = \frac{W}{T} = \frac{e^2 E_x^2 \omega}{4\,m\,\omega_0} \left[\frac{a}{a^2 + (\omega - \omega_0)^2} - \frac{a}{a^2 + (\omega + \omega_0)^2} \right] \qquad \text{(I.10)}$$

A consideration of Eqs. (I.3) and (I.10) is sufficient to show that detailed balance has not been obtained if for no other reason than that the shape factors as given by the brackets in the two equations differ. Van Vleck and Margenau overcame this difficulty by passing from the Lorentz to the Van Vleck–Weisskopf boundary conditions on x and $\dot{x}$ at collision. We have shown in some detail how this transition added the terms as given by Eq. (6.180) to x and subsequently added to the absorption coefficient resulting in Eq. (6.182). It should suffice here to say that a similar computation adds the term:

$$\frac{e^2 E_x a}{4\,m\,\omega^2} \left[\frac{\omega^2 - \omega\,\omega_0}{a^2 + (\omega - \omega_0)^2} + \frac{\omega^2 + \omega\,\omega_0}{a^2 + (\omega + \omega_0)^2} \right] \qquad \text{(I.11)}$$

to Eq. (I.10) to arrive at the power absorbed from the light wave as a function of frequency.

We let $\Upsilon(\omega)\,d\omega$ be the energy density in the field for the frequency interval ω to $\omega + d\omega$. Then this quantity is given by

$$\Upsilon(\omega)\,d\omega = \frac{E^2}{4\pi} = \frac{1}{4\pi} \langle E^2 \rangle \langle \sin^2 \omega\,t \rangle$$

$$= \frac{1}{8\pi} \langle E^2 \rangle$$

when we substitute for the mean value of the sine squared term. E^2 is averaged over the directions of polarization. We agree that:

$$E^2 = E_x^2 + E_y^2 + E_z^2$$

so that:

$$\langle E^2 \rangle = 3 E_x^2$$

which yields:

$$\Upsilon(\omega)\,d\omega = \frac{3}{8\pi} E_x^2$$

Let us diverge for a moment.

We might recall that the principle of equipartition requires that the mean kinetic energy associated with a degree of freedom be $(1/2)\,kT$. Then, since the mean kinetic energy, $\langle(1/2)\,m\omega_0^2 x_0^2 \cos^2 \omega_0 t\rangle$, is the same as the mean potential energy, $\langle(1/2)\,m\omega_0^2 x_0^2 \sin^2 \omega_0 t\rangle$, for the oscillator, we may reasonably expect the mean total energy of oscillation to be kT. Likewise, $(1/2)\,m\omega_0^2 x_0^2 (\sin^2 \omega_0 t + \cos^2 \omega_0 t) = kT$, so that

$$x_0^2 = 2\frac{kT}{m\omega_0^2}$$

Jeans' number, which specifies the allowed number of proper vibrations in a wave number interval, will prove useful here. Let us recall the solution $\sin\left(2\pi k\frac{x}{X}\right)\sin\left(2\pi l\frac{y}{Y}\right)\sin\left(2\pi m\frac{z}{Z}\right)$ to the wave equation $\nabla^2\psi - \frac{1}{c^2}\psi = 0$ when we impose the boundary conditions, standing waves in a box of sides X, Y, and Z. Surely our wave numbers will be:

$$\tilde{\nu}_x = \frac{x}{X}; \quad \tilde{\nu}_y = \frac{l}{Y}; \quad \tilde{\nu}_z = \frac{m}{Z}$$

and we form a three space of these wave number components. It would appear to follow that the number of different proper vibrations in the wave number interval $d\tilde{\nu}_x\, d\tilde{\nu}_y\, d\tilde{\nu}_z$ (Jeans' number) is:

$$d\mathfrak{Z} = dk\, dl\, dm = d\tilde{\nu}_x\, d\tilde{\nu}_y\, d\tilde{\nu}_z\, XYZ = d\tilde{V}\cdot V \tag{I.12a}$$

where $d\tilde{V}$ is a volume element in wave number space and V is, of course, a volume in ordinary space.

Now surely $d\tilde{V} = 4\pi^2\tilde{\nu}^2 d\tilde{\nu}$, so Eq. (I.12a) becomes:

$$d\mathfrak{Z} = V 4\pi^2\tilde{\nu}^2\, d\tilde{\nu} \tag{I.12b}$$

We, of course, recall that $\nu = \tilde{\nu}c$ and also that electromagnetic radiation is possessed of two transverse components. Utilizing these two facts we obtain from Eq. (I.12b):

$$d\mathfrak{Z} = V\frac{8\pi^2\nu^2}{c^3}\, d\nu \tag{I.12c}$$

Eq. (I.12c) gives us Jeans' Number for the number of proper vibrations in the frequency interval $d\nu$, but we still desire an expression for the energy density of the radiation. It would certainly appear reasonable that we could obtain this quantity by multiplying $d\mathfrak{Z}$ by the mean energy for each proper vibration, which we may recall is kT. Thus:

$$\mathfrak{T}(\nu)\, d\nu = \frac{d\mathfrak{Z}}{V}kT = \frac{8\pi\nu^2}{c^3}kT\, d\nu$$

or:

$$\mathfrak{T}(\omega)\, d\omega = \frac{\omega^2 kT}{\pi^2 c^3}\, d\omega \tag{I.13}$$

Eq. (I.13) is the well known Rayleigh–Jeans radiation law.

Thus, we replace E_x^2 by,

$$E_x^2 = \frac{4}{3} \frac{m \omega^2 \omega_0^2}{\pi c^3} x_0^2 \tag{I.14}$$

in the sum of Eqs. (I.10) and (I.11) to obtain:

$$P_A(\omega) = \frac{c^2 x_0^2}{3\pi c^3} \omega^4 \left[\frac{a}{a^2 + (\omega_0 - \omega)^2} + \frac{a}{a^2 + (\omega_0 + \omega)^2} \right] \tag{I.15}$$

A comparison of Eqs. (I.3) and (I.15) shows that a detailed balance condition between absorption and emission has indeed been obtained by considering, in absorption, the work done by the field (a) between collisions and (b) at collisions and by including the Maxwell–Boltzmann boundary conditions. In essence, a choice of methods was available for obtaining x for Eq. (I.2a). We could have (a) taken ω^2 times the Fourier analysis of x (as we did) or (b) taken the Fourier analysis of $\ddot{x}$ for x. As van Vleck and Margenau note, method (b) would not have implicitly included the infinite accelerations at the time boundaries, and, as a consequence, detailed balancing would not have been obtained.

We have remarked that the boundary conditions especially are an approximation. In an actual molecular system, of course, finite collision times would have to be considered. In addition, the Planck radiation law would more logically replace the Rayleigh–Jeans law. In the limit of low frequency, however, the former reduces to the latter, so that the results are particularly applicable to the microwave region.